6
Plant Cell Monographs

Series Editor: David G. Robinson

Plant Cell Monographs

Recently Published Titles

Nitric Oxide in Plant Growth,
Development and Stress Physiology
Volume Editors: Lamattina, L., Polacco, J.
Vol. 6, 2007

The Expanding Cell
Volume Editors: Verbelen, J.-P., Vissenberg, K.
Vol. 5, 2007

The Plant Endoplasmic Reticulum
Volume Editor: Robinson, D. G.
Vol. 4, 2006

The Pollen Tube
A Cellular and Molecular Perspective
Volume Editor: Malhó, R.
Vol. 3, 2006

Somatic Embryogenesis
Volume Editors: Mujib, A., Šamaj, J.
Vol. 2, 2006

Plant Endocytosis
Volume Editors:
Šamaj, J., Baluška, F., Menzel, D.
Vol. 1, 2005

Nitric Oxide in Plant Growth, Development and Stress Physiology

Volume Editors: Lorenzo Lamattina, Joseph C. Polacco

With 39 Figures and 5 Tables

Springer

Volume Editors:

Professor Dr. Lorenzo Lamattina
Instituto de Investigaciones Biológicas
Facultad de Ciencias Exactas y Naturales
Universidad Nacional de Mar del Plata
CC 1245, (7600) Mar del Plata
Argentina

Professor Joseph C. Polacco
Department of Biochemistry
112 Schweitzer Hall
University of Missouri-Columbia
Columbia, MO 65211
USA

Series Editor:

Professor Dr. David G. Robinson
Ruprecht-Karls-University of Heidelberg
Heidelberger Institute for Plant Sciences (HIP)
Department Cell Biology
Im Neuenheimer Feld 230
D-69120 Heidelberg
Germany

ISSN 1861-1370
ISBN 978-3-642-07954-2 e-ISBN 978-3-540-45131-0

DOI 10.1007/11563280

Springer is a part of Springer Science+Business Media

springer.com

Editor: Dr. Christina Eckey, Heidelberg
Desk Editor: Anette Lindqvist, Heidelberg
Cover design: WMXDesign GmbH, Heidelberg

Printed on acid-free paper 149/3100 YL – 5 4 3 2 1 0

Editors

Lorenzo Lamattina was born in Argentina, where he studied agronomy and biology and obtained his Ph.D. from the University of Mar del Plata. He did postdoctoral research at the Institut de Biologie Moleculaire des Plantes (IBMP-CNRS) at Strasbourg, France, where he and his co-workers discovered RNA editing in plant mitochondria. In 1991, he joined the Instituto de Investigaciones Biologicas (IIB), at the Faculty of Exact and Natural Sciences, University of Mar del Plata, as professor of molecular biology. He chaired the IIB from 2000 to 2004 and since 1999 has been director of the Molecular and Integrative Physiology Lab. He has taught molecular biology and plant molecular biology to undergraduate and graduate students. His research focuses on pathways and molecules involved in nitric oxide (NO)-mediated processes in plants. Initially focused on the antioxidant properties of NO and its protection against the deleterious effects of reactive oxygen species (ROS), his research has recently dealt with the NO regulation of stomatal closure, adventitious and lateral root formation as well as root hair development. More recently, the studies related to the NO involvement in plant iron metabolism and transport is finding application in several areas.
Lorenzo Lamattina is a member of several learned societies. He is a Guggenheim Fellow (2005–2006) and he obtained the 2006 award from the Bunge and Born Foundation for his research regarding NO effects in plant biology.

Joe Polacco, a native of Brooklyn NY, obtained his B.S. at Cornell University and Ph.D. at Duke University, both in biochemistry with emphases in chemistry and genetics, respectively. After two years as an assistant professor at the Universidad del Valle in Cali, Colombia, he held a six-month postdoctoral position at Brookhaven National Laboratories, where he entered the field of plant science under the tutelage of Peter Carlson. Subsequently he spent five years as a staff geneticist at the Connecticut Agricultural Experiment Station – the oldest in the U.S. and internationally second only to that at Wageningen. Since 1979 he has been a member of the Biochemistry Department of the University of Missouri-Columbia. He has taught undergraduates and in the school of medicine and graduate school. He introduced biotech outreach courses for non-science majors, such as "Biotechnology in Society." His research has touched on mitochondrial function, lipoxygenase roles, Ni metabolism and plant interactions with methylotrophic commensal bacteria. The main focus of his research is the assimilation of fixed nitrogen in the form of ureides and on mobilization of N reserves. His interest in nitrogen metabolism led directly to studies on nitric oxide (NO) production and function in plants.
Joe Polacco, twice a Fulbright Fellow, has extensive international experience in science. His collaboration with Lorenzo Lamattina involved a five-month stay in his lab, followed by hosting Dr. Lamattina as a Guggenheim Fellow. He is a member of several scientific societies and an award-winning teacher.

Preface

Nitric oxide (NO) has definitively emerged as a ubiquitous molecule in life and its study has contributed to a better knowledge of many mechanisms and functions that were not well understood until very recently. NO is probably the inorganic molecule with the best characterized influence on biological processes. It has been proposed that at earlier stages of evolution, NO could have acted as an antioxidant. As complexity evolved, new and specialized antioxidant activities coevolved to detoxify reactive oxygen species (ROS). Thus, NO was acquiring new functions to contribute to the current exquisite network of cellular signaling. We now know that NO is involved in disease resistance, abiotic stress, growth and development processes, cell division and cell death, nervous system responses, immune responses, etc.

There are several sources of NO. However, its chemical nature and structure make NO even more fascinating. NO is a gas, a free radical with an unpaired electron, but without charge. These properties enable NO to move within cells, cross membranes and be removed efficiently by different cellular components. Because of its rapid movement and removal, the behavior of NO in signaling is optimal (Neil et al. 2003; Lamattina et al. 2003). NO is able to switch on and switch off, in a precise manner, a number of cellular functions. NO can also interact with other signaling molecules contributing to the generation of amplified signaling cascades. NO behaves like a hormone. Its versatility confers to this molecule the property of acting simultaneously in different cellular compartments and in opposite directions. This ability is essential to accomplish a housekeeping role in homeostatic cell processes and to act, at the same time, as a synchronizer of cellular metabolism.

The NO source in plants constitutes a complex panorama. NO can be generated through enzymatic and non-enzymatic pathways, each of which is finely tuned and regulated (Yamasaki et al. 1999, Guo et al. 2003, Corpas et al. 2004, Rockel et al. 2002, Bethke et al. 2004). This complexity challenges genetic approaches to unravel plant NO functionality. In addition to the ability of NO to cross cell membranes and hence act at “a distance”, the intracellular localization of the different NO sources might be linked to different targets and to the regulation of different physiological processes.

No DNA sequence has been found to bind NO directly. However, microarrays, northern blots and AFLP analyses have revealed that NO is implicated in the regulation of the level of numerous transcripts in *Arabidopsis*. The results indicate that NO-regulated genes are involved in every aspect of plant growth and development, as well as in stress responses (Polverari et al. 2003, Parani et al. 2004, He et al. 2005). In addition, post-translational modifications induced by NO through S-nitrosylation have revealed an impressive number of proteins as potential NO targets (Lindermayr et al. 2005). These proteins are thought to be involved in primary and secondary cell metabolism. Furthermore, the ability of NO to be one of the cellular messengers involved in the regulation of cytosolic Ca^{2+} concentration while, nitric oxide synthase (NOS) is itself a Ca^{2+}-dependent activity, make NO an excellent candidate as a signaling molecule in every Ca^{2+}-modulated cell response (Garcia-Mata et al. 2003; Lamotte et al. 2004; Lanteri et al. 2006).

Figure A shows the exponential growth in the number of papers that have appeared in a 10 years period (1995-2004) concerning NO action in plant biology. In comparison to the number of papers on auxin over the same period, it can be observed that NO is rapidly catching up this in an exciting era of worldwide auxin research. Figure B shows that the ratio between auxin-

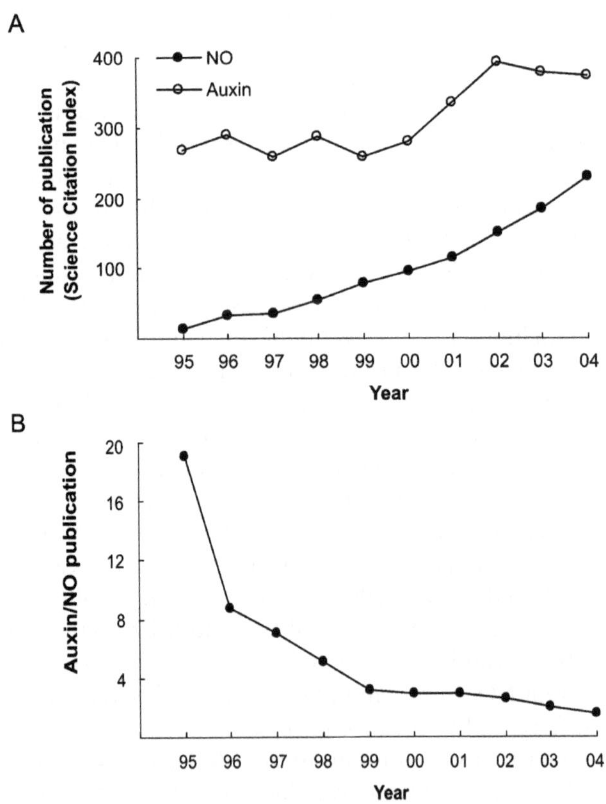

vs NO-related publications was 19 in favor of auxin in 1995, whereas the ratio dropped to 1.4 in 2004.

Finally, the industrial revolution and its subsequent exponential growth led to both environmental pollution and a parallel increase of NO levels. Increased NO levels in the troposphere bring about a natural pressure on NO-regulated processes in biosphere. We have no idea, at the moment, about the real significance of that pressure and its evolutionary side-effects. However, we would like to cite at least two examples that touch on this question: (i) pharmacologists at the Free University of Berlin obtained unexpected results in a measurements of the NO-regulated guanylate cyclase (Friebe et al. 1996) and (ii) trees in New York's Central Park showed better growth than those of neighboring rural areas as a result of NOx-mediated depletion of urban growth-inhibiting ozone (Gregg et al. 2003). Thus, at least these particular examples possibly corroborate the popular saying "What doesn't kill you, makes you stronger". They also point to the need for interactive molecular and ecological approaches to understand the adaptive changes occurring in a very dynamic equilibrium.

The revolution of genomics, proteomics and metabolomics has infused a flood of data in biology. Nevertheless, basic ecological, biochemical, physiological and cellular biological approaches are being revisited to validate and integrate the voluminous bioinformatics and *in silico* data. This book compiles chapters each of which provides a balanced treatment of new results and prospects in every field of NO biology in plants, from physiological, biochemical and molecular points of view. We intend this book to be provocative, but at the same time, a valuable source of information and a tool to inspire scientists to think about how NO could be a variable in their own fields of plant research. Considering that the majority of the data reporting NO action in plant biology has appeared during the last 8 years, this book will be absolutely necessary for teachers and students to update "standard" knowledge in plant biology courses. All the chapters of this volume were written by actively working plant scientists. The editors would like to express their gratitude for their valuable contributions and we hope that they spur further advances by you the reader. We are certainly embarking on a century that will bring us unexpected and exciting discoveries involving NO actions in plants.

Mar del Plata, Argentina — Lorenzo Lamattina
Columbia, MO, USA — Joseph C. Polacco

Finally, we wish to remember with affection our dear colleague Dr. Radomir Konjević (corresponding author, chapter 6) who died on July 22nd, 2006.

Note added in Proof

Durner and coworkers (Zemotjel et al., 2006) have called to question the actual activity of the protein encoded by AtNOS1. They have not been able to detect NOS activity in the recombinant protein using both an [^{3}H]-arginine and a Griess reagent-based NOS assay. Crawford and colleagues, in their response to Durner's letter (Crawford et al., 2006), have accepted that the new experiments show no detectable citrulline in the analysis of the products produced from [^{14}C]-arginine in presence of the recombinant AtNOS1. Moreover, the attempts to reproduce the reported production of NO using the Griess reagent have also failed.

That *Atnos1* produces much less basal NO than wild type is indisputable (Guo et al., 2003), hence the proposed renaming of the gene as NOA1 (NO-Associated 1). However, since the molecular basis of the *Atnoa1* phenotype is unknown, its use as a genetic tool in manipulating endogenous NO is now strongly diminished. Science is a self-correcting mode of inquiry, and erroneous results must be accepted, rapidly communicated to the scientific community, and then withdrawn. This new turn of events will undoubtedly help deter us from making wrong turns in the continuing quest for NO sources in plants.

References

Bethke PC, Badger MR, Jones RL (2004) Apoplastic synthesis of nitric oxide by plant tissues. Plant Cell 16:332–341

Crawford NM, Galli M, Tischner R, Heimer YM, Okamoto M, Mack A (2006) Response to Zemojtel et al.: Plant nitric oxide synthase: back to square one. Trends in plant Sci (November issue)

Corpas FJ, Barroso JB, del Río LA (2004c) Enzymatic sources of nitric oxide in plant cells – beyond one protein-one function. New Phytologist 162:246–248

Friebe A, Malkewitz J, Schultz G, Koesling D (1996) Positive effects of pollution? Nature 382:120

Garcia-Mata C, Gay R, Sokolovski S, Hills A, Lamattina L, Blatt MR (2003) Nitric oxide regulates K^+ and Cl^- channels in guard cells through a subset of abscisic acid-evoked signaling pathways. Proc Natl Acad Sci USA 100:11116–11121

Gregg JW, Jones CG, Dawson TE (2003) Urbanization effects on tree growth in the vicinity of New York City. Nature 424:183–187

Guo FQ, Okamoto M, Crawford NM (2003) Identification of a plant nitric oxide synthase gene involved in hormonal signaling. Science 302:100–103

He J-M, Xu H, She X-P, Song X-G, Zhao W-M (2005) The role and the interrelationship of hydrogen peroxide and nitric oxide in the UV-B-induced stomatal closure in broad bean. Function Plant Biol 32:237–247

Lamattina L, Garcia-Mata C, Graziano M, Pagnussat G (2003) Nitric oxide: The versatility of an extensive signal molecule. Ann Rev Plant Biol 54:109–136

Lamotte O, Gould K, Lecourieux D, Sequeira-Legrand A, Lebrun-Garcia A, Durner J, Pugin A, Wendehenne D (2004) Analysis of nitric oxide signaling functions in tobacco cells challenged by the elicitor cryptogein. Plant Physiol 135:516–529

Lanteri ML, Pagnussat GC, Lamattina L (2006) Calcium and calcium-dependent protein kinases are involved in nitric oxide- and auxin-induced adventitious root formation in cucumber. J Exp Bot 57:1341–1351
Lindermayr C, Saalbach G, Durner J (2005) Proteomic identification of S-nitrosylated proteins in Arabidopsis. Plant Physiol 137:921–930
Neill SJ, Desikan R, Hancock JT (2003) Nitric oxide signaling in plants. New Phytol 159:11–35
Parani M, Rudrabhatla S, Myers R, Weirich H, Smith B, Leaman DW, Goldman SL (2004) Microarray analysis of nitric oxide responsive transcripts in *Arabidopsis*. Plant Biotechnol J 2:359–366
Polverari A, Molesini B, Pezzotti M, Buonaurio R, Marte M, Delledonne M (2003) Nitric oxide-mediated transcriptional changes in Arabidopsis thaliana. Mol Plant Microbe Interact 16:1094–1105
Rockel P, Strube F, Rockel A, Wildt J, Kaiser W (2002) Regulation of nitric oxide (NO) production by plant nitrate reductase in vivo and in vitro. J Exp Bot 53:103–110
Yamasaki H, Sakihama Y, Takahashi S (1999) Alternative pathway for nitric oxide production in plants: new features of an old enzyme. Trends Plant Sci 4:12
Zemojtel T, Frohlich A, Palmieri MC, Kolanczyk M, Mikula I, Wyrwicz LS, Wanker EE, Mundlos S, Vingron M, Martasek P, Durner J (2006) Plant nitric oxide synthase: a never-ending story? Trends in Plant Sci (November issue)

Contents

Plant Cell Monogr (6)
L. Lamattina · J. C. Polacco: Nitric Oxide in Plant Growth
DOI 10.1007/7089_2006_081/Published online: 14 October 2006

Higher Plant Mitochondria as a Source for NO

Werner M. Kaiser (✉) · Kapuganti J. Gupta · Elisabeth Planchet

Julius-von-Sachs Institute for Biosciences, Molecular Plant Physiology and Biophysics, University of Würzburg, Julius-von-Sachs Platz 2, 97082 Würzburg, Germany
kaiser@botanik.uni-wuerzburg.de

Abstract Higher plant mitochondria produce nitric oxide (NO) by two separate systems. One is a mitochondrial nitric oxide synthase (NOS), which catalyzes the synthesis of NO and L-citrulline from L-arginine using NAD(P)H. The other one is the respiratory electron transport chain, with the terminal oxidases, CytOx and AOX, which both reduce nitrite to NO. While oxygen is obligatory for the former reaction, the latter activity appears very low in air but high under oxygen deficiency. However, even under anoxia, the rate of nitrite:NO reduction rarely reaches $\pm 1\%$ of respiratory electron transport. For as yet unknown reasons, nitrite:NO reduction appears absent in mitochondria from green leaves. The contribution of NOS and of nitrite reduction to overall NO production, and possible functions of nirite:NO reduction under hypoxia/anoxia are discussed.

1 Introduction

Mitochondria are known as the organelles housing the citric acid cycle and membrane systems for respiratory electron transport and oxidative phosphorylation. The basic structural similarity of eukaryotic mitochondria reflects an overall similarity in these basic functions. However, in addition, mitochondria serve many more important purposes, such as synthesis of vitamins, cofactors, nucleotides, metabolism of organic acids, amino acids, lipids, and partial reactions of the photorespiratory cycle (Rebeille et al. 1997; Bartoli et al. 2000; Gueguen et al. 2000; Kowaltkowski 2000). Further, mitochondria appear to play an important role in programmed cell death (Balk et al. 1999; Balk and Leaver 2001; Lam et al. 2001; also compare the contributions in this book) and are a major source for reactive oxygen species (ROS) (for review see Møller 2001). These multiple functions are not expected to occur to the same extent in all plant organs, but are more or less expressed in specific organs, cell types, and developmental stages. Recently, evidence has been accumulating that mitochondria are also involved in two apparently opposed processes, namely the production, but also the consumption of nitric oxide (NO). As NO has gained increasing attention in plants as a signaling compound as well as a highly reactive modifying agent for biomolecules, including proteins, this new mitochondrial function appears potentially important. Thus, we will briefly summarize present knowledge on NO in context with mitochondria.

2 NO Production

Plants appear to produce NO by two basically different pathways (also compare Fig. 1):

1. The L-arginine-dependent pathway uses NAD(P)H and O_2 as cosubstrates and is catalyzed by nitric oxide synthase (NOS) according to reaction (1):

$$\text{L-Arg} + \text{NAD(P)H} + \text{H}^+ + \text{O}_2 \rightarrow \text{L-Citr} + \text{NAD(P)}^+ + \text{NO} . \quad (1)$$

2. The nitrite-dependent pathway uses NADH or "electrons" as reductands and is catalyzed by a number of different enzymes according to reaction (2):

$$\text{NO}_2{}^- + \text{e}^- + 2\text{H}^+ \rightarrow 2\text{NO} + \text{H}_2\text{O} . \quad (2)$$

In plants, reduction of nitrite to NO was originally thought to be only catalyzed by nitrate reductases (NR). Xanthine oxidase/dehydrogenase (XDH)

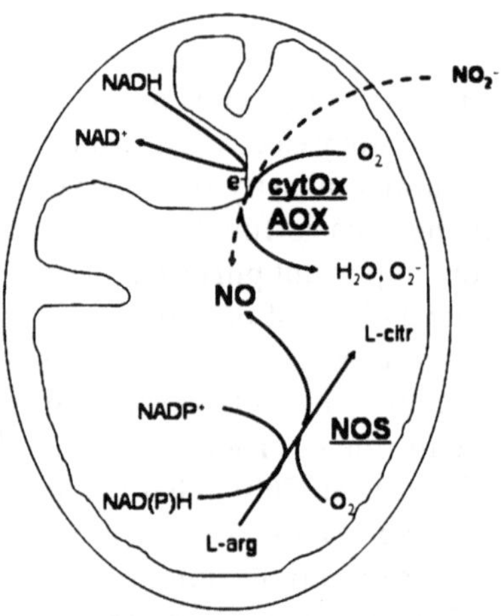

Fig. 1 Reactions producing NO in plant mitochondria. Nitric oxide synthase (NOS) has been shown in *Arabidopsis* to produce NO from L-arginine imported from the cytosol. NOS (AtNOS1) is probably integrated into the inner membrane. NO can be also produced by mitochondrial electron transport, reducing nitrite to NO (the formation of NO from NO_2^- does not require oxygen and is represented by a *dotted arrow*). The reaction is hardly detectable in air. One possible explanation is competition with oxygen. For further details see text. *Cyt ox* cytochrome oxidase, *AOX* alternative oxidase

has also been occasionally suggested as a source for NO using nitrite and xanthine as a substrate (Millar et al. 1998; Godber et al. 2000). However, our own experiments, using recombinant XDH, gave no evidence for NO production by the enzyme itself (Mendel and Kaiser, unpublished). Non-enzymatic NO production from nitrous acid according to Eq. 3 should occur at significant rates only at pH values below pH 5 (pK_a 3.2 of nitrous acid).

$$2HNO_2 \leftrightarrow NO + NO_2 + H_2O \leftrightarrow 2NO + 1/2O_2 + H_2O\,. \qquad (3)$$

Conditions favoring non-enzymatic NO formation are probably rare, but occasionally they may be met in the apoplast of plant cells (Bethke et al. 2004) and perhaps also in the vacuoles.

Overall, the contribution of XDH and of non-enzymatic NO formation to overall plant NO production seems negligible compared to the contribution of NR. However, according to more recent research, mitochondria are another important source for NO in plants and indeed both reactions (1) and (2) appear to occur inside plant mitochondria, as will be shown.

2.1
Nitric Oxide Synthase is Located in the Mitochondria

The above-mentioned nitric oxide synthase (NOS) reaction was suggested as a source for NO in plants, mainly based on pharmacological evidence. Inhibition of NO formation or of NO-dependent reactions by chemical analogs of L-arginine is usually taken as an indication that the reaction was triggered by NOS-derived NO. Immunological evidence for NOS in plants was obtained with antibodies against animal NOS (Kuo et al. 1995; Sen and Chema 1995; Barroso et al. 1999; Ribiero et al. 1999), but those antibodies proved to be rather unspecific (Lo et al. 2000; Butt et al. 2003). As no *Arabidopsis* gene or protein homolog to the large and complex animal protein has yet been found, the existence of NOS in plants is still an enigma.

More recently, a breakthrough in NO research was achieved by the finding of the Crawford group (Guo et al. 2003; Crawford and Guo 2005) that *Arabidopsis* contains a gene with sequence similarity to a gene from *Helix pomatia* that is implicated in NO synthesis. The gene encodes a 60 kDa protein, which, when expressed in *E. coli*, increased NO synthesis in cell extracts. When the corresponding gene (*AtNOS1*) was knocked out in *Arabidopsis*, the resulting mutant had reduced NO production in roots (measured with DAF-2DA). Contrary to animal NOS (about 140 kDA), the much smaller AtNOS1 requires no flavin or tetrahydrobiopterin, but only Ca^{2+}, CaM and NADPH. AtNOS1 seems constitutively expressed. It has been suggested to be part of the signaling pathway involved in ABA-induced stomatal closure, germination, root and shoot growth, seed fertility (for review Crawford and Guo 2005), control of flower timing (He et al. 2004), senescence and protection against oxidative damage (Guo and Crawford 2005), and seems also involved in NO production during plant–pathogen interactions, as derived from experiments with DAF-FM DA and EPR (Zeidler et al. 2004; Guo and Crawford 2005). Also, in the *atnos1* knock out mutant, induction of defence-related genes by *Pseudomonas syringae* was suppressed compared to the wild type (Zeidler et al. 2004). All these data suggest that AtNOS1, despite its different molecular properties, has functions analogous to animal NOS, but without the requirement for tetrahydrobiopterin as cofactor.

The first report on mitochondrial localization of NOS was by Giulivi et al. (1998), who detected NOS activity in purified animal mitochondria, mitochondrial homogenates, and submitochondrial particles, using EPR and oxyhemoglobin to detect NO. Indeed, NOS activity of animal mitochondria appears located in the inner mitochondrial membrane (Ghafourifar and Richter 1997).

Very recent work by Crawford's group indicates that plant AtNOS1, like the animal enzyme, is also located in the mitochondria (compare Fig. 1). This view was based on the following lines of evidence:

- Computational analysis of the NOS1 protein sequence reported a high probability of being targeted to the mitochondria
- Fluorescence from a p35S-NOS1cDNA-GFP construct strongly overlapped with MitoTracker fluorescence in mitochondria of roots and root hairs examined by confocal microscopy
- NO production in mitochondria isolated from *Arabidopsis* WT and AtNOS1 mutant plants was detected using DAF-fluorescence (Guo and Crawford 2005)

Whether AtNOS1, or (yet unknown) isoforms may be also located in other plant cell organelles, is not totally clear. Using an immunological approach, NOS-like activity in pea plants has been reported to be localized in both peroxisomes and chloroplasts (Barroso et al. 1999). The specificity of anti-NOS antibodies used for the experiments, however, has been questioned (Lo et al. 2000; Butt et al. 2003). Thus, at present it seems most probable that NOS-like activity in plants is exclusively located in the mitochondria. Sufficient supply of reductant in the mitochondria is assured by the citric acid cycle, and the second NOS substrate, L-arginine, may pass the mitochondrial membranes via a recently identified translocator for basic amino acids (Catoni et al. 2003; Hoyos et al. 2003). At this point it is also unknown whether AtNOS1 is actually exposed to the matrix side, or to the intermembrane space, as in animal mitochondria (Ghafourifar and Richter 1997). It is also not clear whether AtNOS1 can use NADH, as well as NAD(P)H, as substrate.

In the above-mentioned experiments with mitochondria purified from *Arabidopsis* leaves, Guo and Crawford (2005) surprisingly detected DAF-fluorescence indicative for NOS-dependent NO production without any reductand addition (NAD(P)H or others). However, plant mitochondria may contain 0.2–0.7 nmol NADP mg^{-1} protein, of which up to 40% may be in the reduced state, at least in vivo or in the presence of added substrate (Møller

2001 and literature cited). The NOS activity reported by Guo and Crawford (2005) for purified mitochondria was very low (0.1 pmol mg^{-1} protein min^{-1}), and thus the above NAD(P)H concentration might be sufficient to support NOS activity for some time.

A similar problem arises when considering the suggested protection against oxidative damage by NO. Reported rates of O_2^- formation in intact plant mitochondria vary considerably from about 100–1000 nmol mg^{-1} protein h^{-1} (Møller 2001 and literature cited). But in any case they appear much higher than even the maximum rates of NO production by mitochondria under anoxia, which are 5–10 nmol mg^{-1} protein h^{-1} (Gupta et al. 2005), and which may be much lower in aerobic conditions. If mitochondrial protein were 10% of the total leaf protein, the above mentioned NOS activity in leaf extracts from *Arabidopsis* WT plants (0.1 pmol mg^{-1} protein min^{-1}) would correspond to a NOS activity based on mitochondrial protein of 60 pmol mg^{-1} protein h^{-1}. This is far below estimated rates of ROS formation in the mitochondria. It is therefore not completely clear how intramitochondrial NO could contribute to ROS scavenging, as has been suggested (Millar et al. 2002; Guo and Crawford 2005; Crawford and Guo 2005).

2.2 Mitochondria also Produce NO by Reduction of Nitrite

Cytosolic nitrate reductase (NR) has been known for some time to reduce nitrite to NO with NADH as reductant, although with only a small fraction (about 1%) of its normal nitrate reducing capacity (Rockel et al. 2002; Planchet et al. 2005). In addition, in plant roots a PM-bound nitrite:NO reductase appears to catalyze a similar reaction, in close association with a PM-NR (Stöhr et al. 2001, also see the chapter by Stöhr in this volume). For a number of years, NR, together with the PM-bound enzyme, appeared to be the only source for nitrite-derived NO.

In 1999, Kozlov et al. demonstrated that animal mitochondria are able to produce NO from added nitrite in the absence of oxygen, and the reaction was abolished by the complex III inhibitor myxothiazol, indicating that the respiratory electron transport was donating electrons for nitrite reduction. In plants, a first hint on an involvement of the respiratory chain in nitrite-dependent NO production came from experiments with a *nia* mutant of the unicellular green alga, *Chlorella sorokiniana*. These mutant algal cells are not able to reduce nitrate to NO, and usually did not produce NO (measured as NO emission into the gas phase by chemiluminescence) when supplied with nitrate. When nitrite was added, however, they emitted NO under anoxia, but much less in air. Obviously, the algae could reduce nitrite to NO by means other than NR. The myxothiazol-sensitivity of the reaction (compare Fig. 2) was a hint that respiratory electron transport was the electron source (Tischner et al. 2004).

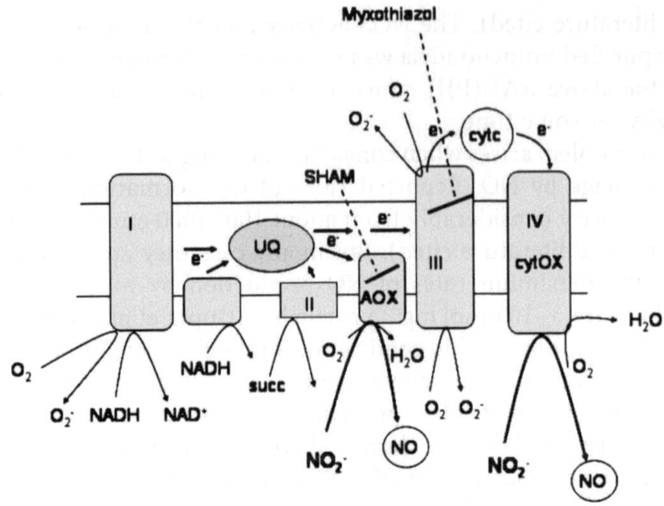

Fig. 2 Reduction of nitrite to NO by the electron transport chain of plant mitochondria. For reasons of simplicity, only one NADH dehydrogenase is shown in the diagram. According to the inhibition sites of myxothiazol (complex III) and salicylhydroxamic acid (AOX), both terminal oxidases appear to posses nitrite:NO reducing activity, which would also explain a possible competition of nitrite and O_2, leading to a very low NO formation rate in air. It is also shown that superoxide (O_2^-) can be produced at several different sites and in close neighborhood to NO, which would facilitate formation of oxidized NO species. ⊥ inhibition, *UQ* ubiquinone, *succ* succinate, *cyt c* cytochrome c, *AOX* alternative oxidase, *Cyt ox* cytochrome oxidase

Similarly, NR-free tobacco suspension cells, either WT cells grown in the absence of nitrate or ammonium, or on ammonium plus tungstate, or cells of a NR-free *nia* double mutant, never produced NO when supplied with nitrate. However, they emitted NO at low rates in air, and at up to 100-fold higher rates under nitrogen, when supplied with nitrite. The reaction was partially inhibited by myxothiazol, and further inhibited by salicyl hydroxamic acid (SHAM), an inhibitor of the alternative oxidase (AOX) (Planchet et al. 2005; Gupta et al. 2005; also compare Fig. 2).

This was confirmed with purified mitochondria from various plant sources. Preparations (suspensions) of mitochondria from roots of tobacco, pea, barley, and *Arabidopsis*, produced NO (detected and quantified by chemiluminescence) under nitrogen, when supplied with NADH and nitrite (Gupta et al. 2005). NO production rates under anoxia were 1–10 nmol NO mg^{-1} protein h^{-1}, which is only 1/1000 of the respiratory electron transport capacity of these preparations (5–8 μmol O_2 mg^{-1} protein h^{-1}). Mitochondria purified from cell suspensions also produced NO from nitrite, although (on a protein basis) less than root mitochondria. Surprisingly, NO production was hardly detectable in mitochondria purified from leaves of all the above-mentioned plant species.

To guard against an artefact of the purification procedure, experiments were carried out with leaf slices in solution. Leaf slices from WT leaves expressing NR readily emitted NO when supplied with nitrite, but this NO emission was almost insensitive to myxothiazol. Further, NO emission from NR-free leaf slices was very low (Planchet et al. 2005; Gupta et al. 2005). In contrast, NR-free root segments fed with nitrite under nitrogen showed almost the same NO emission as normal NR-containing roots (Gupta et al. 2005) and as root segments in solution (compare Fig. 3). Thus it seems that

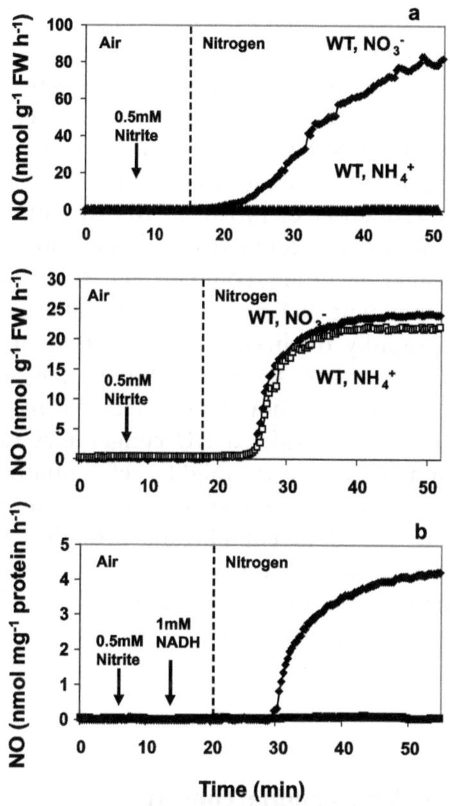

Fig. 3 **a** NO emission (measured by chemiluminescence detection) from suspensions of leaf slices or root segments of pea plants grown either with nitrate or with ammonium as N source. Only the former expressed functional NR, whereas the latter had less than 5% of the normal NR activity. All samples contained 0.5 mM nitrite. Lack of NR prevented reduction of nitrite to NO in the leaf slices, but not in the root segments. Accordingly, roots must be able to reduce nitrite to NO by means other than NR. **b** Mitochondria purified from roots, but not those from leaves of pea plants (grown on nitrate) are able to reduce nitrite to NO, thus explaining the in vivo observations in (**a**). Similar results (not shown) were obtained with mitochondria from tobacco, barley, and *Arabidopsis*. Nitrite (0.5 mM) and NADH (1 mM) were added as indicated. (From Gupta et al. 2005, modified)

green leaf cells produce NO from nitrite only via NR, which is myxothiazol-insensitive, whereas roots reduced nitrite to NO even in the complete absence of NR, indicating that practically all NO originated from the mitochondria. It should not be ignored, however, that Modolo et al. (2005) using EPR, showed that mitochondrial electron transport does also produce NO in leaves (see also Salgado's contribution in this volume). The cause for this discrepancy is not yet known.

2.3 Do Mitochondria Produce NO in Air?

As pointed out above, L-arginine-dependent NO production via NOS should require oxygen, if the reaction occurs in the same way as with NOS from animal origin. However, using chemiluminescence detection, purified mitochondria from tobacco roots gave no NO emission in air with L-arginine (unpublished results), nor with nitrite plus NADH. There are several possible explanations: Either, reduction of nitrite to NO by mitochondrial electron transport is very sensitive to (competitive) inhibition by oxygen and accordingly does not function in air (compare Fig. 3). This would, of course, not hold for NOS-dependent NO production. Or alternatively, NO is produced, but becomes rapidly oxidized (e.g., to NO_2 or to N_2O_3), either by O_2 itself or by reactive oxygen species (ROS). In the latter case, at least part of the oxidation products should be detectable by DAF-fluorescence, which appears to depend on oxidized NO congenitors rather than on NO itself. Direct oxidation of NO by O_2 might preferentially occur within the membrane lipid phase where the low water solubility of NO would provide higher concentrations of NO in equilibrium with the water phase (Shiva et al. 2001). On the other hand, mitochondrial electron transport complexes are one of the major sources for ROS in plant cells (for review see Møller 2001), and a large part of NO oxidative scavenging might occur by reaction with ROS.

3 Functions of NO in the Mitochondrial Context

Using DAF-2DA preloaded mitochondria, some fluorescence increase was observed in air, which was slightly stimulated by L-arginine addition, and which was decreased by the NO scavenger cPTIO (Fig. 4). At first sight this seems to indicate that some NOS activity was present, which produced a small amount of NO in air. However, addition of all NOS cofactors together with L-arginine, which should stimulate NO emission, almost completely prevented the increase in DAF-fluorescence. Thus, at this stage, we consider the data derived from DAF-fluorescence to be inconclusive.

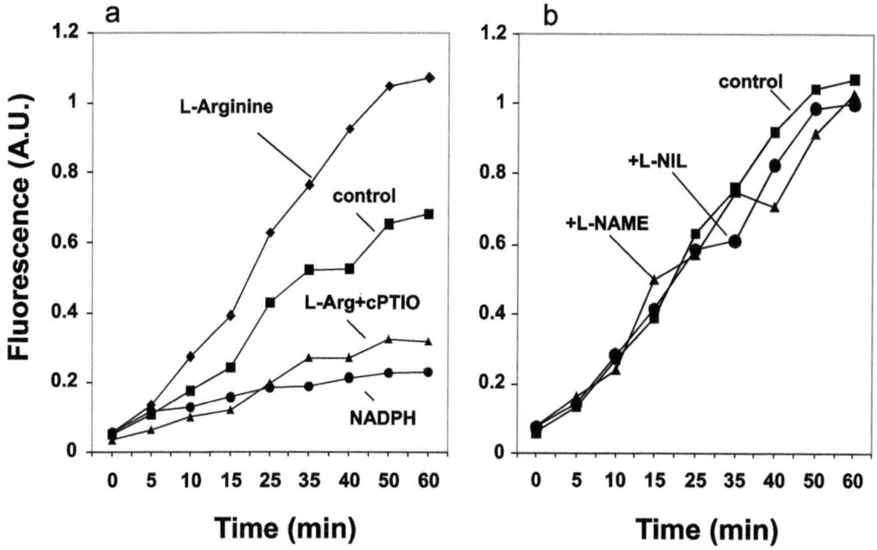

Fig. 4 L-Arginine-dependent NO production of purified mitochondria as indicated by DAF-fluorescence. **a** Mitochondria purified from barley roots were preincubated on ice for 30 min with L-arginine (2.5 mM) alone or with 200 μM cPTIO, or with 0.5 mM NADPH. **b** With 5 mM L-NIL or 5 mM L-NAME. After preincubation, mitochondria were loaded with 10 μM DAF2-DA. Aliquots were used to measure fluorescence (495 nm excitation, 515 nm emission, band width 2 nm) at the indicated times. L-Arginine causes a more rapid increase in DAF-fluorescence, which might indicate NO production by NOS. Consistent with that, the NO scavenger cPTIO (2-(4-carboxyphenyl)-4,4,5,5-tetramethylimidazoline-1-oxyl-3-oxide) almost completely prevented the fluorescence increase. On the other hand, addition of NADPH (expected to increase NOS activity) actually decreased fluorescence. Addition of the NOS inhibitors L-NAME and L-NIL (expected to abolish NO formation) had no effect on L-arginine-stimulated DAF-fluorescence. Thus, it is not yet clear whether the above DAF-fluorescence really indicates NO production by NOS

NO and its coproducts can react with many soluble and membrane-bound constituents of the mitochondria. Generally, NO reacts with transition metals, the biologically most relevant one being iron. In bacteria and in animal cells, hemoglobin play an important role in regulating NO levels. In animal cells, NO reacts with oxyhemoglobin to produce nitrate and methemoglobin. NO may also bind to the heme-Fe^{2+} of deoxyhemoglobin to produce iron-nitrosylhemoglobin. In addition, NO can react with thiols to produce *S*-nitroso-thiols. As plants contain the so-called non-symbiotic hemoglobins (nsHb), these reactions should also occur in plants. Hb-NO complexes are relatively stable and might therefore serve as "buffers" for NO, thereby extending the half-life of NO in plant cells. NsHbs appear to be located in the cytosol, because the DNA sequence of nsHbs does not have a transit signal peptide (Arredondo-Peter et al. 1997; Taylor et al. 1994;

Trevaskis et al. 1997). Therefore, in order to react with Hb, NO produced in the mitochondria has to diffuse out to the cytosol.

Aconitase is a soluble enzyme that catalyzes the reversible isomerization of citrate and isocitrate. It exists in a cytosolic and a mitochondrial form, and its iron-sulfur center has been described as a target for NO in mammals (Hentze and Kuhn 1996). NO has been shown to inhibit isocitrate to citrate conversion (Navarre et al. 2000) and may thereby affect the turnover of the citric acid cycle. NO is also a reversible inhibitor of mitochondrial electron transport and phosphorylation through its interaction with the terminal cytochrome c oxidase (cytOX) (Yamasaki et al. 2001; Millar et al. 2002 and literature cited). The inhibition appears to be competitive with oxygen (Brown and Cooper 1994). In contrast to animals, the plant mitochondrial electron transport chain contains a second terminal oxidase, the "alternative oxidase" AOX, which is NO insensitive. Electron transport from ubiquinol to AOX is not coupled and releases energy as heat. While the cytOX pathway is sensitive to myxothiazol, the AOX pathway is blocked by salicyl hydroxamate (SHAM). As described above, myxothiazol and SHAM each cause a partial inhibition of anoxic NO formation, and in combination inhibit more strongly, though not completely. These results indicate that both terminal oxidases contribute to the reduction of nitrite to NO, while only cytOX can be inhibited by NO. Thus, under hypoxia or anoxia, which induce the AOX pathway, NO may play an important role in regulating the flow of electrons through cytOX, thereby acting as amplifier for AOX induction. The above-mentioned inhibition of mitochondrial aconitase would also contribute to that effect, since it would increase levels of citrate, which is known to induce AOX (Vanlerberghe and McIntosh 1996). The resulting decrease in ATP synthesis may also have far-reaching consequences for other metabolic events outside the mitochondria. In addition, the inhibition of aconitase would decrease the availability of oxoglutarate, which is the unique substrate for ammonia assimilation (Millar et al. 2002).

Another interesting question is whether reduction of nitrite to NO by mitochondrial electron transport might provide an alternative to fermentation under anoxia. Ethanol and lactic acid are the major end products of fermentation in roots, and both are toxic if they accumulate to high concentrations. Reduction of nitrate to nitrite by cytosolic NR, and further reduction of nitrite to NO by NR or by mitochondria, might represent an alternative to fermentative NAD^+ regeneration. Indeed, we could show that roots expressing NR, which accumulate nitrite under anoxia and emit NO, produce much less ethanol and lactate and acidify their cytosol less than NR-deficient roots, which do not form nitrite and, therefore, emit no NO (Stoimenova et al. 2003). However, the measured rates of anoxic NO emission, though 1000-fold higher than NO emission in air, are still only in the range of 10–20 nmol g^{-1} FW h^{-1}, whereas fermentation rates (in terms of NADH oxidized) are several μmol g^{-1} FW h^{-1} (Stoimenova et al. 2003; Gupta et al. 2005). Thus, the role of mitochondrial electron transport from NADH to nitrite as an alternative NADH sink under anoxia seems doubtful. Under hypoxia, however, the situation might be different: it has been proposed that at low oxygen partial pressures, the "hemoglobin cycle" might serve to oxidize NO to nitrate while consuming NADH (for details see the contribution by Hill in this volume). In that case, electron flow through nitrite/NO could be much higher than expected from the measured rates of NO emission under anoxia, where the hemoglobin cycle should not work.

4
Conclusions and Perspectives

More work is required in order to find out whether the extremely low aerobic NO emission rates really reflect low NO production due to competition with oxygen at the terminal oxidases, or whether they are due to oxidative scavenging of NO. It should be noted that with purified mitochondria, the hemoglobin cycle cannot contribute to NO oxidation because non-symbiotic Hb is located in the cytosol. Thus, in mitochondria oxygen/NO competition may seem probable, but the possibility of a quick reaction of NO with ROS cannot be neglected. However, addition of catalase and SOD to purified mitochondria did not improve NO emission (Kaiser, unpublished results). In intact cells, however, the hemoglobin cycle might scavenge NO to the very low aerobic NO emission usually found. Also, it is completely unknown why leaf mitochondria, most probably having the same terminal oxidases as root mitochondria, are not able to produce NO.

The functions of mitochondrial NO are also far from being clear. The above suggestion that NO may serve to regulate respiratory electron flow through cytOX and AOX is indeed fascinating, but its physiological relevance will depend on the NO concentrations required for inhibition and those that are really reached within mitochondria (which are not yet known). Estimates of in vivo NO concentrations in plants vary widely. While our own estimates of in vivo NO concentrations (tobacco leaves) based on chemiluminescence measurements were in the picomolar or low nanomolar range (Planchet et al. 2005), previous studies using other methods gave NO concentrations that were several orders of magnitude higher (0.1–2 μM). Yamasaki et al. (2001) used NO concentrations around 50 nM to cause an inhibition of the steady state membrane potential. The $K_{0.5}$ for cytOX inhibition was also quite high, approximately 0.1–0.3 μM NO (Millar et al. 2002 and literature cited). Thus, for final conclusions on the possible functions of NO in mitochondria and elsewhere in the cell it is crucial to know the real and potential concentrations of NO in plant tissues, cells, and organelles.

Acknowledgements This work was supported by the Deutsche Forschungsgemeinschaft (DFG), SFB 567 and Ka 456-15/1-3.

References

Arredondo-Peter R, Moran JF, Sarath G, Luan P, Klucas RV (1997) Molecular cloning of the cowpea leghemoglobin II gene and expression of its cDNA in Escherichia coli. Purification and characterization of the recombinant protein. Plant Physiol 114: 493–500

Balk J, Leaver CJ, McCabe PF (1999) Translocation of cytochrome c from the mitochondria to the cytosol occurs during heat-induced programmed cell death in cucumber plants. FEBS Lett 463:151–154

Balk J, Leaver CJ (2001) The PET1-CMS mitochondrial mutation in sunflower is associated with premature programmed cell death and cytochrome c release. Plant Cell 13:1803–1818

Barroso JB, Corpas FJ, Carreras A, Sandalio LM, Valderrama R, Palma JM, Lupianez JA, del Rio LA (1999) Localization of nitric-oxide synthase in plant peroxisomes. J Biol Chem 274:36729–36733

Bartoli CG, Pastori GM, Foyer CH (2000) Ascorbate biosynthesis in mitochondria is linked to the electron transport chain between complexes III and IV. Plant Physiol 123:335–344

Bethke PC, Badger MR, Jones RL (2004) Apoplastic synthesis of nitric oxide by plant tissues. Plant Cell 16:332–341

Brown GC, Cooper CE (1994) Nanomolar concentrations of nitric oxide reversibly inhibit synaptosomal respiration by competing with oxygen at cytochrome oxidase. FEBS Lett 356:295–298

Butt YKC, Lum JHK, Lo DCL (2003) Proteomic identification of plant proteins probed by mammalian nitric oxide synthase antibodies. Planta 216:762–771

Catoni E, Desimone M, Hilpert M, Wipf D, Kunze R, Schneider A, Flügge UI, Schumacher K, Frommer WB (2003) Expression pattern of a nuclear encoded mitochondrial arginine-ornithine translocator gene from Arabidopsis. BMC Plant Biology 3:1

Crawford NM, Guo FQ (2005) New insights into nitric oxide metabolism and functions. Trends Plant Sci 10:195–200

Ghafourifar P, Richter C (1997) Nitric oxide synthase activity in mitochondria. FEBS Lett 418:291–296

Godber BLJ, Doel JJ, Sapkota GP, Blake DR, Stevens CR, Eisenthal R, Harrison R (2000) Reduction of nitrite to nitric oxide catalyzed by xanthine oxidoreductase. J Biol Chem 275:7757–7763

Gueguen V, Macherel D, Jaquinod M, Douce R, Bourguignon J (2000) Fatty acid and lipoic acid biosynthesis in higher plant mitochondria. J Biol Chem 275:5016–5025

Giulivi C, Poderoso JJ, Boveris A (1998) Production of nitric oxide by mitochondria. J Biol Chem 273:11038–11043

Guo FQ, Crawford NM (2005) Arabidopsis nitric oxide synthase1 is targeted to mitochondria and protects against oxidative damage and dark-induced senescence. Plant Cell 17:3436–3450

Guo FQ, Okamoto M, Crawford NM (2003) Identification of a plant nitric oxide synthase gene involved in hormonal signaling. Science 302:100–104

Gupta KJ, Stoimenova M, Kaiser WM (2005) In higher plants, only root mitochondria, but not leaf mitochondria reduce nitrite to NO, in vitro and in situ. J Exp Bot 56:2601–2609

He Y, Tang RH, Hao Y, Stevens RD, Cook CW, Ahn SM, Jing L, Yang Z, Chen L, Guo F, Fiorani F, Jackson RB, Crawford NM, Pei ZM (2004) Nitric oxide represses the Arabidopsis floral transition. Science 305:1968–1971

Hentze MW, Kuhn LC (1996) Molecular control of vertebrate iron metabolism: mRNA-based regulatory circuits operated by iron, nitric oxide, and oxidative stress. Proc Natl Acad Sci USA 93:8175–8182

Hill RD, Igamberdiev AU, Baron KN (2007) Nitric Oxide as an Alternative Electron Carrier During Oxygen Deprivation (in this volume). Springer, Berlin Heidelberg New York

Hoyos ME, Palmieri L, Wertin T, Arrigoni R, Polacco J, Palmieri F (2003) Identification of a mitochondrial transporter for basic amino acids in Arabidopsis thaliana by functional reconstitution into liposomes and complementation in yeast. Plant J 33:1027–1033

Kowaltowski AJ (2000) Alternative mitochondrial functions in cell physiopathology: beyond ATP production. Brazilian J Med Biol Res 33:241–250

Kozlov AV, Staniek K, Nohl H (1999) Nitrite reductase activity is a novel function of mammalian mitochondria. FEBS Lett 454:127–130

Kuo WN, Ku TW, Jones DL, Baptiste J (1995) Nitric oxide synthase immunoreactivity in baker's yeasts, lobster and wheat germ. Biochem Arch 11:73–78

Lam E, Kato N, Lawton M (2001) Programmed cell death, mitochondria and the plant hypersensitive response. Nature 411:848–853

Lo DCL, Butt YKC, Chan YSG (2000) False nitric oxide synthase immunoreactivity in *Asparagus* Bean (*Vigna sesquipdalis*). Nitric Oxide: Biol Chem 4:175

Meyer C, Stöhr C (2003) Soluble and plasma membrane-bound enzymes involved in nitrate and nitrite metabolism. In: Foyer C, Noctor G (eds) Photosynthetic nitrogen assimilation and associated carbon and respiratory metabolism. Advances in photosynthesis and respiration, vol 12. Kluwer, Dordrecht, The Netherlands, pp 49–62

Millar AH, Day DA, Mathieu C (2002) Nitric oxide synthesis by plants and its potential impact on nitrogen and respiratory metabolism. In: Foyer C, Noctor G (eds) Photosynthetic nitrogen assimilation and associated carbon and respiratory metabolism. Advances in photosynthesis and respiration, vol 12. Kluwer, Dordrecht, The Netherlands, pp 193–204

Millar TM, Stevens CR, Benjamin N, Eisenthal R, Harrison R, Blake DR (1998) Xanthine oxidoreductase catalyses the reduction of nitrate and nitrite to nitric oxide under hypoxic conditions. FEBS Lett 427:225–228

Modolo LV, Augusto O, Almeida IMG, Magalhaes JR, Salgado I (2005) Nitrite as the major source of nitric oxide production by *Arabidopsis thaliana* in response to *Pseudomonas syringae*. FEBS Lett 579:3814–3820

Møller IM (2001) Plant mitochondria and oxidative stress: electron transport, NADPH turnover and metabolism of reactive oxygen species. Annu Rev Plant Physiol Plant Mol Biol 52:561–591

Navarre DA, Wendehenne D, Durner J, Noad R, Klessig DF (2000) Nitric oxide modulates the activity of tobacco aconitase. Plant Physiol 122:573–582

Planchet E, Gupta KJ, Sonoda M, Kaiser WM (2005) Nitric oxide emission from tobacco leaves and cell suspensions: rate limiting factors and evidence for the involvement of mitochondrial electron transport. Plant J 41:732–743

Rébeillé F, Macherel D, Mouillon JM, Garin J, Douce R (1997) Folate biosynthesis in higher plants: purification and molecular cloning of a bifunctional 6-hydroxymethyl-7,8-dihydropterin pyrophosphokinase/7,8-dihydropteroate synthase localized in mitochondria. EMBO J 16:947–957

Ribiero EA, Cunha FQ, Tamashino WMSC, Martins IS (1999) Growth phase-dependent subcellular localization of nitric oxide synthase in maize cells. FEBS Lett 445:283–286

Rockel P, Strube F, Rockel A, Wildt J, Kaiser WM (2002) Regulation of nitric oxide (NO) production by plant nitrate reductase in vivo and in vitro. J Exp Bot 53:103–110

Salgado I, Modolo LV, Augusto O, Braga MR, Oliveira HC (2007) Mitochondrial Nitric Oxide Synthesis During Plant–Pathogen Interactions: Role of Nitrate Reductase in Providing Substrates (in this volume). Springer, Berlin Heidelberg New York

Sen S, Cheema IR (1995) Nitric oxide synthase and calmodulin immunoreactivity in plant embryonic tissue. Biochem Arch 11:221–227

Shiva S, Brookes PS, Patel RP, Anderson PG, Darley-Usmar VM (2001) Nitric oxide partitioning into mitochondrial membranes and the control of respiration at cytochrome c oxidase. Proc Natl Acad Sci USA 98:7212–7217

Stoimenova M, Libourel IGL, Ratcliffe RG, Kaiser WM (2003) The role of nitrate reduction in the anoxic metabolism of roots. II. Anoxic metabolism of tobacco roots with or without nitrate reductase activity. Plant Soil 253:155–167

Stöhr C (2007) Nitric oxide – a product of plant nitrogen assimilation. In: Lamattina L, Polacco JC (eds) Nitric oxide in plant growth. Plant Cell Monographs, vol 6. Springer, Berlin Heidelberg New York (in press)

Stöhr C, Strube F, Marx G, Ullrich WR, Rockel P (2001) A plasma membrane-bound enzyme of tobacco roots catalyses the formation of nitric oxide from nitrite. Planta 212:835–841

Taylor ER, Nie XZ, MacGregor AW, Hill RD (1994) A cereal haemoglobin gene is expressed in seed and root tissues under anaerobic conditions. Plant Mol Biol 24:853–862

Tischner R, Planchet E, Kaiser WM (2004) Mitochondrial electron transport as a source for nitric oxide in the unicellular green alga *Chlorella sorokiniana*. FEBS Lett 576:151–155

Trevaskis B, Watts RA, Andersson C, Llewellyn D, Hargrove MS, Olson JS, Dennis ES, Peacock WJ (1997) Two hemoglobin genes in *Arabidopsis thaliana*: the evolutionary origins of leghemoglobins. Proc Natl Acad Sci USA 94:12230–12234

Vanlerberghe GC, McIntosh L (1996) Signals regulating the expression of the nuclear gene encoding alternative oxidase of plant mitochondria. Plant Physiol 111:589–595

Yamasaki H, Shimoji H, Ohshiro Y, Sakihama Y (2001) Inhibitory effects of nitric oxide on oxidative phosphorylation in plant mitochondria. Nitric Oxide: Biol Biochem 3:261–270

Zeidler D, Zahringer U, Gerber I, Dubery I, Hartung T, Bors W, Hutzler P, Durner J (2004) Innate immunity in *Arabidopsis thaliana*: Lipopolysaccharides activate nitric oxide synthase (NOS) and induce defense genes. Proc Natl Acad Sci USA 101:15811–15816

Plant Cell Monogr (6)
L. Lamattina · J. C. Polacco: Nitric Oxide in Plant Growth
DOI 10.1007/7089_2006_082/Published online: 29 November 2006

Nitric Oxide – A Product of Plant Nitrogen Metabolism

Christine Stöhr

Institut für Botanik, Ernst-Moritz-Arndt-Universität, Grimmer Strasse 88,
17487 Greifswald, Germany
stoehr@uni-greifswald.de

Abstract Nitric oxide is an intermediate product of inorganic nitrogen assimilation. In plants, it can be formed either by reducing inorganic nitrogen by the nitrite-dependent pathway or by oxidation of organic nitrogen by the arginine-dependent pathway. Both pathways require adequate nitrogen supply to the plant and may not operate under nitrogen deficiency. However, the pathways are differently regulated in relation to oxygen availability and, therefore, have a different importance for underground organs like roots, than for above-ground organs like the shoot.

1
Introduction

A confusing bulk of information is available about possible functions and synthesis of nitric oxide (NO) in plants. New roles played by NO in plant systems are identified constantly. It has been suggested that NO plays important roles in such diverse physiological processes as growth and development, plant disease resistance, abiotic stress, and signal transduction in above and underground plant organs (see selected reviews: Lamattina et al. 2003; Neill et al. 2003; Wendehenne et al. 2004; Shapiro 2005; Crawford and Guo 2005; Lamotte et al. 2005; Crawford 2006). On the cellular level NO was proven to have definite roles in various compartments, such as cytoplasm, mitochondria, peroxisomes, and chloroplasts and as an "extracellular compartment" the apoplast. Specificity demands subcellular targeting or generation of NO.

Formation of NO by plants is necessarily closely linked to nitrogen assimilation and metabolism, since it is produced from inorganic or organic nitrogen sources. Either reduction of the oxidized form forms nitrate via nitrite, or a five-electron oxidation of reduced nitrogen in the form of the guanidine nitrogen of the amino acid L-arginine can lead to release of nitric oxide. Nitrogen is the mineral nutrient required in the highest amounts by plants and is most frequently limiting to growth and yield.

For most plants, nitrate is the inorganic nitrogen source available to roots, especially in temperate agricultural soils (Cookson et al. 2005). Once taken up by the root system, nitrate can be reduced to nitrite in the cytosol, stored in the vacuole, or transported to the shoot. Nitrate incorporation into biological molecules such as amino acids and amino acid-derived molecules (Fig. 1) involves reduction of nitrate by nitrate reductase (cNR) in the cytosol. Under normal growing conditions the resulting nitrite is further reduced to

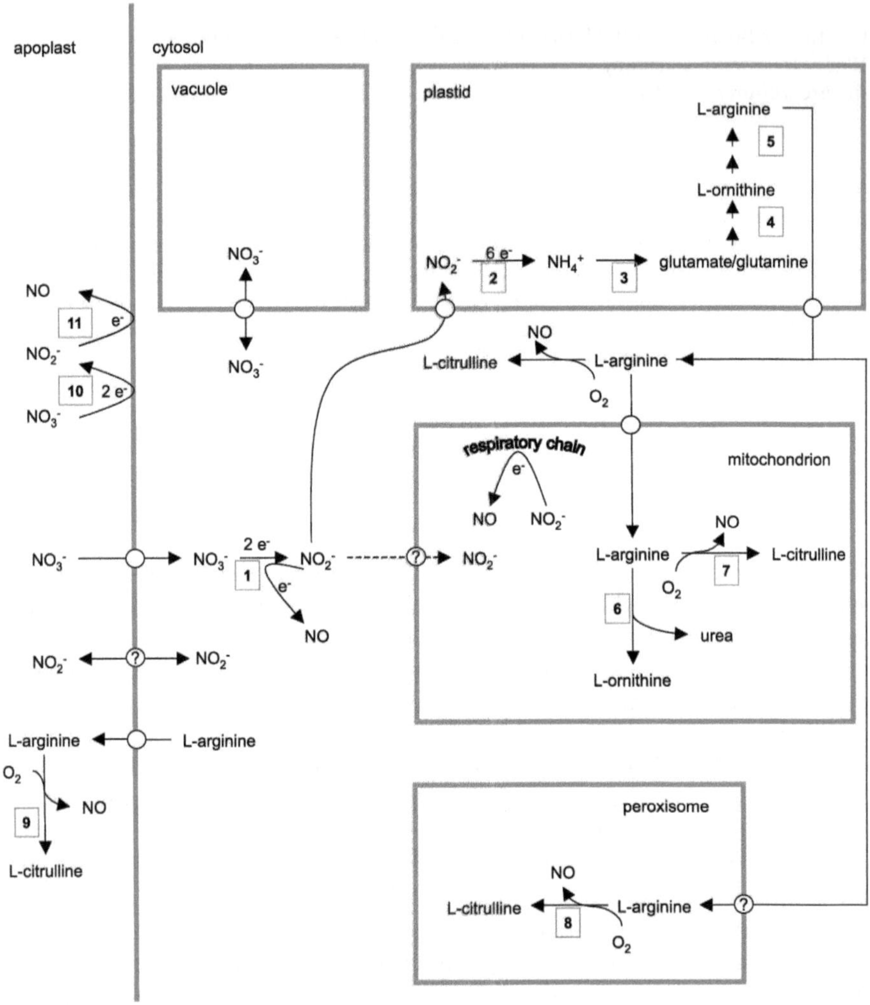

Fig. 1 NO-forming pathway in a root cell in relation to nitrogen assimilation. Enzymes that catalyze the indicated reactions are: *1* cytosolic nitrate reductase; *2* nitrite reductase; *3* glutamine synthetase and glutamine-2-oxoglutarate aminotransferase; *4* the "ornithine pathway": *N*-acetylglutamate synthase, *N*-acetylglutamate kinase, *N*-acetylglutamate-5-P ductase, *N*2-acetylornithine aminotransferase, *N*2-acetylornithine:glutamate acetyltransferase; *5* the "arginine-pathway": ornithine transcarbamoylase, argininosuccinate synthase, argininosuccinate lyase; *6* arginase; *7* mitochondrial nitric oxide synthasel; *8* peroxisomal nitric oxide synthase; *9* apoplastic nitric oxide synthase; *10* plasma membrane-bound nitrate reductase; *11* nitrite:NO reductase

ammonia in the plastid by nitrite reductase (NiR). In both the chloroplasts and the non-photosynthetic plastids, reduced ferredoxins supply the necessary six electrons (Matsumura et al. 1997; Emes and Neuhaus 1997). Only if the toxic nitrite accumulates may it serve as a second substrate for cNR, to become reduced to NO. Cytosolic NR holds a key position in the nitrate assimilation pathway and is under complex regulation at both transcriptional and posttranscriptional level. Yet, nitrate is not only a nutrient. It serves also as signal for rapid changes in metabolism, which include the induction of the synthesis of nitrate assimilatory enzymes and the shift from starch biosynthesis to the production of organic acids to assimilate ammonium (for review see Crawford 1995; Stitt 1999; Foyer et al. 2003). More than 1000 genes are found to respond to low levels of nitrate after only 20 min (Wang et al. 2003). As a direct descendant of nitrate, NO may even trigger some of its effects.

This report does not review the characteristics of the different NO-producing enzymes found in plants nor summarize all of the putative functions of NO (the reader is referred to the excellent reviews available [see above selection] and to the other chapters in this book). This review will focus exclusively on NO production and specificity in regard to nitrogen assimilation and metabolism as well as to oxygen availability.

2
Nitrite as Substrate for Plant NO Formation

Nitrite and NO are intermediates of nitrate assimilation in plants and are also products in bacterial nitrification and denitrification processes (Stewart 1988). Nitrite accumulation in the soil may occur while microbial nitrite oxidation is inhibited by nitrifying and denitrifying bacteria (Burns et al. 1996). In fact, much of the natural NO emission from soil was originally deduced to be of microbial origin and the possible contributions of plants were largely neglected (reviewed by Stöhr and Ullrich 2002).

Nitrite serves as substrate for plant cell NO formation in apoplast, cytosol, and mitochondria (Fig. 1). To reduce nitrite to NO an accumulation of nitrite is necessary, relative to the various enzyme affinities. Generally, cell nitrite concentration is kept very low because of its toxic properties (Sinclair 1987). It is a strong oxidant in neutral and acidic solutions (Hinze and Holzer 1985) and is withdrawn by immediate reduction by nitrite reductase to ammonia in the plastid. Two possibilities allow transport of cytoplasmically formed nitrite to the plastid. The nitrite anion exists in equilibrium with the protonated form, nitrous acid (pK_a 3.1 – 3.5), setting up a small concentration of HNO_2 under physiological conditions (Yamasaki 2000). This would allow a free diffusion across membranes, as proposed by Shingles and coauthors (1996). Alternatively, nitrite might be transported as an anion by a saturable nitrite transporter (Brunswick and Cresswell 1988a,b). Only in *Chlamydomonas* has the plastid transport protein Nar1 been identified (Rexach et al. 2000), but homologs have been found in *Arabidopsis*. Yet, the function of Nar1 orthologs as nitrite transporters in higher plants has to be established. Due to nitrite reductase (e.g., K_m of 0.3 mM for spinach NiR, Bellissimoa and Privalle 1995) the plastidic nitrite concentration is estimated to be in the submillimolar range (Yamasaki 2000).

For the most part, nitrite accumulation in plant cells has been only noted in plants supplied abundantly with nitrate, when nitrite reduction to ammonia is limited. Nitrite accumulates in the cytosol if reduction equivalents are not available due to inhibited photosynthesis or to reduced respiration combined with a delayed negative regulation of nitrate reductase, usually via phosphorylation (Kaiser and Huber 2001). In leaves, nitrate reductase is rapidly inactivated in the dark or when CO_2 is removed (for a recent review the reader is referred to Meyer et al. 2005) leading to a transient nitrite accumulation. Under natural conditions a sudden CO_2 limitation in photosynthesis caused by stomatal closure or reduced light conditions due to clouding may be events causing short-term nitrite build-up in leaves.

In roots, nitrite accumulates upon anaerobiosis (Botrel et al. 1996), whereas hypoxia of leaves has been only reported in aquatic plants (Schlüter and Crawford 2001). Roots of higher plants might be frequently exposed to fluctuations in oxygen availability in their local environment. These variations can range from 21 kPa, which is the value of the pO_2 in air, and represents a maximum probably never met even in well aerated soils, to values close to zero in flooded soils (Saglio et al. 1984). In contrast to marsh plants, internal O_2 transport in non-adapted mesophytes plays only a limited role (Vartapetian et al. 1978; Saglio et al. 1983) and cannot meet the respiratory requirements of buried organs, which draw most of their O_2 from the rhizosphere (Saglio et al. 1984). Under such conditions nitrite accumulates in root cells and is secreted into the rhizosphere (Botrel and Kaiser 1997), probably by-passing nitrate transporters in the plasma membrane. In *C. reinhardtii*, four high affinity nitrate/nitrite transporters have been described (Rexach et al. 1999 and references therein).

In higher plants, nitrite has been found to inhibit nitrate influx in a competitive manner, which suggests that both ions share at least some transport systems (Siddiqi et al. 1992). Nitrite influx and efflux across the plasma membrane may involve a combination of nitrous acid and nitrite ions (Meyer et al. 2005). In addition, nitrite can be formed enzymatically in root apoplast. A plasma membrane-bound nitrate reductase (PM-NR) reduces apoplastic nitrate with succinate as electron donor (Stöhr and Ullrich 1997) in the root apoplast. This enzyme activity is diurnally regulated (Stöhr and Mäck 2001) and highly influenced by external nitrate availability (Stöhr 1999). So far, transport or formation of nitrite in mitochondria remains unclear. Nitrite as a source for nitric oxide might be important mainly for roots in transient hypoxic environments under sufficient nitrate nutrition.

3
L-Arginine as Substrate for NO Formation

L-Arginine-dependent NO formation is catalyzed by NO synthases in various locations in the cell, using NADPH and molecular oxygen as cosubstrates and employing calmodulin as cofactor (see review by Crawford 2006). Besides mitochondrial *At*NOS1 (Guo and Crawford 2005) the protein nature of the enzymes is unknown and they are mainly identified by their activity (see review by Crawford 2006). The biosynthesis of L-arginine (Fig. 1) is primarily dependent on ammonia and, ultimately, on the inorganic nitrogen supply of the plant. Ammonia is assimilated by glutamine synthetase and glutamine-2-oxoglutarate aminotransferase (GS/GOGAT cycle) into the organic forms, glutamine and glutamate serving as nitrogen donors in the biosynthesis of essentially all amino acids (Coruzzi 2003). Glutamine and glutamate can then be used to form aspartate and asparagine, and these four amino acids are used to translocate organic nitrogen from sources to sinks (Lea and Miflin 1980; Peoples and Gifford 1993).

For L-arginine synthesis, two distinct processes are necessary (reviewed by Slocum 2005) leading first to the synthesis of L-ornithine from glutamate, and second to the synthesis of L-arginine from L-ornithine via the urea cycle in plants. Prediction and subcellular fractionation data indicate plastid localization for all arginine pathway enzymes (Slocum 2005). Plastid-localized basic amino acid transporters permit exchange of arginine from the plastid to the cytosol. Catabolism of arginine seems to occur in the mitochondria (Goldraij and Polacco 2000), where two amino acid transporters have been proven to be involved in the exchange of L-arginine (and also L-lysine, L-ornithine, and L-histidine in order of decreasing affinity) between cytoplasm and mitochondria (Catoni et al. 2003; Hoyos et al. 2003). These data point out that arginine might be present in mitochondria where NO formation from L-arginine has been observed. Yet, it remains unclear whether and how L-arginine is delivered to peroxisomes.

As an essential amino acid for protein synthesis and substrate for NOS, arginine has even more roles in plant metabolism. It is the precursor of the diamine putrescine from which polyamines and many important plant alkaloids are derived (reviewed by Slcocum 2005). In turn, polyamines induce NO-biosynthesis in *Arabidopsis* seedlings (Tun et al. 2006). The NO-generating activity of added compounds increased from arginine to putrescine, spermidine, and spermine with an as-yet unknown mechanism (Tun et al. 2006). Most interestingly, the presence of free arginine might also inform about the global nitrogen status of plant tissues. A universal mechanism has been detected by which the activity of *N*-acetylglutamate kinase, the second enzyme of the ornithine pathway and therefore also in arginine synthesis (Fig. 1), may be modulated in response to arginine availability in concert with PII proteins (Chen et al. 2006). These are highly conserved signal transduction proteins involved in sensing the carbon and nitrogen status of cells. PII targets include transcriptional regulators and enzymes of nitrogen metabolism in bacteria and plants (Burillo et al. 2004; Maheswaran et al. 2004).

The concentration of L-arginine in plant cells obviously depends on the developmental state. Its function as a storage amino acid is well known. Since it contains four N atoms per six C atoms, arginine may represent as much as 30% of the total nitrogen in seed storage proteins (Van Etten et al. 1963). It is the most abundant free amino acid in the cotyledons of pea seeds, where it is catabolized as a nitrogen source during the early stages of germination (de Ruiter and Kollöffel 1985). Under environmental conditions, significant changes in the content of various free amino acids in all examined plant parts during the course of a year can be followed. Their content rises in autumn, remains stable during winter and declines quickly at the beginning of spring. The most abundant amino acids in the end of winter storage period – asparagine, arginine and glutamine – made up about 90% of nitrogen in the fraction of free amino acids (Gloser 2002). This also includes a high availability of arginine for NO production under conditions of germination and primary developmental stages.

Arginine as a signaling molecule for high nitrogen availability might be only available for NO production in plants well supplied with nitrogen or under certain developmental conditions when storage proteins are degraded. Together with the view on nitrite, this points to NO as a signal molecule only for plants with good nitrogen nutrition, which is not always the case in a natural environment.

4 Subcellular Location of NO

Cellular formation of NO in plant cells has been followed using fluorescent dyes such as the compounds of the diaminofluorescein (DAF) group. Formation of the fluorescent dyes is not reversible and only reflects a build-up of NO over time. It cannot reveal fluctuations in NO concentrations. Although control experiments were mostly run, these observations have to be judged carefully since DAF compounds are rather unspecific (Stöhr and Stremlau 2006). Moreover they react strongly with ascorbate present in plant tissue. In photosynthetic tissue, intracellular concentrations of ascorbate were estimated at 1–10 mmol g^{-1} FW (Noctor 2006) facilitating chemical reduction of nitrite by ascorbate to yield NO. Beside fluorescent dyes, subcellular synthesis was indicated by the presence of NO-forming enzymes in various organelles. Quantitative measurements of NO concentrations in plant cells or even organelles are still lacking.

Nitric oxide as a non-polar molecule is supposed to cross membranes without restriction. Yet, by unlimited diffusion of NO passing all membranes, the questions arise: how can the specificity of the NO signal be achieved and, moreover, how are the dangerous and toxic reactions of NO restricted? Toxicity of NO is a consequence of its reactivity with transition metal proteins and oxygen and of its ability to form adducts with amines and thiols of varying stability (Van der Vliet et al. 1998). Considering the behavior of NO in membranous environment it has to be taken into account that membranes are not homogenous lipid layers. Moreover, they consist of different areas caused mainly by varying lipid compositions (Meder and Simons 2005; Bérczi and Horvath 2003). Because of their physical properties, NO cannot easily traverse membranes as often assumed. When membranes are in the gel phase, no significant membrane penetration was observed for NO and nor for O_2. In the fluid phase, the transmembrane profiles of NO and O_2 are similar, but that of NO is less steep and shifted towards the center of the membrane, relative to O_2 (Nedeianu et al. 2004). NO seems to be trapped in the hydrophobic core of the membrane. This might explain the observation that fluid-phase membranes were also strong barriers to NO transport, whereas sterols significantly increased NO diffusion (Subczynski et al. 1996). This points to the diffusion coefficient of NO being highly dependent on membrane composition, which is flexible and dynamic (Meder and Simons 2005).

Also, NO reacts differently within membranes in comparison to the aqueous environment. Autoxidation proceeds about 240-fold faster in membranes than in aqueous phases (Lancaster 2000; Shapiro 2005). Whereas NO terminates lipid peroxidation in aqueous medium, it induces lipid peroxidation in a non-aqueous environment (Hiramoto et al. 2003). This might explain conflicting observations on the effects of NO on lipid peroxidation. Protective effects of NO on lipid peroxidation were explained by terminating the radical chain reaction by the reaction of NO with the lipid peroxyl radical (O'Donnell et al. 1997, 1999). However, NO itself can drive lipid peroxidation (Hiramoto et al. 2003).

Possible trapping of NO in the hydrophobic core of membranes, as well as probable chemical reactions with lipids or proteins, raises the question of whether NO is as mobile as usually assumed. Regarding the well-ordered subcellular organization of NO formation (Fig. 1) it seems that a certain local available amount of NO is responsible for its correct function.

4.1 Cytosol

Ascertaining the cytosolic location of proteins is difficult, since the cytoplasm cannot be isolated as easily as chloroplast or mitochondria. Visualized by fluorescence microscopy, NO production in the cytosol has been assumed to be mediated by the cytosolic nitrate reductase. In vitro NO production by purified cNR with NADH as electron source was measured by Yamasaki et al. (1999) and confirmed for different species (reviewed by Meyer et al. 2005). *In planta* NO production by cNR is dependent on enzyme activity and the availability of nitrite and reduction equivalents. The expression of cNR is induced by small amounts of nitrate, and the activity of the protein is altered in response to changes in environmental conditions, such as light, dark, anoxia, pH, and carbon dioxide concentration (for review see Kaiser et al. 1999, 2001; Kaiser and Huber 2001; Meyer and Stitt 2001; Stitt et al. 2002).

As mentioned before, nitrite accumulation in leaves can be achieved during hypoxia or abrupt darkness, resulting in a more immediate cessation of photosynthetic electron transport than in down-regulation of nitrate reductase activity (Kaiser et al. 2002). The K_m of cytosolic nitrate reductase for nitrite ranges from 100 to 300 μM. Besides, nitrate competitively inhibits nitrite reduction by cNR with a K_i of 50 μM for nitrate (Rockel et al. 2002). This implies that only under conditions when nitrite accumulates to concentrations far above those of nitrate, will enzymatic reduction of nitrite to NO occur. However, cytosolic nitrate concentration is maintained at a constant level under many environmental conditions (Miller and Smith 1996; Cookson et al. 2005). A large range of values have been reported for cytosolic nitrate concentrations, but microelectrode measurements suggest that in mature root cells this parameter is regulated at a value independent of changes in the external concentration (Miller and Smith 1996). The vacuolar nitrate pools change with the external nitrate supply and this store is remobilized to maintain the cytosolic concentration of nitrate (van der Leij et al. 1998). In *Arabidopsis* mesophyll cells, cytosolic nitrate was maintained at approximately 1.5 and 2.0 mM during the light and dark treatment, respectively (Cookson et al. 2005). Different from leaves, the roots of many plant species commonly meet hypoxic environmental conditions. Then, cNR activity increases (Stoimenova et al. 2003) and as a consequence, the usually low cytosolic nitrite content builds up and nitrite is excreted into the apoplast (Botrel et al. 1996). Under these conditions NO is formed by cNR.

4.2 Apoplast

The apoplast is an important space for storage, mineral nutrition, certain enzyme activities, stress reactions, and defence responses (Sattelmacher 2001). It consists of an aqueous phase, which (apart from the xylem) is usually a rather thin film adjacent to or within the cell walls. Whereas most transport in and out of cells has to cross this aqueous film and is influenced by its ionic milieu, in the non-aqueous part gas exchange takes place (Felle 2005). Besides secretion from the cytosol, nitrite in the root apoplast can also be formed by local nitrate reduction by plasma membrane-bound nitrate reductase (Stöhr and Ullrich 1997). A root-specific form of PM-NR uses succinate as electron donor and is highly regulated by external nitrate availability (Stöhr 1999) and light (Stöhr and Mäck 2001) in a manner different from cytosolic NR. It is still under discussion whether this enzyme may act as nitrate sensor in combination with exudated succinate, reflecting the carbon status of a cell (Meyer and Stöhr 2002).

Nitrite as substrate is used by root-specific plasma membrane-bound nitrite:NO reductase (NI-NOR), whose nitrite-reducing activity markedly differs from that of cytosolic NR (Stöhr et al. 2001). It does not use reduced nicotine adenine nucleotides; instead reduced cytochrome c can serve as electron donor in vitro. However, a participation of cytochrome c at the plasma membrane in vivo seems unlikely and the physiological electron donor has not been identified. One of the most prominent differences from cNR is its high affinity for nitrite, ranging between 2 and 30 μM (unpublished data). Results of solubilization studies suggest a tight association between NI-NOR and PM-NR (Meyer and Stöhr 2002) indicating a step-by-step reduction of nitrate via nitrite to NO.

Under certain conditions, as an acidic pH and the presence of reduced antioxidants (ascorbate) as are met in barley aleurone layers, NO might be formed via non-enzymatic reduction of apoplastic nitrite (Bethke et al. 2004).

NOS activity in the apoplast has been reported by Zhang and coauthors (2003) and can be blocked by an arginine substrate analog. The authors report a putative activating lipid signal for apoplastic NO production, which might derive from enzymatic breakdown of plasma membrane lipids during the HR response, and probably also during wounding (Shapiro 2005). The question arises: to what extent are the substrates arginine and NADPH available in the apoplast? It is known that amino acids are secreted in the rhizosphere and are important for root colonization by mircoorganisms or fungi (Simons et al. 1997). Arginine was frequently detected in root exudates of various species and may therefore also serve as substrate for NOS in the apoplast (Simons et al. 1997; Phillips et al. 2006). Cell-wall-bound malate dehydrogenase has been early identified as providing NADH at the expense of secreted malate (McNeil et al. 1984) and recently confirmed by cell wall proteome analysis (Zhu et al. 2006). Also, NADH has been detected in solutions obtained from the cell wall free space by a low-speed centrifugation technique (Shinkle et al. 1992) and may serve as electron donor. However, the apoplastic presence and use of NADPH for NO production has to be demonstrated.

4.3 Mitochondria

Nitric oxide affects mitochondrial functionality in plant cells and reduces total cell respiration due to strong inhibition of the cytochrome pathway. The residual respiration depends on the alternative pathway and de novo synthesis of alternative oxidase. These adjustments are associated with depolarization of the mitochondrial membrane potential and release of cytochrome c from mitochondria and will finally lead to induction of about 20% of cell death (Zottini et al. 2002). Thus, it is surprising that mitochondria of all organelles seem to be the major NO producing site in plant cells.

NO formation by mitochondria is both oxygen dependent and independent, with arginine or nitrite, respectively, as substrates. Enzymatic reduction of nitrite was first observed by mitochondria of mammals (Kozlov et al. 1999). It is now proved that the unicellular green alga *Chlorella sorokiniana* (Tischner et al. 2004) and tobacco suspension cells reduce nitrite at the expense of NADH under anoxic conditions, but not in air (Planchet et al. 2005). Obviously, in all higher plants, only root mitochondria, but not leaf mitochondria are able to reduce nitrite to NO, both in vitro (isolated mitochondria) and in situ (Gupta et al. 2005). From inhibitor experiments, it was concluded that electrons from the mitochondrial electron transport chain are transferred to nitrite by cytochrome c oxidase as well as by alternative oxidase. However, the authors cannot exclude the participation of further enzymes.

Production of NO from L-arginine by mitochondrial NOS has been reported in roots and leaves. So far *At*NOS1 is the only protein that has been discovered as NOS in plants (Guo and Crawford 2005). It is targeted to mitochondria where it is required for arginine-dependent NO synthesis. Analysis of a knockout mutant demonstrated that it is responsible for 75% of the basal NOS activity in *Arabidopsis* leaf extracts and most of the ABA-induced NOS activity in roots.

4.4 Peroxisomes

As with the plastid, the role of peroxisomes is different in leaves and roots. Functions of peroxisomes in plant cells are photorespiration, β-oxidation of fatty acids, glyoxylate cycle, metabolism or ureides, and metabolism of ROS (del Río et al. 2002). Peroxisomes contain NOS activity with L-arginine, oxygen, and NADPH as substrates (Barroso et al. 1999; Corpas et al. 2001, 2004; Prado et al. 2004). It was originally discovered in pea seedlings and requires all the cofactors and cosubstrates, as they are necessary for the mammalian enzyme (Barroso et al. 1999). A regulatory role in the glyoxylate cycle, photorespiration, or oxidative catabolism is assumed, but it has not yet been confirmed (Corpas et al. 2001). Beside mitochondria, peroxisomes could be a major site of L-arginine-dependent NO synthesis in plants (Corpas et al. 2001); however, so far no data are available concerning the presence and transport of L-arginine in peroxisomes.

4.5 Plastid

Despite the fact that all various substrates (nitrite, arginine, reduction equivalents, oxygen) for NO formation are present in the plastid, it seems not be a major source of NO in plant cells, maybe because it is particularly sensitive towards NO. With less NO concentrations than required for inhibition of mitochondrial respiration, decline of net photosynthesis is achieved (Saxe 1986; Wodala et al. 2005). Yet, NO formation in the chloroplast has been observed with DAF (Foissner et al. 2000; Gould et al. 2003) and the effect of NOS-inhibitors points to arginine-derived activity in chloroplasts. As mentioned above, the use of DAF has to be judged carefully. Different studies following lipopolysaccharide stimulation of NO formation state clearly that NO was absent from the chloroplast (Zeidler et al. 2004). The involvement of plastidic nitrite reductase (NiR) in NO_x formation was studied in transgenic tobacco plants that express an antisense NiR construct (Vaucheret et al. 1992) and have very low NiR activities, and hence accumulate nitrite (Goshima et al. 1999). In these plants emission of N_2O occurred, but not in the wild type or in transgenic plants grown on ammonium. When NR activity was blocked, no evolution of N_2O was found (Goshima et al. 1999). In vivo NO production by NiR seems to be negligible.

Plastids function not only as the site of photosynthesis and carbon fixation but also as the location of many metabolic biosynthetic pathways, including assimilation of NO_2^- into organic forms, and synthesis of nucleotides and amino acids. As such, the chloroplast is the primary site for the interface of carbon and nitrogen metabolism and is also the compartment where plant PII resides (see above). With that, it might not be a major source of NO formation, but it contains the major prerequisites.

5
Oxygen – The Critical Factor in NO Formation and Action

Regarding the different enzymes for NO formation in plants and their substrates, nitrite together with various reduction equivalents or L-arginine together with oxygen and NADPH, it becomes obvious that NO formation seems to be regulated differently according to the oxygen availability. This may not necessarily concern the final subcellular NO concentration rather than the kinetics of formation, eventually leading to specific "NO signatures" (following the term used for Ca^{2+} signaling). Although discussed controversially (McAinsh and Hetherington 1998; Plieth 2005), specific Ca^{2+} signatures exist for various environmental signals, regarding length, amplitude, and frequency of Ca^{2+} oscillation in the cytosol. To achieve these fluctuations, Ca^{2+} buffers are required that quickly bind free Ca^{2+}. Similarly NO buffer substances seem to exist in the form of plant hemoglobins, as they have been found in the nucleus and cytoplasm (Seregélyes et al. 2000; reviewed by Perazzolli et al. 2006). Together with further reaction partners, cell NO concentration is probably highly controlled. These chemical reactions differ in their dependence on oxygen partial pressure.

Nitric oxide itself is not a particularly reactive molecule nor it is highly toxic (Brunelli et al. 1995; Beckman 1996). The reactivity of NO is comparable with that of molecular oxygen, and like molecular oxygen it becomes toxic by conversion to more strongly oxidizing species. The most pronounced chemical property of NO is an unpaired electron, leading to a high reactivity with O_2 and O_2^- and with several nitrogen compounds. The rate of reaction of NO with O_2 is second-order in NO and first-order in O_2, independent of pH (between pH 4.9 and 7.4) and leads to nitrite; no nitrate was detected (Lewis and Deen 1994). Shapiro (2005) estimated a high diffusion rate for 1 μM NO in air-saturated water. It was 3400 $\mu m^{2\ -1}$ for plant leaves demonstrating that autoxidation would hardly limit NO action within one single cell. However, the combined unpaired radical electrons on NO and the reactive oxygen species superoxide form a stable bond to produce the peroxynitrite anion.

Peroxynitrite is not a free radical and is stable in alkaline solution. It has a pK_a of 6.8 and can decay to produce hydroxylradical, nitrogen dioxide, and bicarbonate radical. The hydroxylradical is a very strong oxidizing species and can rapidly attack biological membranes and all types of biomolecules, such as DNA and proteins, leading to irreparable damage, metabolic dysfunction, and cell death (del Rio et al. 2003). By itself, peroxynitrite is responsible for tyrosine nitration and oxidation of thiol residues to sulphenic and sulphonic acids (Lamattina et al. 2003). Radical oxygen species are produced in mitochondria and peroxisomes, and there, reaction with NO is likely to occur (Vanin et al. 2004; del Río et al. 2002). Yamasaki (2005) included sulphur radicals in this view of oxygen and nitrogen radicals in the combined "ONS" hypothesis to demonstrate that only the balance of all different radicals facilitate the correct operation of plant systems.

The NO chemistry changes in absence of oxygen. NO reacts directly with metal complexes or with thiols in an alkaline environment to yield disulfide and N_2O (Williams 2004). The reaction of NO with secondary amines leads to the formation of nitrosamines, and that with aromatic amines leads to deamination. An example is the irreversible deamination of deoxynucleotides in DNA, thereby causing point mutations. Yet, it also leads to S-nitrosylation of proteins (also in the presence of oxygen), which are discussed as a posttranslational modification to activate or inhibit protein activity (Lindermayr et al. 2005).

Since the NOS-type enzymes (L-arginine-dependent) need oxygen for NO production, maybe more NO-producing enzymes (nitrite-reducing-type) that remain active under oxygen deprivation had to be developed in plant roots. Because of the higher oxygen concentration in green tissue originated by photosynthesis and by direct contact to ambient air, the NOS-type enzymes may function preferentially in the shoot (Fig. 2). The nitrite-reducing-type in mitochondria (Gupta et al. 2005) and that present in apoplast (Stöhr and Stremlau 2006) are both root-specific. They are not only independent of oxygen but rather they are reversibly inhibited by oxygen in a dose-

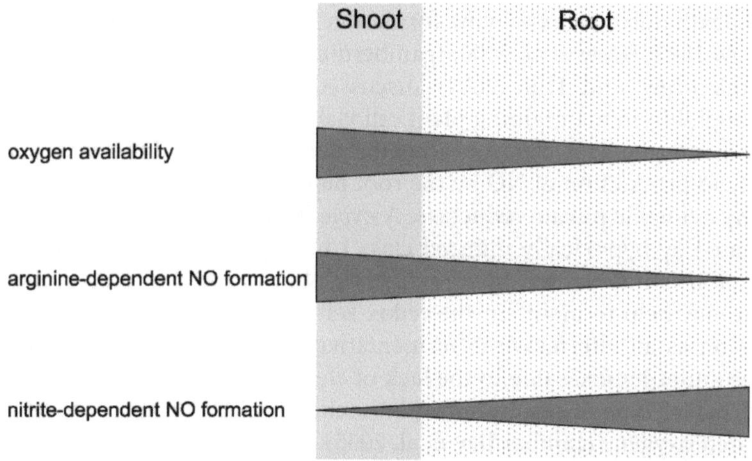

Fig. 2 Dependence of nitrite- and L-arginine-dependent NO production on oxygen availability. High oxygen availability, as mainly found in green tissue, leads to NO formation by NOS-type enzymes that use L-arginine, oxygen, NADPH, and cofactors. Oxygen deprivation, which mainly occurs in root tissue, leads to activation of the nitrite-dependent NO-forming enzymes in the apoplast, cytosol, and mitochondria.

dependent manner, indicating a strong regulation of NO formation by oxygen availability. NO formed by cNR, which is present in roots and leaves, may play a role when the cells are in transition to the unfavorable anaerobic conditions.

Under low oxygen concentrations, plant cells cannot sustain respiration due to lack of an electron acceptor for the terminal oxidases (Geigenberger 2003) and possible functions of NO formed during plant hypoxia have been proposed:

1. Nitrate and nitrite may act as alternative electron acceptors for regeneration of NAD^+ to maintain glycolysis under hypoxic conditions, as an alternative to the use of alcohol dehydrogenase (Crawford 1978). The hereby formed NO might be detoxified by stress-induced hemoglobins (Dordas et al. 2003), which have also been implicated in regeneration of NAD^+ during hypoxia (Hill 1998). In the same context, the induction of stress-induced hemoglobin in *Arabidopsis* by elevated nitrate (Wang et al. 2000) has been also related to modulation of NO levels (Igamberdiev et al. 2004).
2. NO may either accelerate or inhibit programmed cell death depending on its concentration and the combination of additional factors (Kim et al. 2001; Pedroso et al. 2000; Zhang et al. 2003). As observed during pathogen defence, the effects may be either local by inducing cell necrosis or systemic by regulatory pathways. It may also be selective in relation to the responding cells. A similar type of reaction could be responsible for selected

cell death during aerenchyma formation in roots exposed to waterlogging (Drew 1997; Drew et al. 2000; Igamberdiev et al. 2005).

3. Igamberdiev and Hill (2004) discussed the role of nitrate, NO, and hemoglobin in maintaining plant cell viability under anoxic stress. Nitrate as an intermediate electron acceptor during oxygen deficiency leads finally to production of NO by the root plasma membrane-bound enzymes or by cytosolic nitrate reductase. A cycle is proposed whereby NO is oxygenated by hypoxically induced class 1 hemoglobin. The turnover of this reaction is maintained by a methemoglobin reductase. This cyclic reaction may help to maintain the redox status of the cell at very low oxygen tension as an alternative to fermentative pathways. NADH that accumulates under hypoxia due to the lack of electron acceptors is oxidized in an alternative type of respiration to mitochondrial electron transport under limited oxygen (Igamberdiev et al. 2005).
4. NO production as signal for oxygen deficiency was postulated for apoplastic (Stremlau and Stöhr 2006) and intracellular NO by binding to hemoglobin, and to regulate the citric acid cycle and respiration (Millar et al. 2002).

6 Conclusions

Several NO-producing systems operate in plants and are differentially influenced by oxygen partial pressure. Subcellular location of NO-forming enzymes, the chemistry of NO, and the presence of NO scavengers (e.g. hemoglobins) indicate a narrow field of NO activity within a cell. Probably, the NO formation pathways operate in an organ-specific manner and are dependent on plant nitrogen supply. Further data have to be collected to prove whether various kinetic features could lead to specific NO signatures as postulated for the Ca^{2+} signaling pathways.

References

Barroso JB, Corpas FJ, Carreras A, Sandalio LM, Valderrama R, Palma JM, Lupianez JA, del Río LA (1999). Localization of nitric-oxide synthase in plant peroxisomes. J Biol Chem 274:36729–36733

Beckman JS (1996) The physiological and pathological chemistry of nitric oxide. In: Lancaster JR (ed) Nitric oxide: principles and actions. Academic, pp 1–82

Bellissimoa DB, Privalle LS (1995) Expression of spinach nitrite reductase in *Escherichia coli*: Site-directed mutagenesis of predicted active site amino acids. Arch Biochem Biophys 323:155–163

Bérczi A, Horvath G (2003) Lipid rafts in the plant plasma membrane? Acta Biol Szegediensis 47:7–10

Bethke PC, Badger MR, Jones RL (2004) Apoplastic synthesis of nitric oxide by plant tissues. Plant Cell 16:332–341
Botrel A, Kaiser WM (1997) Nitrate reductase activation state in roots in relation to the energy and carbohydrate status. Planta 201:496–501
Botrel A, Magné C, Kaiser WM (1996) Nitrate reduction, nitrite reduction and ammonium assimilation in barley roots in response to anoxia. Plant Physiol Biochem 34:645–652
Brunellei L, Crow JP, Beckman JS (1995) The comparative toxicity of nitric oxide and peroxynitrite to *Escherichia coli*. Arch Biochem Biophys 316:327–334
Brunswick P, Cresswell CF (1988a) Nitrite uptake into intact pea chloroplasts. I. Kinetics and relationship with nitrite assimilation. Plant Physiol 86:378–383
Brunswick P, Cresswell CF (1988b) Nitrite uptake into intact pea chloroplasts. II. Influence of electron transport regulators, uncouplers, ATPase and anion uptake inhibitors and protein binding reagents. Plant Physiol 86:384–389
Burillo S, Luque I, Fuentes I, Contreras A (2004) Interactions between the nitrogen signal transduction protein PII and *N*-acetyl glutamate kinase in organisms that perform oxygenic photosynthesis. J Bacteriol 186:3346–3354
Burns LC, Stevens RJ, Laughlin RJ (1996) Production of nitrite in soil by simultaneous nitrification and denitrification. Soil Biol Biochem 27:47–59
Catoni E, Desimone M, Hilpert M, Wipf D, Kunze R, Schneider A, Flügge UI, Schumacher K, Frommer WB (2003) Expression pattern of a nuclear encoded mitochondrial arginine-ornithine translocator gene from *Arabidopsis*. BMC Plant Biol 3:1
Chen YM, Ferrar TS, Lohmeir-Vogel E, Morrice N, Mizuno Y, Berenger B, Ng KK, Muench DG, Moorhead GB (2006) The PII signal transduction protein of *Arabidopsis thaliana* forms an arginine-regulated complex with plastid N-acetyl glutamate kinase. J Biol Chem 281:5726–5733
Cookson SJ, Williams LE, Miller AJ (2004) Light-dark changes in cytosolic nitrate pools depend on nitrate reductase activity in *Arabidopsis* leaf cells. Plant Physiol 138:1097–1105
Corpas FJ, Barroso JB, del Río LA (2001) Peroxisomes as a source of reactive oxygen species and nitric oxide signal molecules in plant cells. Trends Plant Sci 6:145–150
Coruzzi GM (2003) Primary N-assimilation into amino acids in *Arabidopsis*. In: Somerville C, Meyerowitz E (eds) The *Arabidopsis* book, vol 31. American Society of Plant Biologists, Rockville, pp 1–17
Crawford NM (1995) Nitrate: nutrient and signal for plant growth. Plant Cell 7:859–868
Crawford NM (2006) Mechanisms for nitric oxide synthesis in plants. J Exp Bot 57:471–478
Crawford NM, Guo FQ (2005) New insights into nitric oxide metabolism and regulatory functions. Trends Plant Sci 10:195–200
Crawford RMM (1978) Metabolic adaptations to anoxia. In: Hook DD, Crawford RMM (eds) Plant life in anaerobic environments. Ann Arbor Science, Ann Arbor MI, pp 119–136
del Río LA, Corpas FJ, Sandalio LM, Palma JM, Gómez M, Barroso JB (2002) Reactive oxygen species, antioxidant systems and nitric oxide in peroxisomes. J Exp Bot 53:1255–1272
deRuiter H, Kollöffel C (1983) Arginine catabolism in the cotyledons of developing and germinating pea seeds. Plant Physiol 73:525–528
Dordas C, Rivoal J, Hill RD (2003) Plant haemoglobins, nitric oxide and hypoxic stress. Ann Bot 91:173–178

Drew MC (1997) Oxygen deficiency and root metabolism: injury and acclimation under hypoxia. Ann Rev Plant Physiol Plant Mol Biol 48:223–250

Drew MC, He I, Morgan PW (2000) Programmed cell death and aerenchyma formation in roots. Trends Plant Sci 5:123–127

Emes MJ, Neuhaus HE (1997) Metabolism and transport in non-photosynthetic plastids. J Exp Bot 48:1995–2005

Felle HH (2005) pH regulation in anoxic plants. Ann Bot 96:519–532

Foissner I, Wendehenne D, Langebartels C, Durner J (2000). In vivo imaging of an elicitor-induced nitric oxide burst in tobacco. Plant J 23:817–824

Foyer CH, Parry M, Noctor G (2003) Markers and signals associated with nitrogen assimilation in higher plants. J Exp Bot 54:585–593

Geigenberger P (2003) Response of plant metabolism to too little oxygen. Curr Opin Plant Biol 6:247–256

Gloser V (2002) Seasonal changes of nitrogen storage compounds in a rhizomatous grass *Calamagrostis epigeios*. Biol Plant 45:563–568

Goldraij A, Polacco JC (2000) Arginine degradation by arginase in mitochondria of soybean seedling cotyledons. Planta 210:652–658

Goshima N, Mukai T, Suemori M, Takahashi M, Caboche M, Morikawa H (1999) Emission of nitrous oxide (N_2O) from transgenic tobacco expressing antisense NiR mRNA. Plant J 19:75–80

Gould KS, Lamotte O, Klinguer A, Pugin A, Wendehenne D (2003) Nitric oxide production in tobacco leaf cells: A generalized stress response? Plant Cell Environ 26:1851–1862

Guo FQ, Crawford NM (2005) *Arabidopsis* nitric oxide synthase1 is targeted to mitochondria and protects against oxidative damage and dark-induced senescence. Plant Cell 17:3436–3450

Gupta KJ, Stoimenova M, Kaiser WM (2005) In higher plants, only root mitochondria, but not leaf mitochondria reduce nitrite to NO, in vitro and in situ. J Exp Bot 56:2601–2609

Hill RD (1998) What are hemoglobins doing in plants? Canad J Bot 76:707–712

Hinze H, Holzer H (1985) Accumulation of nitrite and sulfite in yeast cells and synergistic depletion of the intracellular ATP content. Z Lebensm Unters Forsch 180:117–120

Hiramoto K, Ohkawa T, Oikawa N, Kikugawa K (2003) Is nitric oxide (NO) an antioxidant or a prooxidant for lipid peroxidation? Chem Pharm Bull 51:1046–1050

Hoyos ME, Palmieri L, Wertin T, Arrigoni R, Polacco JC, Palmieri F (2003) Identification of mitochondrial transporter for basic amino acids in Arabidopsis thaliana by functional reconstitution into liposomes and complementation in yeast. Plant J 33:1027–1035

Hsieh MH, Lam HM, van de Loo FJ, Coruzzi G (1998) A PII-like protein in *Arabidopsis*: putative role in nitrogen sensing. Proc Natl Acad Sci USA 95:13965–13970

Igamberdiev AU, Baron K, Manac'h-Little N, Stoimenova M, Hill RD (2005) The haemoglobin/nitric oxide cycle: Involvement in flooding stress and effects on hormone signaling. Ann Bot 96:557–564

Igamberdiev AU, Hill RD (2004) Nitrate NO and haemoglobin in plant adaption to hypoxia: an alternative to classic fermentation pathways. J Exp Bot 55:2473–2482

Kaiser WM, Huber SC (2001) Post-translational regulation of nitrate reductase: mechanism, physiological relevance and environmental triggers. J Exp Bot 52:1981–1989

Kaiser WM, Weiner H, Huber SC (1999) Nitrate reductase in higher plants: a case study for transduction of environmental stimuli into control of catalytic activity. Physiol Plant 105:385–390

Kaiser WM, Weiner H, Kandlbinder A, Tsai CB, Rockel P, Sonoda M, Planchet E (2002) Modulation of nitrate reductase: some new insights, an unusual case and a potentially important side reaction. J Exp Bot 53:1–8
Kim PK, Zamora R, Petrosko P, Billiar TR (2001) The regulatory role of nitric oxide in apoptosis. Intern Immunopharmacol 1:1421–1441
Kozlov AV, Staniek K, Nohl H (1999) Nitrite reductase activity is a novel function of mammalian mitochondria. FEBS Lett 454:127–130
Lamattina L, García-Mata C, Graziano M, Pagnussat G (2003) Nitric oxide: The versatility of an extensive signal molecule. Ann Rev Plant Biol 54:109–136
Lamotte O, Coutois C, Barnavon L, Pugin A, Wendehenne D (2005) Nitric oxide in plants: the biosynthesis and cell signaling properties of a fascinating molecule. Planta 221:1–4
Lancaster JRJ (2000) The physical properties of nitric oxide: determinants of the dynamics of NO in tissue. In: Ignarro LJ (ed) Nitric oxide: biology and pathobiology. Academic, San Diego, pp 209–224
Lea PJ, Miflin BJ (1980) Transport and metabolism of asparagine and other nitrogen compounds within the plant. In: Stumpf PK, Conn EE (eds) The biochemistry of plants, vol 5. Academic, New York, pp 569–607
Lewis RS, Deen WM (1994) Kinetics of the reaction of nitric oxide with oxygen in aqueous solutions. Chem Res Toxicol 7:568–574
Lindermayr C, Saalbach G, Durner J (2005) Proteomic identification of S-nitrosylated proteins in *Arabidopsis*. Plant Physiol 137:921–930
Maheswaran M, Urbanke C, Forchhammer K (2004) Complex formation and catalytic activation by the PII signaling protein of *N*-acetyl-L-glutamate kinase from Synechococcus elongatus strain PCC 7942. J Biol Chem 279:55202–55210
Matsumura T, Sakakibara H, Nakano R, Kimata Y, Sugiyama T, Hase T (1997) A nitrate-inducible ferredoxin in maize roots – genomic organization and differential expression of two nonphotosynthetic ferredoxin isoproteins. Plant Physiol 114:653–660
McAinsh MR, Hetherington AM (1998) Encoding specificity in Ca^{2+} signaling systems. Trends Plant Sci 3:32–36
McNeil M, Darvill AG, Fry SC, Albersheim P (1984) Structure and function of the primary cell walls of plants. Annu Rev Biochem 53:625–663
Meder D, Simons K (2006) Lipid rafts, caveolae, and membrane traffic. In: Fielding CJ (ed) Lipid rafts and caveolae. Wiley, Weinheim, pp 1–17
Meyer C, Lea US, Provan F, Kaiser WM, Lillo C (2005) Is nitrate reductase a major player in the plant NO (nitric oxide) game? Photosyn Res 83:181–189
Meyer C, Stöhr C (2002) Soluble and plasma membrane-bound enzymes involved in nitrate and nitrite metabolism. In: Foyer C, Noctor G (eds) Photosynthetic nitrogen assimilation and associated carbon and respiratory metabolism. Advances in photosynthesis and respiration, vol 12. Kluwer, Dordrecht, The Netherlands, pp 49–62
Fryer MJ, Baker NR, Coope CE, Vartapetian BB, Andreeva I, Nuritdinov N (1978) Plant cells under oxygen stress. In: Hook DD, Crawford RMM (eds) Plant life in anaerobic environments. Ann Arbor Science, Ann Arbor MI, pp 13–88
Millar AH, Day DA, Mathieu C (2002) Nitric oxide synthesis by plants and its potential impact on nitrogen and respiratory metabolism. In: Foyer CH, Noctor G (eds) Photosynthetic nitrogen assimilation and associated carbon and respiratory metabolism. Advances in photosynthesis and respiration, vol 12. Kluwer, Dordrecht, The Netherlands, pp 193–204
Miller AJ, Smith SJ (1996) Nitrate transport and compartmentation in cereal root cells. J Exp Bot 47:843–854

Nedeianu S, Pali T, Marsh D (2004) Membrane penetration of nitric oxide and its donor *S*-nitroso-*N*-acetylpenicillamine: a spin-label electron paramagnetic resonance spectroscopic study. Biochim Biophys Acta Biomem 1661:135–143
Neill SJ, Desikan R, Hancock JT (2003) Nitric oxide signaling in plants. New Phytol 159:11–35
Noctor G (2006) Metabolic signaling in defence and stress: the central roles of soluble redox couples. Plant Cell Environ 29:409–425
O'Donnell VB, Chumley PH, Hogg N, Bloodsworth A, Darley-Usmar VM, Freeman BA (1997) Nitric oxide inhibition of lipid peroxidation: Kinetics of reaction with lipid peroxyl radicals and comparison with alpha-tocopherol. Biochem 36:15216–15223
O'Donnell VB, Eiserich JP, Bloodsworth A, Chumley PH, Kirk M, Barnes S, Darley-Usmar VM, Freeman BA (1999) Nitration of unsaturated fatty acids by nitric oxide-derived reactive species. Methods Enzymol 301:454–470
Peoples MB, Gifford RM (1993) Long-distance transport of carbon and nitrogen from sources to sinks in higher plants. In: Dennis DT, Turpin DH (eds) Plant physiology, biochemistry and molecular biology. Wiley, New York, pp 434–447
Perazzolli M, Romero-Puertas MC, Delledonne M (2006) Modulation of nitric oxide bioactivity by plant haemoglobins. J Exp Bot 57:479–488
Pedroso MC, Magalhaes JR, Durzanc D (2000) Nitric oxide induces cell death in *Taxus* cells. Plant Sci 157:173–180
Phillips DA, Fox TC, Six J (2006) Root exudation (net efflux of amino acids) may increase rhizodeposition under elevated CO_2. Global Change Biol 12:561–567
Planchet E, Kapuganti JG, Sonoda M, Kaiser WM (2005) Nitric oxide emission from tobacco leaves and cell suspensions: rate limiting factors and evidence for the involvement of mitochondrial electron transport. Plant J 41:732–743
Plieth C (2005) Calcium: just another regulator in the machinery of life? Ann Bot 96:1–8
Prado AM, Porterfield DM, Feijo JA (2004) Nitric oxide is involved in growth regulation and re-orientation of pollen tubes. Develop 131:2707–2714
Rexach J, Fernández E, Galván A (2000) The *Chlamydomonas reinhardtii Nar1* gene encodes a chloroplast membrane protein involved in nitrite transport. Plant Cell 12:1441–1453
Rexach J, Montero B, Fernández E, Galván A (1999) Differential regulation of the high affinity nitrite transport systems III and IV in *Chlamydomonas reinhardtii.* J Biol Chem 274:27801–27806
Rockel P, Strube F, Rockel A, Wildt J, Kaiser WM (2002) Regulation of nitric oxide (NO) production by plant nitrate reductase in vivo and in vitro. J Exp Bot 53:103–110
Saglio PH, Rancillac M, Bruzan F, Pradet A (1984) Critical oxygen pressure for growth and respiration of excised and intact roots. Plant Physiol 76:151–154
Sattelmacher B (2001) The apoplast and its significance for plant mineral nutrition. New Phytol 149:167–192
Saxe H (1986) Effects of NO, NO_2 and CO_2 on net photosynthesis, dark respiration and transpiration of pot plants. New Phytol 103:185–197
Schlüter U, Crawford MM (2001) Long-term anoxia tolerance in leaves of *Acorus calamus* L. and *Iris pseudacorus* L. J Exp Bot 52:2213–2225
Seregélyes C, Igamberdiev AU, Maassen A, Hennig J, Dudits D, Hill RD (2004) NO-degradation by alfalfa class-1 haemoglobin (Mhb1): a possible link to *PR-1a* gene expression in Mhb1-overproducing tobacco plants. FEBS Lett 571:61–66
Shapiro AD (2005) Nitric oxide signaling in plants. Vitam Horm 72:339–398
Shingles R, Roh MH, McCarty RE (1996) Nitrite transport in chloroplast inner envelope vesicles. I. Direct measurement of proton-linked transport. Plant Physiol 112:1375–1381

Shinkle JR, Swoap SJ, Simon P, Jones RL (1992) Cell wall free space of *Cucumis* hypocotyls contains NAD and a blue light-regulated peroxidase activity. Plant Physiol 98:1336–1341

Siddiqi MY, King BJ, Glass ADM (1992) Effects of nitrate, chlorate and chlorite on nitrate uptake and nitrate reductase activity. Plant Physiol 100:644–650

Simons M, Permentier HP, de Weger LA, Wijffelman CA, Lugtenberg BJJ (1997) Amino acid synthesis is necessary for tomato root colonization by Pseudomonas fluorescens strain WCS365. Mol Plant-Microbe Interact 10:102–106

Sinclair J (1987) Changes in spinach thylakoid activity due to nitrite ions. Photosyn Res 12:255–263

Slocum RD (2005) Genes, enzymes and regulation of arginine biosynthesis in plants. Plant Physiol Biochem 43:729–745

Stewart V (1988) Nitrate respiration in relation to facultative metabolism in enterobacteria. Microbiol Rev 52:190–232

Stitt M (1999) Nitrate regulation of metabolism and growth. Curr Opin Plant Biol 2:178–186

Stitt M, Muller C, Matt P, Gibon Y, Carillo P, Morcuende R, Scheible WR, Krapp A (2002) Steps towards an integrated view of nitrogen metabolism. J Exp Bot 53:959–970

Stöhr C (1999) Relationship of nitrate supply with growth rate plasma membrane-bound and cytosolic nitrate reductase, and tissue nitrate content in tobacco plants. Plant Cell Environ 22:169–177

Stöhr C, Mäck G (2001) Diurnal changes in nitrogen metabolism of tobacco roots. J Exp Bot 52:1283–1289

Stöhr C, Stremlau S (2006) Formation and possible roles of nitric oxide in plant roots. J Exp Bot 57:463–470

Stöhr C, Ullrich WR (1997) A succinate-oxidising nitrate reductase is located at the plasma membrane of plant roots. Planta 203:129–132

Stöhr C, Ullrich WR (2002) Generation and possible role of NO in plant roots and their apoplastic space. J Exp Bot 53:2203–2303

Stöhr C, Strube F, Marx G, Ullrich WR, Rockel P (2001) A plasma-membrane-bound enzyme of tobacco roots catalyzes the formation of nitric oxide from nitrite. Planta 212:835–841

Stoimenova M, Libourel IGL, Ratcliffe RG, Kaiser WM (2003) The role of nitrate reduction in the anoxic metabolism of roots. II. Anoxic metabolism of tobacco roots with or without nitrate reductase activity. Plant Soil 253:155–167

Subczynski WK, Lomnicka M, Hyde JS (1996) Permeability of nitric oxide through lipid bilayer membranes. Free Radic Res 24:343–349

Taylor AA, Stewart GR (1981) Tissue and subcellular localization of enzymes of arginine metabolism in Pisum sativum. Biochem Biophys Res Commun 101:1281–1289

Tischner R, Planchet E, Kaiser WM (2004) Mitochondrial electron transport as a source for nitric oxide in the unicellular green alga *Chlorella sorokiniana*. FEBS Lett 576:151–155

Tun NN, Santa-Catarina C, Begum T, Silveira V, Handro W, Floh EIS, Scherer GFE (2006) Polyamines induce rapid biosynthesis of nitric oxide (NO) in *Arabidopsis thaliana* seedlings. Plant Cell Physiol 47:346–354

van der Vliet A, Hoen PA, Wong PS, Bast A, Cross CE (1998) Formation of *S*-nitrosothiols via direct nucleophilic nitrosation of thiols by peroxynitrite with elimination of hydrogen peroxide. J Biol Chem 273:30255–30262

Van Etten CH, Wolff IA, Jones Q (1963). Amino acid composition of seeds from 200 angiospermous plant species. J Agr Food Chem 11:399–410

Vanin AF, Svistunenko DA, Mikoyan VD, Serezhenkov VA (2004) Endogenous superoxide production and the nitrite/nitrate ratio control the concentration of bioavailable free nitric oxide in leaves. J Biol Chem 279:24100–24107

Vaucheret H, Kronenberger J, Lépingle A, Vilaine F, Boutin JP, Caboche M (1992) Inhibition of tobacco nitrite reductase activity by expression of antisense RNA. Plant J 2:559–569

Wang R, Guegler K, LaBrie ST, Crawford NM (2000) Genomic analysis of a nutrient response in *Arabidopsis* reveals diverse expression patterns and novel metabolic and potential regulatory genes induced by nitrate. Plant Cell 12:1491–1510

Wang R, Okamoto M, Xing X, Crawford NM (2003) Microarray analysis of the nitrate response in *Arabidopsis* roots and shoots reveals over 1000 rapidly responding genes and new linkages to glucose, trehalose-6-phosphate, iron, and sulfate metabolism. Plant Physiol 132:556–567

Wendehenne D, Durner J, Klessig DF (2004). Nitric oxide: A new player in plant signaling and defense responses. Curr Opin Plant Biol 7:449–455

Williams DLH (2004) Nitrosation reactions and the chemistry of nitric oxide. Elsevier, Amsterdam

Wink DA, Mitchell JB (1998) Chemical biology of nitric oxide: insights into regulatory cytotoxic, and cytoprotective mechanisms of nitric oxide. Free Radic Biol Med 25:434–456

Wodala B, Deák Z, Vass I, Erdei L, Horváth F (2005) Nitric oxide modifies photosynthetic electron transport in pea leaves. Acta Biol Szegediensis 49:7–8

Yamasaki H (2000) Nitrite-dependent nitric oxide production pathway: implications for involvement of active nitrogen species in photoinhibition in vivo. Phil Trans R Soc Lond B 355:1477–1488

Yamasaki H (2005) The NO world for plants: achieving balance in an open system. Plant Cell Environ 28:78–84

Yamasaki H, Sakihama Y, Takahashi S (1999) An alternative pathway for nitric oxide production: new features of an old enzyme. Trends Plant Sci 4:128–129

Zeidler D, Zähringer U, Gerber I, Dubery I, Hartung T, Bors W, Hutzler P, Durner J (2004) Innate immunity in *Arabidopsis thaliana*: Lipopolysaccharides activate nitric oxide synthase (NOS) and induce defense genes. Proc Natl Acad Sci USA 101:15811–15816

Zhang C, Czymmek KJ, Shapiro AD (2003) Nitric oxide does not trigger early programed cell death events but may contribute to cell-to-cell signaling governing progression of the *Arabidopsis* hypersensitive response. Mol Plant-Microbe Interact 16:962–972

Zhu J, Chen S, Alvarez S, Asirvatham VS, Schachtman DP, Wu Y, Sharp RE (2006) Cell wall proteome in the maize primary root elongation zone. I. Extraction and identification of water-soluble and lightly ionically bound proteins. Plant Physiol 140:311–325

Zottini M, Formentin E, Scattolin M, Carimi F, Lo Schiavo F, Terzi M (2002) Nitric oxide affects plant mitochondrial functionality in vivo. FEBS Lett 515:75–78

Plant Cell Monogr (6)
L. Lamattina · J. C. Polacco: Nitric Oxide in Plant Growth
DOI 10.1007/7089_2006_083/Published online: 7 October 2006

NO-Based Signaling in Plants

David Wendehenne (✉) · Cécile Courtois · Angélique Besson ·
Antoine Gravot · Annie Buchwalter · Alain Pugin · Olivier Lamotte

Plante-Microbe-Environnement, UMR INRA 1088/CNRS 5184/Université de Bourgogne,
BP 86510, 17 rue Sully, 21065 Dijon cedex, France
wendehen@dijon.inra.fr

Abstract In animals, nitric oxide (NO) is an endogenously produced radical involved in cell communication and signal transduction. Its functions in plants are currently being discovered at an unprecedented pace, and insight into NO-derived mechanisms has mainly been gained from research on signal transduction. Numerous studies have firmly placed NO as one component of the signal perception–transduction network that connects plant responses to primary signals, including hormones, elicitors of defence responses or abiotic stresses. Protein kinases and the second messengers Ca^{2+}, cGMP, and cADPR convey part of the NO signal within cells. Furthermore, NO-based protein modifications are emerging as broad-based mechanisms for posttranslational regulation of protein function and might be implied in the regulation of numerous signaling pathways.

1
Introduction: A Lesson from the Past

Nitric oxide (NO) is a simple diatomic free radical gas. Twenty five years ago, it was considered a pollutant and its potentially damaging nature led to the widely held belief that it only served a harmful purpose. When the pioneer studies of Ignarro, Furchgott, and Murad (the 1998 Nobel prizewinners for Medicine) re-evaluated this prejudice and provided evidence that NO was produced physiologically and functioned as an endothelium-derived relaxing factor as well as a host defence mediator, many scientists were deeply sceptical. As a matter of fact, given its chemical and physical properties, it was hardly conceivable that NO could be suited to act as a signaling molecule. It was indeed surprising that a free radical that is hardly detectable in biological contexts, diffuses freely across biological membranes, forms complexes with transition metal ions, and represents a precursor for other harmful N-oxides, might fully participate in physiological functions. Yet, discovering that NO is a signaling agent in mammals led to a major revolution in biomedical research and provided new directions and therapeutic approaches for a wide range of human diseases. Furthermore, it allowed the characterization of new and unexpected mechanisms of signal transduction. Hardly anybody would have expected such a scenario in the 1980s.

By the late 1990s, NO had become an important topic in plant research. Within a few years, we learnt that NO was generated in plants by two or more distinct enzymes and had a large number of physiological functions ranging from stomatal closure to defence responses. Understanding how NO is produced, perceived, and transduced has represented one of the highlights of plant biology research over the past years. In this review, we summarize the mechanisms underlying NO-based signaling in animals and describe recent findings that shed light on the role of NO in plant cellular signal transduction.

2 A Signaling Agent in Animals

2.1 General Principles

The "NO as a signaling agent" concept originates from studies aimed at understanding the mechanisms by which a cell perceives changes in intracellular NO concentrations and conveys this information into downstream physiological effects. One of the most relevant actions of NO is the activation of soluble guanylate cyclase (sGC; Hanafy et al. 2001). The interaction of NO with the sGC heme results in the formation of a nitrosyl-heme complex, which triggers an increase in enzyme activity and a subsequent rise in intracellular cyclic 3′,5′-guanosine monophosphate (cGMP) concentration. According to many studies, this mechanism enables cells to transmit a NO signal to the downstream targets of cGMP, namely cGMP-dependent protein kinases (PKG), cyclic-nucleotide-gated channels (CNGC), and cGMP-regulated phosphodiesterases (Schmidt and Walter 1994). Many good examples of physiological processes controlled by the NO/cGMP pathway have come to light, such as smooth muscle relaxation, peripheral and central neurotransmission, and blood pressure regulation.

Soluble GC is not the exclusive target for NO in animal cells: recent advances highlight the occurrence of NO-dependent mechanisms that do not rely on the production of cGMP. Indeed, in biological systems, NO displays chemical reactivity toward various targets including iron-sulfur and heme centers as well as nucleophilic groups such as thiols, amines, or carboxyls (Hanafy et al. 2001). In this regard, S-nitrosylation is emerging as an important form of posttranslational protein modification (Stamler et al. 2001). It consists of the reversible coupling of a NO moiety to a reactive cysteine thiol to form nitrosothiol. S-nitrosylation can trigger or inhibit protein activity and over the past decade, the reported number of protein targets for S-nitrosylation has increased to over a hundred (Stamler et al. 2001). A substantial body of recent work has implicated S-nitrosylation in the regulation of signaling pathways involving low molecular weight G proteins, protein kinases and Ca^{2+}-permeable channels (see below). Control of the *N*-methyl D-aspartate (NMDA) signaling pathway that mediates postsynaptic Ca^{2+} fluxes is an archetype for S-nitrosylation based-processes (Kone et al. 2003). In this model, the stimulation of the NMDA receptor (NMDAR) leads to a rapid and localized Ca^{2+} influx, which promotes the activation of the calmodulin (CaM)-dependent neuronal nitric oxide synthase (nNOS). Once produced, NO S-nitrosylates the low molecular weight G proteins $p21^{ras}$ and dexras co-localized to nNOS. The signaling cascade is stopped when NMDAR is inhibited by S-nitrosylation of a single cysteine residue. According to several reports, NMDAR desensitization through S-nitrosylation might prevent toxicity due to over-stimulation of the NMDA receptor, which leads to neurone death (Schroeter et al. 2002).

In addition to S-nitrosylation, nitration of tyrosine residues by nitrating agents such as peroxynitrite ($ONOO^-$) is another type of reaction through which NO influences cell responses (Schopfer et al. 2003). Peroxynitrite is formed by the reaction of NO with the superoxide anion ($O_2^{\bullet-}$). It is highly reactive and considered to account for NO toxicity in necrotic diseases and apoptosis (Beck et al. 1999). Several proteins have been identified as in vivo targets for $ONOO^-$, such as manganese superoxide dismutase, prostacyclin synthase and actin. Part of the effects of $ONOO^-$ are likely to involve an inhibition of protein tyrosine phosphorylation via tyrosine nitration (Ischiropoulos 1998). Recent studies assume that tyrosine nitration may also represent an important form of posttranslational protein modification in non-pathological contexts (Schopfer et al. 2003).

2.2 NO and Ca^{2+}: the Tale of Two Messengers

Ca^{2+} is a ubiquitous second messenger. The fluctuations of its intracellular concentrations control an incredible number of processes. The specificity of Ca^{2+} signaling might be related to its extreme compartmentalization and its complex temporal and spatial distribution (Petersen et al. 2005). The molecular systems responsible for producing signaling events, namely Ca^{2+} channels and transporters, respond to an array of physiological stimuli including extracellular or intracellular messengers, membrane depolarization or hyperpolarization, and mechanical or thermal stimuli. Works on various animal cell types indicate that most of the channels and transporters involved in the control of Ca^{2+} homeostasis seem to be modulated by NO (Clementi 1998). NO acts either directly on the channels or transporters, or indirectly via interconnected secondary signaling cascades.

The list of transporters and Ca^{2+} channels directly regulated by NO by the means of S-nitrosylation or oxidation comprises Ca^{2+}-ATPase, store-operated Ca^{2+} channels, voltage-gated P/Q-type and L-type Ca^{2+} channels, CNGC and ryanodine receptors (RyRs; Stamler et al. 2001). RyRs are large, homotetrameric high-conductance Ca^{2+} channels mediating the release of Ca^{2+} from the endoplasmic reticulum (Eu et al. 1999). RyRs are targets for NO because they contain a large number of sulfhydryls. For instance, the skeletal muscle and cardiac RYRs (RyR1 and RyR2, respectively) contain 100 and 89 cysteines per 560-kDa subunit (Eu et al. 1999). Modification of RYR activity by NO might influence key physiological functions such as vasomotor control, muscle contraction, and insulin release. At the molecular level, NO interacts with RyRs in a complex way since both activation and inhibition have been reported (Hart and Dulhunty 2000). At physiological pO_2, RyR1 is activated by S-nitrosylation of a single cysteine residue (cysteine 3635) located within a CaM-binding domain. It is assumed that the S-nitrosylation of cysteine 3635 reverses the inhibitory effect of CaM on RyR1 (Sun et al. 2001).

The indirect modulation of Ca^{2+} channels by NO is mainly mediated by cGMP. cGMP can modulate ion channel activity through two different mechanisms: (1) direct action through its binding to cyclic-nucleotide binding sites of CNGCs, (2) indirect action mediated by PKGs (Ahern et al. 2002; Clementi 1998). PKGs are widespread proteins that control ion-channel function, since virtually every type of ion channel contains PKG consensus phosphorylation sites (White 1999). Accordingly, PKG-mediated phosphorylation of Ca^{2+} channels, including the inositol 1,4,5-triphosphate receptor (IP_3R) and voltage-dependent Ca^{2+} channels, has been proposed to account for the stimulating or inhibitory effect of NO on Ca^{2+} mobilization (Clementi 1998; Yoshimura et al. 2001). There is also increasing evidence that PKGs activate ADP-ribosyl cyclase, the enzyme catalyzing the cyclization of NAD^+ to cyclic ADP-ribose (cADPR; Willmott et al. 1996). cADPR is a nucleotide showing Ca^{2+}-mobilizing activity in a wide variety of cells (Guse 1999). RyRs appear to be the cADPR intracellular target, although it is not clear whether cADPR acts as a direct triggering molecule for RYR-mediated Ca^{2+} release or as a modulating factor (Willmott et al. 1996). Activation of the cGMP/PKG/cADPR cascade is a key pathway used by NO to activate RYRs in sea urchin eggs and in various neuronal cell types (see for example Willmott et al. 1996).

The efficacy of NO in increasing intracellular Ca^{2+} concentration might have deep repercussions on cell metabolism and adaptation. It has been shown that the combined actions of NO and Ca^{2+} lead to the protein kinase A-dependent activation of the CREB transcription factor capable of reprogramming the pattern of gene expression in neurons (Peunova and Enikolopov 1993). NO and Ca^{2+} have to act within a very narrow time lapse for this activation to occur. Another remarkable example of the importance of the cross-talk between NO and Ca^{2+} is provided by Patel et al. (1999) who reported that increases in the frequency of Ca^{2+} oscillations in rat liver cells, induced by endothelial-derived NO, stimulate metabolic activity in hepatocytes together with an increased blood flow resulting from the vasodilating effects of NO. The modulation of Ca^{2+} signals by NO is therefore critical for cellular and, ultimately, organ function.

2.3 Protein Kinases as Targets of NO Action

During the last few years, an increasing number of protein kinases have been shown to be regulated by NO. Nitric oxide activates all three mitogen-activated protein kinase (MAPK) cascades, namely the stress-activated protein kinase/c-Jun N-terminal kinase (SAPK/JNK) cascade, the stress-activated p38 MAPK cascade and the ERK/MAPK cascade (Beck et al. 1999). The activation of the ERK/MAPK module by NO has important implications (Schroeter et al. 2002). For instance, in stress conditions, this process leads to the inhibition (through phosphorylation) of the pro-apoptotic protein Bad and to the expression of anti-apoptotic proteins (such as Bcl-2) and enzymes involved in antioxidant defence. In neuronal cells, NO induces the expression of immediate-early genes through the ERK/MAPK pathway and plays a role in the synaptic plasticity (Lee et al. 2000). NO activates the MAPK cascades via the upstream S-nitrosylation of low molecular weight proteins including $p21^{ras}$, Rac1 and Cdc42 (Lander et al. 1996). It has also been reported that NO can directly suppress JNK activation by S-nitrosylation, supporting pro-survival mechanisms (Beck et al. 1999). Similarly, NO inhibits the autokinase activity of the Janus kinases JAK2 and JAK3, presumably through oxidation of crucial dithiols to disulfide (Duhé et al. 1998). Because JAKs transmit signals from almost all types of cytokine receptors, such inhibition may explain how macrophage-produced NO affects the ability of certain lymphoid cells to proliferate in response to cytokine. Finally, protein phosphatases also represent targets for NO (Stamler et al. 2001). By direct interaction with phosphatases, either activation or inhibition might occur, which will counteract or prolong activation of protein kinase activity (Schroeter et al. 2002). Taken together, these findings establish a role for protein kinase signaling pathways in the cellular response to NO.

3 NO: A Signaling Molecule in Plants?

A large number of physiological responses are regulated by NO in plants including germination, root growth, gravitropic bending, stomatal closure, flowering, orientation of pollen tubes, hypoxia, iron availability, adaptation to biotic and abiotic stresses, and cell death (review by Delledonne 2005). Now, a main question to be answered is how NO regulates these diverse biological processes. Some studies do shed some light on the subject.

3.1
NO Synthesis in Plants: Still So Many Things to Do

In animals, it is thought that the normal biological levels of useful NO are low. Once produced, NO availability depends on the molecular environment: it can be rapidly scavenged and bound to multiple targets within the cells. We should also bear in mind that when used as a signal, NO synthesis is usually rapid, transient, and takes place in subcellular compartments (Kone et al. 2003). Therefore, the properties of NO present considerable problems when attempting to detect it in biological systems and its detection/quantification in animal cells is still a major challenge (NO Gordon Conference, Italy 2005). Strictly speaking, this means that not detecting NO does not necessarily reflect the absence of any production.

In plants, enzymatic as well non-enzymatic sources have been identified including nitrate reductase and AtNOS1, an enzyme displaying NOS-like activity. Excellent reviews have been published recently that describe the aspects of NO synthesis (see for example Crawford 2006; del Rio et al. 2004). However, it should also be noticed that contrasting data have been reported regarding the identity of NO sources, particularly in plant physiological contexts (see for example Desikan et al. 2002 versus Guo et al. 2003; Foissner et al. 2000, and Lamotte et al. 2004 versus Planchet et al. 2006). To our point of view, these inconsistencies highlight the technical difficulties in measuring NO production in plant tissues (see also the chapter by Kaiser in this volume). Furthermore, they show our ignorance about NO-based processes in plants. For example, contrary to what is normally expected (Planchet et al. 2006), the absence of NO synthesis in nitrate-deprived cells does not inevitably point to nitrate reductase as the source of NO (Gauthier et al. submitted). Similarly, do NOS-like enzymes function optimally in mutants impaired in nitrate reductase expression? We should also keep in mind that interpretation of the mechanisms of NO synthesis and action are made more difficult by the limited specificity of pharmacological agents. We simply know nothing, as yet, about the fate of these agents within plant cells. This is true for NOS inhibitors, NO scavengers, and also for their presumed inactive analogs. In short, even if important progress has been made in elucidating the molecular identity of NO sources in plants, many crucial questions remain to be answered.

3.2
Metal Nitrosation, S-Nitrosylation and Tyrosine Nitration

NO has been found to react with several plant heme and non-heme metalloproteins. Lipoxygenase-1 was the first enzyme studied in this regard (Nelson 1987). Lipoxygenase-1 contains a single non-heme iron and catalyzes the hydroperoxydation of linoleic acid. Exposure of ferrous lipoxygenase-1 to NO leads to the formation of a metal-nitrosyl complex. Interestingly, at pH 9, NO and linoleic acid compete for binding to ferrous lipoxygenase-1. Because the substrate binds very close to the iron, it is assumed that NO and linoleic acid occupy overlapping sites. The physiological meaning of this competition remains to be determined. In addition to lipoxygenase-1, both symbiotic and non-symbiotic hemoglobins have been shown to interact with NO. The symbiotic hemoglobin is found in legume root nodules and its function is believed to involve the facilitation of O_2 transport to nitrogen-fixing bacteria. Using electron paramagnetic resonance, Mathieu et al. (1998) detected the presence of a complex (Lb-NO) between NO and Fe_{II} leghemoglobin in root nodules from soybean plants that had been grown in the absence of nitrate. The Lb-NO complex is present at high concentration in the youngest nodules, suggesting the involvement of NO in the regulation of nitrogen fixation. More recently, the possible role for oxy-leghemoglobin (oxyLb), formed in *Medicago truncatula* nodules upon symbiosis, has been deeply investigated (Herold and Puppo 2005). It arises from this study that oxyLb is able to scavenge NO and $ONOO^-$ presumably formed during the nodulation, and thereby may prevent the triggering of NO-mediated defence reactions.

Class-1 non-symbiotic hemoglobin is also emerging as an important target for NO. Class-1 hemoglobin proved to be hexacoordinate and, similarly to animal hexacoordinate hemoglobins, is up-regulated in response to stresses such as hypoxia (Dordas et al. 2003; Perazzolli et al. 2004). In vitro studies indicate that class-1 hemoglobin possesses an NADPH-dependent NO scavenging activity and catalyzes the conversion of NO to nitrate (Perazzolli et al. 2004; Seregélyes et al. 2004). For instance, Perazzolli et al. (2004) demonstrated that the treatment of *A. thaliana* class-1 Fe_{II} hemoglobin (AHb1) with NO results in the formation of a nitrosyl complex and a rapid NO-dependent oxidation of AHb1 to methemoglobin together with nitrate. AHb1 also metabolizes nitrosoglutathione (GSNO) through a similar catalytic cycle, this process being accompanied by S-nitrosylation of the protein. Of importance, AHb1 was also S-nitrosylated in vivo under hypoxia, and genetic approaches indicated that this process might protect plants against nitrosative stress associated with hypoxia by reducing NO emission. Clearly, such experiments highlight the ability of plant cells to tightly control NO availability and NO-dependent effects. An unsettled issue is whether the NO-scavenging activity of hemoglobin is simply a detoxification mechanism or whether it is also a means to channel the effects of NO. Finally, it should be mentioned that other iron-containing proteins, such as the cytosolic and mitochondrial aconitases, have been reported as putative NO targets (Klessig et al. 2000).

Accumulating data suggest that S-nitrosylation is an important posttranslational modification regulating plant protein function (see also the chapter by Lindermayr and Durner in this volume). Using the "biotin switch method", a procedure in which the S-nitrosylated cysteine of nitrosothiols is chemically replaced by a detectable biotinylated cysteine, Lindermayr et al. (2005) identified more than 100 S-nitrosylated proteins from *A. thaliana* cell cultures and leaves exposed to artificially generated NO. These proteins are related to oxidative stress, signaling, cellular architecture, and metabolism and include methionine adenosyltransferase (MAT). MAT catalyzes the synthesis of *S*-adenosylmethionine, a substrate for the biosynthesis of the plant hormone ethylene. MAT1, one of the three MAT isoforms in *A. thaliana*, is inhibited by GSNO in vitro through S-nitrosylation of cysteine 114 (Lindermayr et al. 2006). Accordingly, treatment of *A. thaliana* cell cultures with GSNO was shown to decrease ethylene production, suggesting that NO-induced S-nitrosylation of MAT1 might have a regulatory function on ethylene biosynthesis and influence cross-talk operating between hormone signaling pathways. The *Vicia faba* guard cell outward-rectifying K^+ channel ($I_{K,out}$) is among other NO targets possibly regulated through S-nitrosylation (Sokolovski and Blatt 2004). It was indeed reported that chemical reagents capable of reducing cysteine thiol adducts protect $I_{K,out}$ from NO inactivation. In contrast, the effect of NO on the channel was mimicked by redox reagents able to oxidatively bridge closely spaced thiol groups. Although these experiments do not provide direct proof that $I_{K,out}$ is indeed S-nitrosylated, they point out that NO targets critical cysteine residues of the channel (see more details in the chapter by Sokolovski and Blatt in this volume).

More indirect evidence that NO operates through S-nitrosylation-dependent mechanisms in plants is provided by the analysis of GSNO-reductase (GSNOR) involvement in plant defence responses. *A. thaliana* GSNOR metabolizes GSNO to ammonia and was reported to complement the hypersensitivity to GSNO of a yeast mutant lacking GSNOR activity (Sakamoto et al. 2002). Diaz et al. (2003) reported that the expression of the *A. thaliana AtGSNOR1* gene was regulated by molecular signals related to plant defence. As expected, genetic manipulation of *AtGSNOR1* indicates that the level of nitrosothiols is directly correlated with *AtGSNOR1* function, an increase or a decrease of nitrosothiols concentration being observed in plants exhibiting reduced or increased *AtGSNOR1* activity, respectively (Feechan et al. 2005).

In contrast to the accumulating knowledge of S-nitrosylation-based processes, much less is known about tyrosine nitration as a possible NO signal. As stated above, tyrosine nitration is mediated by $ONOO^-$. Although experiments based on the use of donors or scavengers argue for $ONOO^-$ involvement in plant physiological functions (see for example Alamillo and Garcia-Olmedo 2001), its detection in tissues has not been reported so far. The only detailed demonstration that tyrosine nitration might occur in plants was provided by Morot-Gaudry-Talarmain et al. (2002) who used an immunological approach based on monoclonal antibodies raised against 3-nitrotyrosine ($3-NO_2Tyr$). The authors detected increased protein tyrosine nitration in an antisense nitrite reductase tobacco line displaying high nitrite concentration and enhanced nitrate reductase-dependent NO emission. The identity of the corresponding proteins has not been reported but it is likely that their nitration might account for the reduced development and chlorotic phenotype of the antisense line.

To conclude this list of NO targets, there remains the question of plant sGC. Logically, much attention has been paid to cGMP as a putative NO-induced effector in plants. Artificially generated NO has been found to induce cGMP synthesis in various plant species (for example see Durner et al. 1998; Hu et al. 2005). Works combining complementary pharmacological tools including cGMP analogs, sGC and phosphodiesterase inhibitors, elucidated some of the functions of the NO-cGMP cascade. Examples include growth regulation and re-orientation of pollen tubes (Prado et al. 2004), cell death (Clarke et al. 2000), induction of phenylalanine ammonia lyase expression and activity (Durner et al. 1998), auxin-induced root organogenesis (Pagnussat et al. 2003), and gravitropic curvature (Hu et al. 2005). From this, one can conclude that NO activates a plant enzyme catalyzing an sGC-like activity.

3.3 The Ca^{2+} Link

Several studies indicate that NO can mediate its biological effects in plants through the mobilization of the second messenger Ca^{2+}:

1. Artificially generated NO has the ability to induce: (i) transient increases of cytosolic free Ca^{2+} concentration ($[Ca^{2+}]_{cyt}$) in *Vicia faba* guard cells and tobacco cells (Garcia-Mata et al. 2003; Lamotte et al. 2006; Sokolovski et al. 2005); (ii) fast and transient extracellular Ca^{2+} uptake in tobacco cell cultures (Lamotte et al. 2004). For instance, Lamotte et al. (2006) demonstrated that the treatment of tobacco cells with the NO donor DEA-NONOate resulted in an increase of $[Ca^{2+}]_{cyt}$ which occurred within 5 min and peaked at 350 nM after 8 min before decreasing to the background level within 12 min. By contrast, the same donor did not affect the nuclear free Ca^{2+} concentration in tobacco cells (Lecourieux et al. 2005), suggesting that the effects of NO on Ca^{2+} homeostasis are restricted to specific cellular compartments.
2. Pharmacological analyses point out that the NO-evoked increases in $[Ca^{2+}]_{cyt}$ are sensitive to inhibitors of plasma membrane and/or intracellular Ca^{2+} channels. In particular, these analyses make a strong case for the regulation of RYR-like channels by NO (Garcia-Mata et al. 2003; Lamotte et al. 2004 and 2006).
3. The regulatory effects of exogenously applied NO on stomatal closure (Neill et al. 2002), inactivation of inward-rectifying K^+ channels ($I_{k,in}$, Garcia-Mata et al. 2003), adventitious root formation (Lanteri et al. 2006), and *PR-1* expression (Klessig et al. 2000) are partly reduced by Ca^{2+} channel antagonists.

4. NOS inhibitors and NO scavengers were shown to reduce rises in $[Ca^{2+}]_{cyt}$ induced by elicitors of defence responses (Lamotte et al. 2004; Vandelle et al. 2006) and hyperosmotic stress (Gould et al. 2003; Lamotte et al. 2006).

Pharmacological approaches have provided some insight into the way by which NO promotes $[Ca^{2+}]_{cyt}$ increases. Durner et al. (1998) and Klessig et al. (2000) provided the first results supporting the involvement of cADPR in that process. The authors showed that cADPR mediates the up-regulation of *PR-1* expression by NO in tobacco. Accordingly, cADPR itself is capable of inducing *PR-1* expression in tobacco leaf disks, this mechanism being sensitive to RYR inhibitors. The role of cADPR as a second messenger for NO-triggered Ca^{2+} mobilization was further supported by the demonstration that the NO-evoked increase of $[Ca^{2+}]_{cyt}$ is partly reduced by 8-Br-cADPR, a cADPR antagonist (Lamotte et al. 2006). Cyclic GMP might also be a component of the NO-driven cascade leading to rises in $[Ca^{2+}]_{cyt}$, as suggested by experiments using mammalian sGC inhibitors (Garcia-Mata et al. 2003). However, it is important to mention that up to now, the classical view of cGMP acting on Ca^{2+} release through cADPR has not been proved in plants.

Besides cyclic nucleotides, two recent studies emphasize a role for protein kinases in NO-mediated mobilization of free Ca^{2+} in *Vicia faba* guard cells and tobacco suspension cells (Lamotte et al. 2006, Sokolovski et al. 2005). Both reports show that the NO-mediated elevation of $[Ca^{2+}]_{cyt}$ is reduced by protein kinase inhibitors. This result was further supported by the demonstration that NO induces the activation of a 42-kDa protein kinase which corresponds to *Nicotiana tabacum* osmotic stress-activated protein kinase (NtOSAK, see below, Lamotte et al. 2006). The kinetics of NtOSAK activation parallels the kinetics of $[Ca^{2+}]_{cyt}$ elevation. The precise link between NtOSAK and $[Ca^{2+}]_{cyt}$ is not known. Finally, it has also been assumed that NO, artificially generated or endogenously produced, might also regulate extracellular Ca^{2+} influxes by promoting changes in plasma membrane electrical potential (Lamotte et al. 2006; Vandelle et al. 2006).

Taken together, these studies provide evidence that NO regulates cytosolic Ca^{2+} homeostasis in plant cells through multiple processes. Furthermore, they reveal that NO is required for optimal stimulus-induced $[Ca^{2+}]_{cyt}$ elevation, highlighting its role as a signaling agent in plants. The question that further comes to mind is the incidence of the NO/Ca^{2+} pathway on the cell response to stimulus. The first, and many would say the most important, consequence is the amplification of Ca^{2+} signaling. By favoring Ca^{2+} mobilization, NO is likely to influence the temporal and spatial arrangement of Ca^{2+} signaling and, consequently, the cell response to primary signals including elicitors, hormones, and abiotic stresses. The demonstration that ABA-induced $I_{k,in}$ inactivation by NO requires elevated $[Ca^{2+}]_{cyt}$ is a remarkable example of the impact of the NO/Ca^{2+} pathway on cell response (Garcia-Mata et al. 2003).

Furthermore, the contribution of NO to cell death (Delledonne 2005) might also be explained by its Ca^{2+} mobilizing properties. Supporting this hypothesis, compelling pharmacological, biochemical, and genetic evidence point to a role for Ca^{2+} channels and/or $[Ca^{2+}]_{cyt}$ increases in mediating cell death in plant cells exposed to avirulent pathogens or elicitors (Lam 2004). If verified, future research exploring the function of NO in Ca^{2+}-dependent cell death should address at least two questions: Do cells perceive the NO-dependent Ca^{2+} release as a Ca^{2+} overload causing cytotoxicity? Does the Ca^{2+} mobilized in response to NO act on specific cell death regulators?

3.4 Protein Kinase Activation: A Route for NO Effects

Interest in protein kinase cascades involved as components of NO-induced effects is significant, but only few components of such pathways have been identified. The "in-gel kinase assay" allows the characterization of protein kinases whose activities are rapidly and transiently induced by artificially generated NO in several plant species including tobacco (Klessig et al. 2000; Lamotte et al. 2006), *A. thaliana* (Clarke et al. 2000) and cucumber (Lanteri et al. 2006; Pagnussat et al. 2004). Most of these proteins exhibit MAPK properties. This observation is substantiated by the demonstration that the GSNO-sensitive 48 kDa kinase in tobacco leaves and cell suspensions corresponds to the MAPK salicylic acid-induced protein kinase (SIPK; Klessig et al. 2000). As stated above, in addition to MAPKs, NO was shown to induce the activation of the tobacco serine/threonine protein kinase NtOSAK (Lamotte et al. 2006). NtOSAK belongs to the plant SNF1-related protein kinase type 2 (SnRK2) family, SNF1 being a yeast protein kinase activated by glucose starvation (Mikolajczyk et al. 2000). The SnRK2 group is specific to plants and may be involved in the response to environmental stresses, including hyperosmotic and saline stresses (Mikolajczyk et al. 2000). The possibility that NO could also regulate the activity of Ca^{2+}-dependent protein kinases (CDPK) was recently addressed by Lanteri et al. (2006) who characterized a 50 kDa NO-dependent CDPK in cucumber hypocotyls.

The physiological impact of NO/protein kinase cascades has been poorly investigated and progress on linking protein kinases to endogenously produced NO has come from pharmacological analysis. Through these strategies, it has been proposed that a NO-dependent/cGMP-independent MAPK signaling cascade (Pagnussat et al. 2004) as well as a 50 kDa CDPK (see above, Lanteri et al. 2006) are activated during the auxin-induced adventitious rooting process (see details in the chapter by Correa-Aragunde et al. in this volume). Furthermore, recent work demonstrates that NtOSAK is activated in response to hyperosmotic stress through a NO-dependent process (Lamotte et al. 2006). Interestingly, a tobacco 41 kDa protein kinase showing the same biochemical properties as NtOSAK was shown to be activated by INF1 through a NO-dependent mechanism (Yamamoto et al. 2004). Finally, Ötvos et al. (2005) reported that NO donor-promoted auxin-dependent division of leaf protoplast-derived alfalfa cells was accompanied by an increased amount and activity of the cell cycle regulatory protein kinase *p34*^{cdc2}. Although preliminary, these studies suggest that NO produced by plant cells may well deliver signals into protein kinase cascades.

4
Conclusion

NO-based mechanisms in biological contexts are complex phenomena. In animals, NO effects are not simply a consequence of the amount of NO produced but, more importantly, are determined by the local environment in which NO is released (e.g., the cellular redox state, the bioavailability of NO-generating enzyme substrates, the nature and proximity of molecular targets, as well as NO-metabolizing proteins) and the nature of the generated reactive nitrogen species. In plants, prevailing questions during the last decade have concerned the synthesis and functions of NO. Remarkable advances have been made in addressing these central problems, and the concept that NO might act as a signaling agent is clearly emerging (Fig. 1):

1. Changes in NO concentration in plant cells are apparent during the transduction of a wide variety of biotic and abiotic signals. NO synthesis is

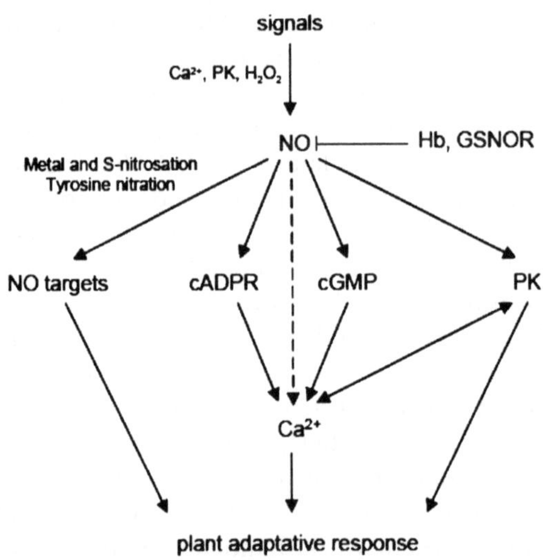

Fig. 1 NO-based signaling in plants. See the text for comments. *Solid arrow* events demonstrated through pharmacological, biochemical, molecular, or genetic approaches; *dashed arrow* supposed events; *PK* protein kinase

dynamically controlled by upstream signaling events including phosphorylation, production of H_2O_2, and increases in intracellular free Ca^{2+} concentration (Bright et al. 2006; Lamotte et al. 2004).
2. NO is integrated in signaling networks in which it regulates Ca^{2+} fluxes, protein kinase activation and cGMP synthesis.
3. NO regulates the expression of several classes of genes (see for example Wendehenne et al. 2004).
4. NO availability is tightly controlled by scavenging proteins including hemoglobins and GSNOR.
5. NO-based protein modification occurs in plants.

One may object that the list of NO targets identified so far does not include typical signaling proteins. Nevertheless, we should bear in mind that NO diffuses freely across cell membranes and might modulate membrane-anchored and transmembrane proteins (including channels) whose identification is still a technical challenge. In addition, signaling proteins are usually present in low abundance, increasing technical difficulties. Finally, in addition to typical signaling proteins, S-nitrosylation of metabolic enzymes such as MAT1 might guide the cells toward specific responses (for instance by promoting SA-dependent pathways). Similarly, regarding the importance of the cytoskeleton in the integration of extracellular signals, S-nitrosylation of tubulin or actin is perhaps a crucial event favoring correct signal transduction.

Clearly, the processes underlying NO effects in plants still remain quite misty and a great deal of work remains to be done to complete our understanding of NO-based signaling. In particular, NO targets need to be better defined and it will be important to ensure that they are physiologically relevant. Also of future interest will be more precise information about NO biochemistry and the natures of the mechanisms controlling NO synthesis and the switching off of the NO signal.

References

Ahern GP, Klyachko VA, Jackson MB (2002) cGMP and S-nitrosylation: two routes for modulation of neuronal excitability by NO. Trends Neurosci 25:510–517

Alamillo JM, Garcia-Olmedo F (2001) Effects of urate, a natural inhibitor of peroxynitrite-mediated toxicity, in the response of *Arabidopsis thaliana* to the bacterial pathogen *Pseudomonas syringae*. Plant J 25:529–540

Beck KF, Eberhardt W, Frank S, Huwiler A, Messmer UK, Mühl H, Pfeilschifter J (1999) Inducible NO synthase: role in cellular signaling. J Exp Biol 202:645–653

Bright J, Desikan R, Hancock JT, Weir IS and Neill SJ (2006) ABA-induced NO generation and stomatal closure in *Arabidopsis* are dependent on H_2O_2 synthesis. Plant J 45:113–122

Clarke A, Desikan R, Hurst RD, Hancock JT, Neill SJ (2000) NO way back: nitric oxide and programmed cell death in *Arabidpsis thaliana* suspension cultures. Plant J 24:667–677

Clementi E (1998) Role of nitric oxide and its intracellular signaling pathways in the control of Ca^{2+} homeostasis. Biochem Pharmacol 55:713–718

Correa-Aragunde N, Lanteri ML, García-Mata C, ten Have A, Laxalt AM, Graziano M, Lamattina L (2007) Nitric oxide functions as intermediate in auxin, abscisic acid, and lipid signaling pathways. In: Lamattina L, Polacco JC (eds) Nitric oxide in plant growth. Plant Cell Monographs, vol 6. Springer, Berlin Heidelberg New York (in press)

Crawford NM (2006) Mechanisms for nitric oxide synthesis in plants. J Exp Bot 57:471–478

Delledonne M (2005) NO news is good news for plants. Curr Opin Plant Biol 8:390–396

del Rio LA, Corpas FJ, Barroso JB (2004) Nitric oxide and nitric oxide synthase activity in plants. Phytochem 65:783–792

Desikan R, Griffiths R, Hancock J, Neill S (2002) A new role for an old enzyme: nitrate reductase-mediated nitric oxide generation is required for abscisic acid-induced stomatal closure in *Arabidopsis thaliana.* Proc Natl Acad Sci USA 99:16314–16318

Diaz M, Achkor H, Titarenko E, Martinez MC (2003) The gene encoding glutathione-dependent formaldehyde dehydrogenase/GSNO reductase in responsive to wounding, jasmonic acid and salicylic acid. FEBS Lett 543:136–139

Dordas C, Hasinoff BB, Igamberdiev AU, Manac'h N, Rivoal J, Hill RD (2003) Expression of a stress-induced haemoglobin affects NO levels produced by alfalfa root cultures under hypoxic stress. Plant J 35:763–770

Duhé RJ, Evans GA, Erwin RA, Kirken RA, Cox GW, Farrar WL (1998) Nitric oxide and thiol redox regulation of Janus kinase activity. Proc Natl Acad Sci USA 95:126–131

Durner J, Wendehenne D, Klessig DF (1998) Defense genes induction in tobacco by nitric oxide, cyclic GMP and cyclic ADP ribose. Proc Natl Acad Sci USA 95:10328–10333

Eu JP, Xu L, Stamler JS, Meissner G (1999) Regulation of ryanodine receptors by reactive nitrogen species. Biochem Pharmacol 57:1079–1084

Feechan A, Kwon E, Yun BW, Wang Y, Pallas JA, Loake GJ (2005) A central role for S-nitrosothiols in plant disease resistance. Proc Natl Acad Sci USA 102:8054–8059

Foissner I, Wendehenne D, Langebartels C, Durner J (2000) In vivo imaging of an elicitor-induced nitric oxide burst in tobacco. Plant J. 23:817–824

Garcia-Mata C, Gay R, Sokolovski S, Hills A, Lamattina L, Blatt MR (2003) Nitric oxide regulates K^+ and Cl^- channels in guard cells through a subset of abscisic acid-evoked signaling pathways. Proc Natl Acad Sci USA 100:11116–11121

Gould KS, Lamotte O, Klinguer A, Pugin A, Wendehenne D (2003) Nitric oxide production in tobacco leaf cells: a generalized stress response? Plant Cell Environ 26:1851–1862

Guo FQ, Okamoto M, Crawford NM (2003) Identification of a plant nitric oxide synthase gene involved in hormonal signaling. Science 302:100–103

Guse AH (1999) Cyclic ADP-ribose: a novel Ca^{2+}-mobilising second messenger. Cell Signal 11:309–316

Hanafy KA, Krumenacker JS, Murad F (2001) NO, nitrotyrosine, and cyclic GMP in signal transduction. Med Sci Monit 7:801–819

Hart JDE, Dulhunty AF (2000) Nitric oxide activates or inhibits skeletal muscle ryanodine receptors depending on its concentration, membrane potential and ligand binding. J Memb Biol 173:227–236

Herold S, Puppo A (2005) Oxyleghemoglobin scavenges nitrogen monoxide and peroxynitrite: a possible role in functioning nodules? J Biol Inorg Chem 2005 10:935–945

Hu X, Neill S, Tang Z, Cai W (2005) Nitric oxide mediates gravitropic bending in soybean roots. Plant Physiol 137:663–670

Ischiropoulos H (1998) Biological tyrosine nitration: a pathophysiological function of nitric oxide and reactive oxygen species. Arch Biochem Biophys 365:1–11

Kaiser WM, Gupta KJ, Planchet E (2007) Higher plant mitochondria as a source for NO. In: Lamattina L, Polacco JC (eds) Nitric oxide in plant growth. Plant Cell Monographs, vol 6. Springer, Berlin Heidelberg New York (in press)

Klessig DF, Durner J, Noad R, Navarre DA, Wendehenne D, Kumar D, Zhou JM, Shah J, Zhang S, Kachroo P, Trifa Y, Pontier D, Lam E, Silva H (2000) Nitric oxide and salicylic acid signaling in plant defense. Proc Natl Acad Sci USA 97:8849–8855

Kone BC, Kuncewicz T, Zhang W, Yu ZY (2003) Protein interactions with nitric oxide synthases: controlling the right time, the right place, and the right amount of nitric oxide. Am J Physiol Renal Physiol 285:178–190

Lam E (2004) Controlled cell death, plant survival and development. Nature Mol Cell Biol 5:305–315

Lander HM, Jacovina AT, Davis RJ, Tauras JM (1996) Differential activation of mitogen-activated protein kinases by nitric oxide-related species. J Biol Chem 271:19705–19709

Lamotte O, Gould K, Lecourieux D, Sequeira-Legrand A, Lebrun-Garcia A, Durner J, Pugin A, Wendehenne D (2004) Analysis of nitric oxide signaling functions in tobacco cells challenged by the elicitor cryptogein. Plant Physiol 135:516–529

Lamotte O, Courtois C, Dobrowolska G, Besson A, Pugin A, Wendehenne D (2006) Mechanisms of nitric oxide-induced increase of free cytosolic Ca^{2+} concentration in *Nicotiana plumbaginifolia* cells. Free Rad Biol Med 40:1369–1376

Lanteri M, Pagnussat GC, Lamattina L (2006) Calcium and calcium-dependent protein kinases are involved in nitric oxide- and auxin-induced adventitious root formation in cucumber. J Exp Bot 57:1341–1351

Lecourieux D, Lamotte O, Bourque S, Mazars C, Wendehenne D, Ranjeva R, Pugin A (2005) Elicitors induce specific changes in nuclear free calcium in tobacco cell suspensions. Cell Calcium 38:527–538

Lee SA, Park JK, Kang EK, Bae HR, Bae KW, Park HT (2000) Calmodulin-dependent activation of p38 and p42/44 mitogen-activated protein kinases contributes to c-fos expression by calcium in PC12 cells: modulation by nitric oxide. Mol Brain Res 75: 16–24

Lindermayr C, Durner J (2007) S-Nitrosylation in plants – spectrum and selectivity. In: Lamattina L, Polacco JC (eds) Nitric oxide in plant growth. Plant Cell Monographs, vol 6. Springer, Berlin Heidelberg New York (in press)

Lindermayr C, Saalbach G, Bahnweg G, Durner J (2006) Differential inhibition of Arabidopsis methionine adenosyltransferase by protein S-nitrosylation. J Biol Chem 281:4285–4291

Lindermayr C, Saalbach G, Durner J (2005) Proteomic identification of S-nitrosylated proteins in Arabidopsis. Plant Physiol 137:921–930

Mathieu C, Moreau S, Frendo P, Puppo A, Davies MJ (1998) Direct detection of radicals in intact soybean nodules: presence of nitric oxide-leghemoglobin complexes. Free Rad Biol Med 24:1242–1249

Mikolajczyk M, Awotunde OS, Muszynska G, Klessig DF, Dobrowolska G (2000) Osmotic stress induces rapid activation of a salicylic acid-induced protein kinase and a homolog of protein kinase ASK1 in tobacco cells. Plant Cell 12:165–178

Morot-Gaudry-Talarmain Y, Rockel P, Moureaux T, Quillere I, Leydecker MT, Kaiser WM, Morot-Gaudry JF (2002) Nitrite accumulation and nitric oxide emission in relation to cellular signaling in nitrite reductase antisense tobacco. Planta 215:708–715

Neill SJ. Desikan R, Clarke A, Hancock JT (2002) Nitric oxide is a novel component of abscisic acid signaling in stomatal guard cells. Plant Physiol 128:13–16

Nelson MJ (1987) The nitric oxide complex of ferrous soybean lipoxygenase-1. J Biol Chem 262:12137–12142

Ötvos K, Pasternak TP, Miskolczi P, Domoki M, Dorjgotov D, Szücs A, Bottka S, Dudits D, Fehér A (2005) Nitric oxide is required for, and promotes auxin-mediated activation of, cell division and embryogenic cell formation but does not influence cell cycle progression in alfalfa cell cultures. Plant J 43:849–860

Pagnussat GC, Lanteri ML, Lamattina L (2003) Nitric oxide and cyclic GMP are messengers in the indole acetic acid-induced adventitious rooting process. Plant Physiol 132:1241–1248

Pagnussat GC, Lanteri ML, Lombardo MC, Lamattina L (2004) Nitric oxide mediates the indole acetic acid induction activation of a mitogen-activated protein kinase cascade involved in adventitious root development. Plant Physiol 135:279–286

Patel S, Robb-Gaspers LD, Stellato KA, Shon M, Thomas AP (1999) Coordination of calcium signaling by endothelial-derived nitric oxide in the intact liver. Nature Cell Biol 1:467–471

Perazzolli M, Dominici P, Romero-Puertas MC, Zago E, Zeier J, Sonoda M, Lamb C, Delledonne M (2004) Arabidopsis nonsymbiotic haemoglobin AHb1 modulates nitric oxide bioactivity. Plant Cell 16:2785–2794

Petersen OH, Michalak M, Verkhratsky A (2005) Calcium signaling: past, present and future. Cell Calcium 38:161–169

Peunova N, Enikolopov G (1993) Amplification of calcium-induced gene transcription by nitric oxide in neuronal cells. Nature 264:450–453

Planchet E, Sonoda M, Zeier J, Kaiser WM (2006) Nitric oxide (NO) as an intermediate in the cryptogein-induced hypersensitive response – a critical re-evaluation. Plant Cell Environ 29:59–69

Prado AM, Porterfield DM, Feijo JA (2004) Nitric oxide is involved in growth regulation and re-orientation of pollen tubes. Development 131:2702–2714

Sakamoto A, Ueda M, Morikawa H (2002) *Arabidopsis* glutathione-dependent formaldehyde dehydrogenase is an S-nitrosoglutathione reductase. FEBS Lett 515:20–24

Schmidt HHHW, Watler U (1994) NO at work. Cell 78:919–925

Schopfer FJ, Baker PRS, Freeman BA (2003) NO-dependent protein nitration: a cell signaling event or an oxidative inflammatory response? Trends Biochem Sci 28:646–654

Schroeter H, Boyd C, Spencer JPE, Williams RJ, Cadenas E, Rice-Evans C (2002) MAPK signaling in neurodegeneration: influences of flavonoids and of nitric oxide. Neurobiol Aging 23:861–880

Seregélyes C, Igamberdiev AU, Maassen A, Henning J, Dudits D, Hill RD (2004) NO-degradation by alfalfa class 1 hemoglobin (Mhb1): a possible link to PR-1a gene expression in Mhb1-overproducing tobacco plants. FEBS Lett 571:61–66

Sokolovski S, Blatt MR (2004) Nitric oxide block of outward-rectifying K^+ channels indicates direct control by protein nitrosylation in guard cells. Plant Physiol 136:4275–4284

Sokolovski S, Blatt MR (2007) Nitric oxide and plant ion channel control. In: Lamattina L, Polacco JC (eds) Nitric oxide in plant growth. Plant Cell Monographs, vol 6. Springer, Berlin Heidelberg New York (in press)

Sokolovski S, Hills A, Gay R, Garcia-Mata C, Lamattina L, Blatt MR (2005) Protein phosphorylation is a prerequisite for intracellular Ca^{2+} release and ion channel control by nitric oxide and abscisic acid in guard cells. Plant J 43:520–529

Stamler JS, Lamas S, Fang FC (2001) Nitrosylation: the prototypic redox-based signaling mechanism. Cell 106:675–678

Sun J, Xin C, Eu JP, Stamler JS and Meissner G (2001) Cysteine-3635 is responsible for skeletal muscle ryanodine receptor modulation by NO. Proc Natl Acad Sci USA 98:11158–11162

Vandelle E, Poinssot B, Wendehenne D, Bentejac M, Pugin A (2006) Integrated signaling network involving calcium, nitric oxide, active oxygen species but not mitogen-activated protein kinases in BcPG1-elicited grapevine defenses. Mol Plant-Microbe Interact 19:429–440

Wendehenne D, Durner J, Klessig D (2004) Nitric oxide: a new player in plant signaling and defence responses. Curr Opin Plant Biol 7:449–455

White RE (1999) Cyclic GMP and ion channel regulation. Adv Second Messenger Phosphoprotein Res 33:251–277

Willmott N, Sethi JK, Walseth TF, Lee H, White AM, Galione A (1996) Nitric oxide-induced mobilization of intracellular calcium via the cyclic ADP-ribose signaling pathway. J Biol Chem 271:3699–3705

Yamamoto A, Katou S, Yoshioka H, Doke N, Kawakita K (2004) Involvement of nitric oxide generation in hypersensitive cell death induced by elicitin in tobacco cell suspension culture. J Gen Plant Pathol 70:85–92

Yoshimura N, Seki S, De Groat W (2001) Nitric oxide modulates Ca^{2+} channels in dorsal root ganglion neuros innervating rat urinary bladder. J Neurophysiol 86:304–311

Plant Cell Monogr (6)
L. Lamattina · J. C. Polacco: Nitric Oxide in Plant Growth
DOI 10.1007/7089_2006_084/Published online: 7 October 2006

S-Nitrosylation in Plants – Spectrum and Selectivity

Christian Lindermayr · Jörg Durner (✉)

Institute for Biochemical Plant Pathology,
GSF-Research Center for Environment and Health,
Ingolstädter Landstrasse 1, 85764 Neuherberg-Munich, Germany
durner@gsf.de

Abstract Nitric oxide (NO) has become recognized as a key signaling molecule in plants over the last few years, but still little is known about the way in which NO regulates different events in plants. Analyses of NO-dependent processes in animal systems have demonstrated protein S-nitrosylation – the covalent attachment of NO to the sulfhydryl group of cysteine residues – to be one of the dominant regulation mechanisms for many animal proteins. This reversible protein modification is an important posttranslational, redox-based regulation mechanism for many proteins of different classes in animals. For plants, however, the importance of protein S-nitrosylation remained to be elucidated.

This chapter will discuss the chemistry of S-nitrosothiol formation and the release of NO from S-nitrosylated cysteine residues, as well as the specificity and regulation of S-nitrosylation. Furthermore, the identification of plant proteins as candidates for this type of protein modification, and the physiological functions of protein S-nitrosylation in plants are described.

1 Introduction

Although a lot of studies have implicated nitric oxide (NO) as a key regulator for many different physiological processes in plants, less is known about how this molecule regulates these different events. As a readily diffusible free radical, NO reacts with a variety of intracellular and extracellular targets and can act as activator or inhibitor of enzymes, ion-channels or transcription factors, as well as modulator of protein function. As a result of these modifications NO regulates specific processes during plant growth and development and abiotic or biotic stress situations. The alteration of protein function/activity by NO is a well-described mechanism for the control of physiological processes in animals.

Different reactions have been described by which NO interacts with proteins. As a precursor of the reactive nitrating species, peroxinitrite and nitrogen dioxide, it modifies proteins by generation of 3-nitrotyrosine, as shown for the tyrosine residues 161 and 357 of alpha-tubulin (Tedeschi et al. 2005). Protein nitration is an irreversible reaction, which is of importance for pathophysiological processes but probably not for signaling. Because of its reactivity with transition metals NO can also bind to metal ions of heme groups. This is reported for bacterial flavohemoglobin (Hausladen et al. 2001) and for the activation of the soluble guanylate cyclase (GC) (Brandish et al. 1998; Russwurm and Koesling 2004). The NO-sensitive GC plays a key role in the NO/cGMP signaling cascade involved in many different physiological processes in animals (Ignarro et al. 1999; Danielewski et al. 2005; Zsombok et al. 2005). Since endogenous cGMP concentration is increased in tobacco leaves and cell suspension cultures after NO treatment, a NO-dependent cGMP signaling pathway has been also proposed for plants (Durner et al. 1998).

Finally, NO can react with the thiol group of cysteine residues to form S-nitrosothiols (S-nitrosylation) as described, e.g., for mammalian methionine adenosyltransferase (Perez-Mato et al. 1999) and ryanodine receptor/ Ca^{2+}-channel (Eu et al. 2000). The majority of all NO-affected proteins seem to be regulated by S-nitrosylation, making this type of protein modification a predominant mechanism in NO signaling. We will focus on this NO-dependent modification of cysteine residues, describing its chemistry/ formation, specificity, and physiological function in plants.

2 Chemistry of S-Nitrosylation/Formation and NO Release

Many different posttranslational modifications of proteins are known, including glycosylation, acetylation, phosporylation, and redox-dependent reactions. Phosphorylation and redox-dependent alterations are especially conserved throughout evolution and play a pivotal role in cellular signaling systems. Redox-dependent modification involves small reactive molecules, which reversibly modify redox-sensitive amino acids or groups such as the iron-sulfur cluster or heme to regulate cellular processes. Particularly, cysteine residues are targets for multiple redox-based modifications, resulting in the formation of sulfenic acid (S-OH), sulfinic acid ($S\text{-}O_2H$), S-nitrosylation (S-NO), glutathionylation (S-SG) or intramolecular disulfides (S-S). All these modifications may potentially alter protein function/activity and therefore influence cellular processes. Within the redox-based modifications, nitrosylation of cysteine residues in target proteins, with corresponding changes in the functional parameters of the modified proteins, seem to be of enormous importance for the regulation of many different physiological processes in microorganisms, animals, and plants.

S-Nitrosylation of proteins is an enzyme-independent process, but the exact chemistry underlying this posttranslational modification is still not fully understood. NO can exist in three different reactive states: nitrosonium cation (NO^+), NO radical ($NO^\bullet$), and nitroxyl anion (NO^-), which all show different reactivities with thiol groups. The most important reactions resulting in the formation of S-nitrosothiols are summarized in Fig. 1. While $NO^\bullet$ does not

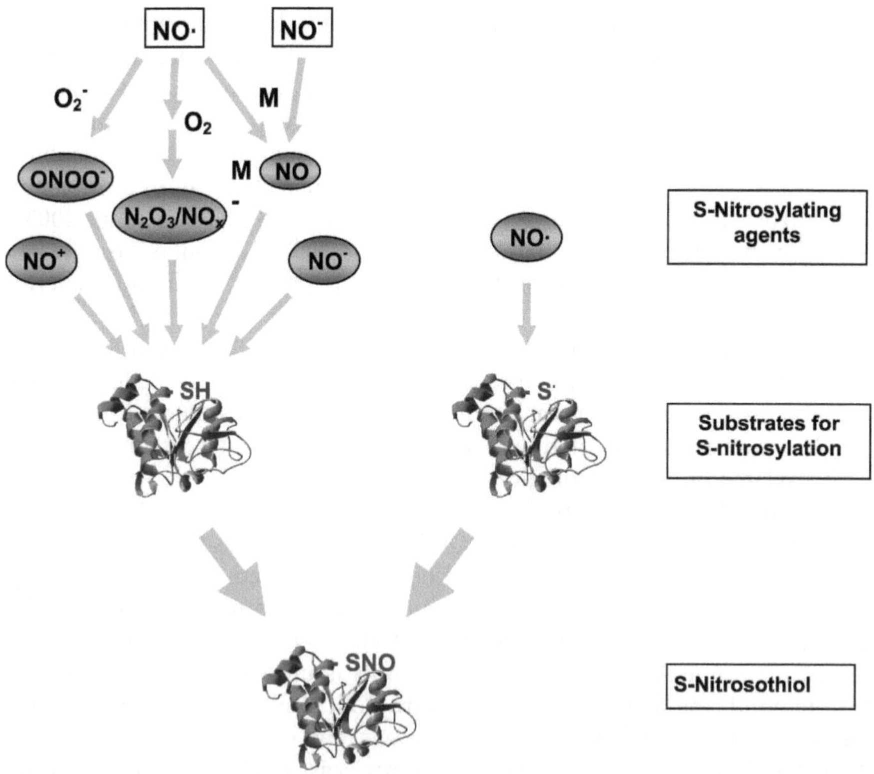

Fig. 1 Summary of the most important reactions of the different NO redox states leading to formation of S-nitrosothiols. NO reacts primarily with superoxide (O_2^-), oxygen (O_2), and redox metals (M) resulting in formation of S-nitrosylating agents, such as peroxynitrite ($ONOO^-$), NO oxides, or metal–NO complexes (M-NO). The latter can also be formed by the reaction of nitroxyl anions with redox metals. Furthermore, the different reactive states of NO can act as S-nitrosylating agents, since the direct reaction of NO^+ and NO^- with thiol groups, as well as the reaction of $NO^•$ with thinyl radicals ($RS^•$), result in the formation of S-nitrosothiols. The thiol groups and thinyl radicals can either be part of a protein or a low molecular weight compound such as glutathione

directly interact with thiols, NO^+ confers strong electrophilicity and reactivity towards most biological R-SH species. But, next to the direct S-nitrosylation activity of NO, it mainly functions as a precursor for several higher nitrogen oxides, which effectively mediate S-nitrosylation of proteins.

The complexity of the S-nitrosylation chemistry resulted in the description of many different reaction mechanisms, whereas most of them begin with the oxidation of $NO^•$ to form $NO_2^•$ and N_2O_3 (Hogg 2002; Gaston et al. 2003). Both are effective S-nitrosylating molecules in biological systems and these higher oxides of nitrogen are likely to be the major nitrosylating species, especially in cells that produce moderate to high levels of NO. Furthermore, the formation of these nitrogen oxides is favored in lipophilic environments, such as membranes or hydrophobic pockets in proteins. Under these conditions, $NO^{\bullet}$ and O_2 may reach higher concentrations, accelerating the reaction of both molecules 300-fold in comparison to the reaction in the surrounding medium (Liu et al. 1998; Martinez-Ruiz and Lamas 2004). Nucleophilic attack of NO^- on relatively electropositive cysteine sulfur groups also results in S-nitrosylation (Lipton et al. 1993; Kim et al. 1999; Nelson et al. 2003). However, the direct reaction between thiols and NO^- is dependent on the energy state of NO^- and occurs only when NO^- is present in the high energy singlet state.

Additionally, thinyl radicals ($RS^{\bullet}$) can react with $NO^{\bullet}$ to build S-nitrosothiols. Thiols, such as glutathione, are necessary to control the cellular redox status and occur in millimolar amounts in cells. It is absolutely imaginable that low concentrations of thinyl radicals are present in cells and become S-nitrosylated under physiological conditions.

The simultaneous production of superoxide (O_2^-) and $NO^{\bullet}$ can result in the formation of peroxynitrite ($ONOO^-$), which is also able to nitrosylate sulfhydryl groups (Viner et al. 1999). $ONOO^-$ can also be formed by the reaction between O_2 and NO^-, provided that the nitroxyl anions exist in their lower energy triplet state (Kim et al. 1999). The mechanism of the peroxynitrite-mediated modification of thiols is unclear and might involve, on one hand, the formation of the electrophilic S-nitrosylating agent N_2O_3 and, on the other hand, nucleophilic pathways (van der Vliet et al. 1998; Goldstein et al. 1999; Viner et al. 1999). In the latter case nucleophilic attack of the thiolate anion on the nitrogen atom in ONOOH results in the formation of corresponding S-nitrosothiol with displacement of HOO^- (van der Vliet et al. 1998).

As mentioned above, at low NO concentrations the production of N_2O_3 would be minimal and a greater fraction of NO would directly react with transition metals, especially iron ions, to form metal–NO complexes. The metals can be present as free ions or bound to other compounds, such as heme groups. The metal–NO complexes are able to transfer their NO moiety to thiol groups of cysteine residues, making them important S-nitrosylating agents operating efficiently in hydrophilic environments, rather than in hydrophobic environments (Lane et al. 2001). Next to $NO^{\bullet}$, NO^- molecules are also able to form metal–NO complexes. Metal-catalyzed S-nitrosylation is described for the hemoglobin of the worm *Ascaris lumbricoides*, a common parasite of human and domestic animals (Minning et al. 1999). The worm tolerates O_2 poorly and has therefore adapted hemoglobin to detoxify it. Worm-produced NO binds ferric heme to produce HbFe(III)NO, and NO is then transferred to a nearby cysteine, followed by a one-electron reduction of iron, yielding ferrous heme, HbFe(II). Oxygen binds to HbFe(II) and reacts with the newly formed SNO to generate nitrate. In this way, the parasite uses a heme-catalyzed S-nitrosylation reaction to eliminate oxygen.

Another type of a S-nitrosylation mechanism is the direct transfer of an NO group from a nitrosylated protein or a low molecular weight S-nitrosothiol, such as S-nitrosoglutathione, to the sulfur group of a cysteine residue (trans-S-nitrosylation) (Scharfstein et al. 1994; Stamler 1994; Gaston et al. 2003). Within a cell exists an equilibrium between high- and low-mass thiols and high- and low-mass S-nitrosothiols, which is accurately regulated by a small set of proteins.

3 Selectivity/Specificity of S-Nitrosylation

Since most proteins possess cysteine residues, substrate specificity is a very important feature of endogenous protein S-nitrosylation. This includes structural factors that influence the susceptibility to S-nitrosylation, like surrounding acidic or basic amino acids and the presence of a hydrophobic environment that enables the formation of S-nitrosylating species via the reaction between oxygen and NO (Stamler et al. 1997). An acid-base consensus motif for S-nitrosylation was derived by Stamler et al. (1997) by comparing amino acid sequences surrounding NO-sensitive cysteine residues ([GSTCYNQ][KRHDE]C[DE]). It is well-established that deprotonation of thiol to the nucleophilic thiolate (RS^-) is suppressed and enhanced, respectively, by neighboring acidic and basic groups. Electrostatic interaction between the basic side chains of these amino acids and the thiol group of cysteine residues increases its nucleophilicity and H^+ release is promoted. Furthermore, the interaction of S-nitrosylated cysteines with acidic side chains results in reduction of the nucleophilicity of the thiol and promotion of NO^+ donation. This seemed to be important for trans-S-nitrosylation reactions from the in vivo NO donor nitrosoglutathione (Perez-Mato et al. 1999; Hess et al. 2005). In Arabidopsis, methionine adenosyltransferase 1 S-nitrosylation of Cys114 is responsible for inhibition of the enzyme activity (Lindermayr et al. 2006). This cysteine residue is surrounded by basic (Lys113) as well as acidic (Glu116, Glu117) amino acids, whose side chains are positioned close to the thiol group of Cys114 and facilitate, on one side, the H^+-release of the SH group of Cys114 and, on the other side, the NO^+-release of GSNO (Lindermayr et al. 2006). But, it should be mentioned that generally not only amino acids directly flanking a NO-sensitive cysteine interact with this residue, but also amino acid side chains, which are brought to such a cysteine by the three-dimensional conformation (Perez-Mato et al. 1999).

However, S-nitrosylation experiments done with peptides containing cysteine residues flanked by different combinations of amino acids revealed that S-nitrosylation appears to be independent of the type of amino acids surrounding the cysteine residue (Taldone et al. 2005). The analyzed model peptides were based on the protein sequence of mammalian eNOS. Several point mutations were introduced to create basic, acidic, hydrophobic, or hydrophilic environments surrounding the cysteine residue. The peptides were S-nitrosylated under different conditions and cysteine modification was analyzed by electrospray ionization mass spectrometry. Unfortunately, the peptides were exposed to an excess of S-nitrosylating agents, which do not reflect physiological conditions and therefore probably overcast the influence of the amino acids neighboring the cysteine residue and promote non-specific S-nitrosylation. Additionally, analyses of the three-dimensional structure of several proteins containing NO-reactive cysteine residues showed that nitrosylation and denitrosylation processes may depend on the cysteine-Sγ atomic structural microenvironment rather than on the surrounding amino acids (Ascenzi et al. 2000).

Another important S-nitrosylation motif promotes binding of GSNO in such a way that the NO group of GSNO is positioned close to the sulfhydryl group of the reactive cysteines. Molecular modeling revealed that GSNO docks to OxyR through hydrogen bonding between the γ-glutamyl amine of GSNO and the γ-carboxylate of Asp202 of OxyR (Hess et al. 2005). The potential motif for GSNO-mediated trans-S-nitrosylation was then identified by database analysis ([H,K,R]C[hydrophobic]X[D,E]), and was shown to be generally present in some copies in about 650 substrates for protein S-nitrosylation, including bacterial transcription factor OxyR, annexin-6, glutathione-S-transferase-mu, and semaphoring-4D (Hess et al. 2005).

4
Protein S-Nitrosylation in Plants

Analyses of NO-dependent processes in animal systems have demonstrated protein S-nitrosylation of cysteine residues to be one of the dominant regulation mechanisms for many animal proteins. Although NO has become recognized as a key signaling molecule in plants over the last few years, less is known about this NO-dependent type of posttranslational modification in plants. To obtain insight into the regulatory function of protein S-nitrosylation, the identification of proteins sensitive to this modification is of enormous importance. However, the major problem for this venture is the lability of S-nitrosylated proteins. Nitrosothiols decompose in the presence of intracellular reducing agents, such as ascorbate or glutathione, and reducing metals, especially Cu(I). This sensitivity to reducing agents results in half-lives of seconds to a few minutes and makes it difficult to isolate S-nitrosylated proteins.

To detect S-nitrosylated proteins a three-step procedure was developed, which converts S-nitrosylated cysteines to biotinylated cysteines (Jaffrey and Snyder 2001). In brief, free thiols were blocked with methyl methanethiosulfonate followed by selective decomposition of the nitrosothiol bonds with ascorbate, resulting in the reduction of nitrosothiols to thiols. In the final step, the newly formed thiols react with the biotinylating agent *N*-[6-biotinamido)hexyl]-3′-(2′-pyridyldithion)propionamide. The biotin-labeled proteins can easily be detected by immunoblotting using anti-biotin antibodies, or purified using immobilized streptavidin. This method was successfully used to identify candidates for S-nitrosylation in mammalian brain lysates, mitochondria, endothelia cells, and mesanglia cells, and to determine the S-nitroso proteome of *Mycobacterium tuberculosis* (Jaffrey et al. 2001; Kuncewicz et al. 2003; Foster and Stamler 2004; Martinez-Ruiz and Lamas 2004; Rhee et al. 2005).

To identify candidates for S-nitrosylation "in plants", extracts of *Arabidopsis thaliana* cell suspension cultures were treated with the NO-donor S-nitrosoglutathione to generate S-nitrosylated proteins. Furthermore, *A. thaliana* plants were treated with gaseous NO to analyze whether S-nitrosylation can occur in the specific redox environment of a plant cell in vivo. S-Nitrosylated proteins were detected by the biotin switch method and biotin-labeled proteins were purified and analyzed using nano liquid chromatography in combination with mass spectrometry. We identified 63 proteins from cell cultures and 52 proteins from leaves that represent candidates for S-nitrosylation including stress-related, redox-related, signaling/regulating, cytoskeleton, and metabolic proteins [(Lindermayr et al. 2005) and selected candidates (Table 1)]. Of the identified proteins from cell cultures and leaves, 33 and 15%, respectively, have already been reported to become S-nitrosylated in animals, suggesting that several NO signaling and regulating mechanisms seem to be conserved within both kingdoms (Fig. 2). Additionally, at least 30% of the identified candidates in both approaches have already been described in contexts associated with redox-regulated processes in animals or plants. Further, 37% of the identified cell culture proteins and 46% of the identified leaf proteins represent new candidates for protein S-nitrosylation (Fig. 2).

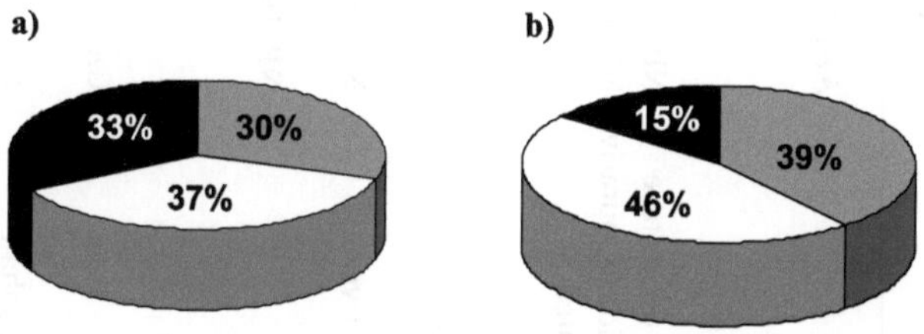

Fig. 2 Candidates identified for protein S-nitrosylation in *Arabidopsis thaliana*. The majority of the identified candidates from cell cultures (**a**) and leaves (**b**) have been already described in the context of S-nitrosylation in animals (*black*) and redox-regulated processes in plants or animals (*gray*). Identified candidates not yet described in either the context of redox-regulated processes or of S-nitrosylation are represented by *white* sections

Table 1 Selected candidates of identified S-nitrosylated proteins from *A. thaliana* cell cultures and leaves

Protein	Acc. no	MW	Identified peptides (score) GSNO/NO	GSH/untreated	Hints to S-nitrosylation
Hsp 90, putative	AAL49788	80036	ADLVNNLGTIAR (50) + 1		Jaffrey et al. 2001; Fratelli et al. 2002, 2003; Lind et al. 2002; Shenton and Grant 2003
Cu/Zn-superoxide dismutase	NP_172360	15088	AVVVHADPDDLGK (55) + 1		Lawler and Song 2002
Glutathione peroxidase, putative	NP_192897	25568	FAPTTSPLSIEK (65) + 1	FAPTTSPLSIEK (51)	Koh et al. 2001
Peroxiredoxin-related	AAF66133	21230	AVNVEEAPSDFK (37)		Motohashi et al. 2001; Fratelli et al. 2002, 2003; Lind et al. 2002
Type 2 peroxiredoxin-related	NP_176773	17417	APIAVGDVVPDGTISFF DENDQLQTASVHSLAAGK (51)		
Glutaredoxin, putative	NP_198853	11749	LVPLLTEAGAIAGK (59)		Klatt et al. 2000; Song and Lee 2003
Actin 2/7	NP_196543	41709	NYELPDGQVITIGAER (114) + 3	AGFAGDDAPR (53) +1	Dalle-Donne et al. 2000, 2003; Jaffrey et al. 2001; Fratelli et al. 2002, 2003; Lind et al. 2002
Annexin	CAA67608	35757	HYNDEDVIR (48) +2	DALLANEATK (48) + 2	Liu et al. 2002; Kuncewicz et al. 2003
Fructose 1,6-biphosphate aldolase, putative	NP_190861	38516	VSPEVIAEHTVR (85) + 4	LASINVENVETNR (64) + 3	Fratelli et al. 2002; Lind et al. 2002; Ito et al. 2003; Shenton and Grant 2003

Table 1 (continued)

Protein	Acc. no	MW	Identified peptides (score) GSNO/NO	GSH/untreated	Hints to S-nitrosylation
Triosephosphate isomerase	T50646	27138	VASPAQAQEVHDELR (109) + 3	VASPAQAQEVHDELR (68) + 3	Fratelli et al. 2002; Ito et al. 2003; Shenton and Grant 2003
GAPDH C subunit	NP_187062	36891	GILGYTEDDVVSTDFVGDNR (92) + 4	AASFNIIPSSTGAAK (46) + 3	Mohr et al. 1996, 1999; Klatt et al. 2000; Motohashi et al. 2001; Kuncewicz et al. 2003; Shenton and Grant 2003
2-Phosphoglycerate hydrolase (enolase)	NP_181192	47689	VTAAVPSGASTGIYEALELR (86) + 3		Fratelli et al. 2002; Lind et al. 2002; Shenton and Grant 2003
Phosphoglycerate kinase	NP_178073	42105	GVTTIIGGGDSVAAVEK (62) + 2		Fratelli et al. 2002
SAM synthetase, putative	AAO11581	42769	KPEEVGAGDQGHMFGYAT + 5 DETPELMPLTHVLATK (70)	FVIGGPHGDAGLTGR (76) + 4	Ruiz et al. 1998; Perez-Mato et al. 1999
Adenosylhomocysteinase	CAB09795	51447	LVGVSEETTTGVK (48)		
Methionine synthase	NP_197294	84304	IPSSEEIADR (82) + 1	FALESFWDGK (53) + 1	Yamazaki et al. 2004
ATP synthase CF1 alpha chain	NP_051044	55294	LIESPAPGIISR (49) + 2		Eaton et al. 2003

Cell culture extracts treated with GSNO or GSH and leaf extracts of NO-treated or untreated *A. thaliana* plants were subjected to the biotin switch method and analyzed by nanoLC/MS/MS after trypic digestion. The MASCOT search engine was used to parse mass spectrometry data to identify proteins from primary sequence databases. The best-matching peptide identifying the protein is given. If there were further peptides found, the number of the peptides is given. References for hints confirming that the identified protein is a candidate for S-nitrosylation are given. The results of two separate experiments for each treatment are summarized

Acc. no accession number

To obtain insight into the physiological and regulatory functions of NO in plants it is of enormous importance to determine the effects of S-nitrosylation on the identified plant proteins, e g., whether enzyme activities are inhibited or enhanced due to S-nitrosylation, or whether structural alterations followed by change of protein function are a consequence of the modification. These results will probably also give us hints about the regulation of cross-talk between plant-specific NO-, salicylic acid-, and jasmonic acid/ethylene-dependent signaling pathways.

5 Regulation of S-Nitrosylation

A very important characteristic for a sensitive signaling mechanism is its fine-tuned regulation, which means in the first instance the fast production as well as the fast degradation of the signaling compounds and the modifications they are responsible for. Since the formation of S-nitrosothiols is an enzyme-independent process, S-nitrosylation depends mainly on the activity of NO-producing systems and the NO accessibility of the target protein. In plants, several different sources of NO are described. A hormone-activated NOS from *A. thaliana* (AtNOS1) was cloned and its sequence showed similarity to a protein implicated in NO synthesis in the snail *Helix pomatia* (Guo et al. 2003). Treatment of Arabidopsis cell suspension cultures with bacterial lipopolysaccharides resulted in a rapid production of NO due to the activation of AtNOS, demonstrating its implication in plant–pathogen interaction (Zeidler et al. 2004). Nitrate reductase (NR) and nitrite reductase (NiR) are also enzymes that are capable of producing NO in plants (Yamasaki 2000; Stöhr et al. 2001; Rockel et al. 2002). Additionally, in tobacco cell suspensions an enzymatic production of NO that is independent of NR and NiR has been described (Planchet et al. 2005). In this NiR activity, electron carriers of the mitochondrial respiratory chain seemed to be involved. Interestingly, in higher plants only mitochondria of roots, but not of leaves, are able to reduce nitrite to NO (Gupta et al. 2005; see also the chapter by Kaiser et al. in this volume). NO production may involve not only other currently unidentified enzyme activities but also non-enzymatic principles (del Rio et al. 2003). For instance, recent work has shown that a non-enzymatic reduction of nitrite to NO occurs in the apoplast of barley aleurone layers (Bethke et al. 2004). This NO production required an acidic pH and was accelerated by reducing agents such as phenolic compounds. All these different NO sources could potentially regulate S-nitrosylation on the level of NO production and influence the S-nitrosothiol content in cells.

The stability of S-nitrosothiols is subject to different factors, such as the redox status of the cell environment or the activity of S-nitrosothiol-degrading enzymes. A number of enzymes, including GSNO reductase (GSNOR) (Liu et al. 2001), thioredoxin reductase (Nikitovic and Holmgren 1996; Haendeler et al. 2002), copper zinc superoxide dismutase (Johnson et al. 2001), and γ-glutamyl transpeptidase (Hogg et al. 1997), are shown to decompose GSNO and have an important physiological role in regulating endogenous S-nitrosothiol content and modulating S-nitrosothiol signaling. S-Nitrosylated proteins can be denitrosylated by the reducing action of glutathione or thioredoxin. The consequence of this reaction is the S-nitrosylation of the reducing molecules, which both can be reduced enzymatically by GSNO reductase or thioredoxin reductase, respectively (Nikitovic and Holmgren 1996). GSNO reductase is conserved from bacteria to animals and plants (Liu et al. 2001; Sakamoto et al. 2002; Diaz et al. 2003) and was previously identified as glutathione-dependent formaldehyde dehydrogenase (Fliegmann and Sandermann 1997; Liu et al. 2001; Achkor et al. 2003). The main products of the enzymatic reduction of GSNO are oxidized glutathione and ammonia. The GSNOR has been shown to play a pivotal role in controlling the S-nitrosothiol content in cells, and Arabidopsis plants deficient in GSNOR show higher basal S-nitrosothiol levels than wild-type plants (Feechan et al. 2005).

S-Nitrosothiols can also be reduced by thioredoxin due to transference of the NO from the S-nitrosothiol to Cys69 of thioredoxin (Haendeler et al. 2002). Subsequently, the S-nitrosylated thioredoxin can be denitrosylated by thioredoxin reductase. Both the GSH/GSNOR and the thioredoxin/thioredoxin reductase seemed to be important for protecting the intracellular environment from excessive nitrosative stress. A number of other enzymes were shown to decompose S-nitrosothiols in in-vitro experiments, including glutathione peroxidase (Kuo and Kocis 2002; Kuo et al. 2002), γ-glutamyl transpeptidase (Hogg et al. 1997; Henson et al. 1999), and xanthine oxidase (Trujillo et al. 1998), but none of these enzymes has been demonstrated to regulate endogenous S-nitrosothiol content.

Next to these enzymatic systems, the intracellular redox status is crucial for the S-nitrosothiol content in cells (Beltran et al. 2000). Ascorbate and glutathione are of paramount importance for the antioxidative system. They protect cells from oxidative and nitrosative stress and are therefore important regulators of the intracellular redox status (Noctor and Foyer 1998). Glutathione is the major low molecular weight non-protein thiol in cells and reacts with NO_x to form GSNO, which is known to be capable of transnitrosylating proteins. Since glutathione is present in millimolar amounts, the formation of GSNO due to nitrosative stress is likely to predominate. Depletion of GSH was shown to increase protein S-nitrosylation, while increased GSH levels may lower levels of S-nitrosylated proteins via an S-transnitrosylation reaction that forms GSNO (Haqqani et al. 2003). Ascorbate is also a reductive agent, which effectively releases NO from GSNO (Smith and Dasgupta 2000). The reduction of GSNO by ascorbate proceeds in one state, which involves reduction of the $S-N$ bond followed by release of the NO group.

6 Physiological Function of S-Nitrosothiols in Plants

NO is an important redox-active signaling molecule acting through two principle ways of signaling: (i) through activation of guanylate cyclase, which produces the second messenger cyclic GMP, or (ii) through S-nitrosylation of cysteine residues of redox-sensitive proteins. In animals, protein S-nitrosylation regulates many different processes such as protein–protein interactions (Matsumoto et al. 2003), apoptosis (Haendeler et al. 2002), signal transduction through Src (Akhand et al. 1999), and protein ubiquitynation (Hess et al. 2005). In contrast, investigation of the regulatory functions of protein S-nitrosylation in plants is just beginning. Until now a physiological function of S-nitrosylation has only been described for *Arabidopsis* non-symbiotic class 1-hemoglobin (AHb1). Hemoglobins are well-characterized regulators of NO homeostasis in microorganisms and animals and it has been speculated that the function of ancestral hemoglobin is to protect cells from nitrosative stress by converting NO to NO_3^- (Durner et al. 1999). In plants, non-symbiotic AHb1 is demonstrated to react with GSNO, resulting in the formation of S-nitrosylated AHb1 (Perazzolli et al. 2004). In addition, increased nitrate reductase-mediated NO production rèsulted in the formation of S-nitrosylated AHb1 and significantly lower NO emission in AHb1-overexpressing Arabidopsis plants than wild-type plants. Moreover, AHb1 is S-nitrosylated under hypoxic stress conditions, which are known to stimulate NO production via nitrate reductase (Rockel et al. 2002; Perazzolli et al. 2004). Thus, this protein is able to scavenge NO through formation of S-nitrosohemoglobin, suggesting a role in NO detoxification.

The participation of NO in abscisic acid-induced stomatal closure is well established (Mata and Lamattina 2001; Neill et al. 2002). However, neither the target(s) nor the way of action of this signaling molecule in guard cells has yet been resolved. Recent work provides some evidence that outward-rectifying K^+-channels of *Vicia faba* guard cells are regulated by S-nitrosylation of cysteine sulfhydryl groups, influencing stomatal closure (Sokolovski and Blatt 2004). Patch clamp experiments with outward-rectifying K^+ channels revealed that exposure to submicromolar amounts of NO reduced the amplitude of K^+ current ($I_{K,out}$). The $I_{K,out}$ could be recovered with reducing agents such as dithiothreitol, suggesting a redox-mediated inactivation of $I_{K,out}$ by NO. In addition, the action of NO could be mimicked by the SH-modifying membrane-permeable reagent NEM, confirming an important role for thiol groups in regulation the K^+ channels. Furthermore, the membrane-impermeable thiol-modifying compound 5,5′-dithiobis-2-nitrozoic acid did not inhibit K^+ current, suggesting that the regulatory cysteine residue(s) are not accessible at the external site of the membrane. In summary, these results indicate that the K^+ channel, or a closely associated regulatory protein, is regulated by S-nitrosylation of cysteine sulfhydryl groups.

Plants respond to the attack of pathogens with NO production, emphasizing the important role of this molecule in plant defense (Delledonne et al. 1998; Durner et al. 1998). The increased NO formation, however, also supposes the involvement of S-nitrosothiols in plant defense reactions. As already mentioned above, GSNO reductase controls S-nitrosothiol levels in living cells. Therefore, Arabidopsis mutants with altered GSNO reductase activities were analyzed for their ability to fight against plant pathogens (Feechan et al. 2005). The results of these experiments demonstrated the importance of S-nitrosothiols for R-gene-mediated resistance, basal disease resistance, and non-host resistance. Surprisingly, enhanced levels of S-nitrosothiols did not result in resistance against microbial infections, but led to enhanced susceptibility, suggesting that NO and S-nitrosothiols might have different functions in plant disease resistance. Disease susceptibility in S-nitrosothiol-enriched plants was shown to be associated with reduced and delayed expression of PR1, a marker gene for salicylic acid-dependent signaling (Feechan et al. 2005). However, accumulation of PR1 transcripts could not be restored by exogenous application of SA, concluding a role for S-nitrosothiols coincident with or downstream of SA. Additionally, these plants produced lower amounts of salicylic acid in response to virulent, avirulent, and non-host pathogens in comparison to wild-type plants, suggesting that S-nitrosothiols also control the biosynthesis of SA. It is not surprising that S-nitrosothiols act at many distinct sites within the cell, because S-nitrosothiols is a collective term for hundreds of S-nitrosylated low molecular and high molecular weight thiols, which may all have different functions. From this point of view it is very important to know which set of potential targets for S-nitrosylation is present at the time of treatment.

A reduction in GSNO reductase at the transcript and protein level was observed after SA treatment of Arabidopsis plantlets, underlining an interaction between SA signaling and GSNO reductase/S-nitrosothiols (Diaz et al. 2003). Furthermore, wound-induced reduction of GSNO reductase transcripts and proteins in Arabidopsis, as well as jasmonic acid-dependent down-regulation of GSNO reductase transcripts in tobacco BY-2 cells, emphasize the multiple roles of this enzyme and/or S-nitrosothiols in plant signaling. In summary, GSNO reductase is responsible for controlling the intracellular thiol/S-nitrosothiol ratio and it seems to be involved in regulation of SA, JA, NO, and GSNO signaling.

In-vitro experiments suggested that glyceraldehyde 3-phosphate dehydrogenase (GAPDH) is regulated by NO via S-nitrosylation (Perazzolli et al. 2004; Sokolovski and Blatt 2004; Lindermayr et al. 2005), but a possible physiological function of S-nitrosylated GAPDH in plants is unknown. In mammals, this glycolytic enzyme participates in nuclear processes such as gene transcription and RNA transport. In addition, DNA replication and nuclear translocation of GAPDH during apoptotic cell death is described for many cell systems. Hara et al. (2005) revealed an NO-dependent signaling pathway resulting in S-nitrosylation of GAPDH and triggering of its binding to an E3 ubiquitin ligase (Siah1) followed by translocation to the nucleus. In the nucleus, S-nitrosylated GAPDH stabilizes Siah1, enhancing its E3 ubiquitin ligase activity and causing cell death. Probably, S-nitrosylation of plant GAPDH also results in a new function of this glycolytic enzyme.

Furthermore, Arabidopsis methionine adenosyltransferase (MAT) is differentially inhibited by S-nitrosylation (Lindermayr et al. 2006). Arabidopsis has four isoforms of this enzyme, but only the activity of isoform 1 is significantly reduced after treatment with GSNO. Cysteine 114 is located directly next to the putative catalytic center as part of the active site loop and its S-nitrosylation is responsible for inhibition of the enzyme activity. Since MAT catalyzes the biosynthesis of S-adenosylmethionine, which is a substrate for the biosynthesis of the plant hormone ethylene, inhibition of MAT1 will result in reduction of the ethylene biosynthesis. A linkage of NO and ethylene emission in senescing foliage and in fruit, flower, and vegetable ripening was previously reported by Leshem and colleagues (Leshem and Haramaty 1996; Leshem et al. 1998), who demonstrated the effect of NO in extending the postharvest life of fresh horticultural products by reducing ethylene synthesis. Additionally, ethylene and NO monitoring of both vegetative and generative plant organs revealed that temporal progress of maturation and senescence go hand in hand with a significant decrease in endogenous NO emission and an increase in ethylene emission indicating that NO has opposite effects to ethylene (Leshem 2000). Furthermore, GSNO markedly reduced ethylene emission of Arabidopsis cell cultures (Lindermayr et al. 2006). These observations suggest that NO directly acts by down-regulating ethylene synthesis and that MAT1 could be a switch point for regulating cross-talk between NO and ethylene signaling.

Although we are just at the beginning of the characterization of protein S-nitrosylation in plants, it is assumed that this type of protein modification plays a relevant role in the regulation of many different physiological processes in plants similar to that already demonstrated in animals.

References

Achkor H, Diaz M, Fernandez MR, Biosca JA, Pares X, Martinez MC (2003) Enhanced formaldehyde detoxification by overexpression of glutathione-dependent formaldehyde dehydrogenase from Arabidopsis. Plant Physiol 132:2248–2255

Akhand AA, Pu M, Senga T, Kato M, Suzuki H, Miyata T, Hamaguchi M, Nakashima I (1999) Nitric oxide controls src kinase activity through a sulfhydryl group modification-mediated Tyr-527-independent and Tyr-416-linked mechanism. J Biol Chem 274:25821–25826

Ascenzi P, Colasanti M, Persichini T, Muolo M, Polticelli F, Venturini G, Bordo D, Bolognesi M (2000) Re-evaluation of amino acid sequence and structural consensus rules for cysteine-nitric oxide reactivity. Biol Chem 381:623–627

Beltran B, Orsi A, Clementi E, Moncada S (2000) Oxidative stress and S-nitrosylation of proteins in cells. Br J Pharmacol 129:953–960

Bethke PC, Badger MR, Jones RL (2004) Apoplastic synthesis of nitric oxide by plant tissues. Plant Cell 16:332–341

Brandish PE, Buechler W, Marletta MA (1998) Regeneration of the ferrous heme of soluble guanylate cyclase from the nitric oxide complex: acceleration by thiols and oxyhemoglobin. Biochemistry 37:16898–16907

Dalle-Donne I, Giustarini D, Rossi R, Colombo R, Milzani A (2003) Reversible S-glutathionylation of Cys 374 regulates actin filament formation by inducing structural changes in the actin molecule. Free Radic Biol Med 34:23–32

Dalle-Donne I, Milzani A, Giustarini D, Di Simplicio P, Colombo R, Rossi R (2000) S-NO-actin: S-nitrosylation kinetics and the effect on isolated vascular smooth muscle. J Muscle Res Cell Motil 21:171–181

Danielewski O, Schultess J, Smolenski A (2005) The NO/cGMP pathway inhibits Rap 1 activation in human platelets via cGMP-dependent protein kinase I. Thromb Haemost 93:319–325

del Rio LA, Corpas FJ, Sandalio LM, Palma JM, Barroso JB (2003) Plant peroxisomes, reactive oxygen metabolism and nitric oxide. IUBMB Life 55:71–81

Delledonne M, Xia Y, Dixon RA, Lamb C (1998) Nitric oxide functions as a signal in plant disease resistance. Nature 394:585–588

Diaz M, Achkor H, Titarenko E, Martinez MC (2003) The gene encoding glutathione-dependent formaldehyde dehydrogenase/GSNO reductase is responsive to wounding, jasmonic acid and salicylic acid. FEBS Lett 543:136–139

Durner J, Gow AJ, Stamler JS, Glazebrook J (1999) Ancient origins of nitric oxide signaling in biological systems. Proc Natl Acad Sci USA 96:14206–14207

Durner J, Wendehenne D, Klessig DF (1998) Defense gene induction in tobacco by nitric oxide, cyclic GMP, and cyclic ADP-ribose. Proc Natl Acad Sci USA 95:10328–10333

Eaton P, Jones ME, McGregor E, Dunn MJ, Leeds N, Byers HL, Leung KY, Ward MA, Pratt JR, Shattock MJ (2003) Reversible cysteine-targeted oxidation of proteins during renal oxidative stress. J Am Soc Nephrol 14:S290–296

Eu JP, Sun J, Xu L, Stamler JS, Meissner G (2000) The skeletal muscle calcium release channel: coupled O2 sensor and NO signaling functions. Cell 102:499–509

Feechan A, Kwon E, Yun BW, Wang Y, Pallas JA, Loake GJ (2005) A central role for S-nitrosothiols in plant disease resistance. Proc Natl Acad Sci USA 102:8054–8059

Fliegmann J, Sandermann H Jr (1997) Maize glutathione-dependent formaldehyde dehydrogenase cDNA: a novel plant gene of detoxification. Plant Mol Biol 34:843–854

Foster MW, Stamler JS (2004) New insights into protein S-nitrosylation: Mitochondria as a model system. J Biol Chem 279:25891–25897

Fratelli M, Demol H, Puype M, Casagrande S, Eberini I, Salmona M, Bonetto V, Mengozzi M, Duffieux F, Miclet E, Bachi A, Vandekerckhove J, Gianazza E, Ghezzi P (2002) Identification by redox proteomics of glutathionylated proteins in oxidatively stressed human T lymphocytes. Proc Natl Acad Sci USA 99:3505–3510

Fratelli M, Demol H, Puype M, Casagrande S, Villa P, Eberini I, Vandekerckhove J, Gianazza E, Ghezzi P (2003) Identification of proteins undergoing glutathionylation in oxidatively stressed hepatocytes and hepatoma cells. Proteomics 3:1154–1161

Gaston BM, Carver J, Doctor A, Palmer LA (2003) S-Nitrosylation signaling in cell biology. Mol Interv 3:253–263

Goldstein S, Czapski G, Lind J, Merenyi G (1999) Effect of *NO on the decomposition of peroxynitrite: reaction of N2O3 with ONOO. Chem Res Toxicol 12:132–136

Guo F-Q, Okamoto M, Crawford MJ (2003) Identifcation of a Plant Nitric Oxide Synthase Gene Involved in Hormonal Signaling. Science 302:100–103

Gupta KJ, Stoimenova M, Kaiser WM (2005) In higher plants, only root mitochondria, but not leaf mitochondria reduce nitrite to NO, in vitro and in situ. J Exp Bot 56:2601–2609

Haendeler J, Hoffmann J, Tischler V, Berk BC, Zeiher AM, Dimmeler S (2002) Redox regulatory and anti-apoptotic functions of thioredoxin depend on S-nitrosylation at cysteine 69. Nat Cell Biol 4:743–749

Haqqani AS, Do SK, Birnboim HC (2003) The role of a formaldehyde dehydrogenase-glutathione pathway in protein S-nitrosation in mammalian cells. Nitric Oxide 9:172–181

Hara MR, Agrawal N, Kim SF, Cascio MB, Fujimuro M, Ozeki Y, Takahashi M, Cheah JH, Tankou SK, Hester LD, Ferris CD, Hayward SD, Snyder SH, Sawa A (2005) S-nitrosylated GAPDH initiates apoptotic cell death by nuclear translocation following Siah1 binding. Nat Cell Biol 7:665–674

Hausladen A, Gow A, Stamler JS (2001) Flavohemoglobin denitrosylase catalyzes the reaction of a nitroxyl equivalent with molecular oxygen. Proc Natl Acad Sci USA 98:10108–10112

Henson SE, Nichols TC, Holers VM, Karp DR (1999) The ectoenzyme gamma-glutamyl transpeptidase regulates antiproliferative effects of S-nitrosoglutathione on human T and B lymphocytes. J Immunol 163:1845–1852

Hess DT, Matsumoto A, Kim SO, Marshall HE, Stamler JS (2005) Protein S-nitrosylation: purview and parameters. Nat Rev Mol Cell Biol 6:150–166

Hogg N (2002) The biochemistry and physiology of S-nitrosothiols. Annu Rev Pharmacol Toxicol 42:585–600

Hogg N, Singh RJ, Konorev E, Joseph J, Kalyanaraman B (1997) S-Nitrosoglutathione as a substrate for gamma-glutamyl transpeptidase. Biochem J 323 (Pt 2):477–481

Ignarro LJ, Cirino G, Casini A, Napoli C (1999) Nitric oxide as a signaling molecule in the vascular system: an overview. J Cardiovasc Pharmacol 34:879–886

Ito H, Iwabuchi M, Ogawa K (2003) The sugar-metabolic enzymes aldolase and triose-phosphate isomerase are targets of glutathionylation in Arabidopsis thaliana: detection using biotinylated glutathione. Plant Cell Physiol 44:655–660

Jaffrey SR, Erdjument-Bromage H, Ferris CD, Tempst P, Snyder SH (2001) Protein S-nitrosylation: a physiological signal for neuronal nitric oxide. Nat Cell Biol 3:193–197

Jaffrey SR, Snyder SH (2001) The biotin switch method for the detection of S-nitrosylated proteins. Sci STKE 2001:PL1

Johnson MA, Macdonald TL, Mannick JB, Conaway MR, Gaston B (2001) Accelerated S-nitrosothiol breakdown by amyotrophic lateral sclerosis mutant copper, zinc-superoxide dismutase. J Biol Chem 276:39872–39878

Kaiser WM, Gupta KJ, Planchet E (2007) Higher plant mitochondria as a source for NO. In: Lamattina L, Polacco JC (eds) Nitric oxide in plant growth. Plant Cell Monographs, Vol 6. Springer, Berlin, Heidelberg, New York (in press)

Kim WK, Choi YB, Rayudu PV, Das P, Asaad W, Arnelle DR, Stamler JS, Lipton SA (1999) Attenuation of NMDA receptor activity and neurotoxicity by nitroxyl anion, NO. Neuron 24:461–469

Klatt P, Pineda Molina E, Perez-Sala D, Lamas S (2000) Novel application of S-nitroso-glutathione-Sepharose to identify proteins that are potential targets for S-nitrosoglutathione-induced mixed-disulphide formation. Biochem J 349:567–578

Koh YH, Suzuki K, Che W, Park YS, Miyamoto Y, Higashiyama S, Taniguchi N (2001) Inactivation of glutathione peroxidase by NO leads to the accumulation of H_2O_2 and the induction of HB-EGF via c-Jun NH2-terminal kinase in rat aortic smooth muscle cells. Faseb J 15:1472–1474

Kuncewicz T, Sheta EA, Goldknopf IL, Kone BC (2003) Proteomic analysis of S-nitrosylated proteins in mesangial cells. Mol Cell Proteomics 2:156–163

Kuo WN, Kocis JM (2002) Nitration/S-nitrosation of proteins by peroxynitrite-treatment and subsequent modification by glutathione S-transferase and glutathione peroxidase. Mol Cell Biochem 233:57–63

Kuo WN, Kocis JM, Mewar M (2002) Protein denitration/modification by glutathione-S-transferase and glutathione peroxidase. J Biochem Mol Biol Biophys 6:143–146

Lane P, Hao G, Gross SS (2001) S-nitrosylation is emerging as a specific and fundamental posttranslational protein modification: head-to-head comparison with O-phosphorylation. Sci STKE 2001:RE1

Lawler JM, Song W (2002) Specificity of antioxidant enzyme inhibition in skeletal muscle to reactive nitrogen species donors. Biochem Biophys Res Commun 294: 1093–1100

Leshem Y, Haramaty E (1996) The characterization and contrasting effects of the nitric oxide free radical in vegetative stress and senescence of Pisum sativum Linn. foliage. Journal of Plant Physiology 148:258–263

Leshem YY (2000) Nitric Oxide in Plants – Occurrence, Function and Use. Dordrecht, Netherlands

Leshem YY, Wills RBH, Ku VV (1998) Evidence for the function of the free reduced gas – nitric oxide ($NO^{\cdot}$) as an endogenous maturation and senescence regulating factor in higher plants. Plant Physiol Biochem 36:825–826

Lind C, Gerdes R, Hamnell Y, Schuppe-Koistinen I, von Lowenhielm HB, Holmgren A, Cotgreave IA (2002) Identification of S-glutathionylated cellular proteins during oxidative stress and constitutive metabolism by affinity purification and proteomic analysis. Arch Biochem Biophys 406:229–240

Lindermayr C, Saalbach G, Bahnweg G, Durner J (2006) Differential inhibition of Arabidopsis methionine adenosyltransferases by protein S-nitrosylation. J Biol Chem 281:4285–4291

Lindermayr C, Saalbach G, Durner J (2005) Proteomic identification of S-nitrosylated proteins in Arabidopsis. Plant Physiol 137:921–930

Lipton SA, Choi YB, Pan ZH, Lei SZ, Chen HS, Sucher NJ, Loscalzo J, Singel DJ, Stamler JS (1993) A redox-based mechanism for the neuroprotective and neurodestructive effects of nitric oxide and related nitroso-compounds. Nature 364:626–632

Liu L, Enright E, Sun P, Tsai SY, Mehta P, Beckman DL, Terrian DM (2002) Inactivation of annexin II tetramer by S-nitrosoglutathione. Eur J Biochem 269:4277–4286

Liu L, Hausladen A, Zeng M, Que L, Heitman J, Stamler JS (2001) A metabolic enzyme for S-nitrosothiol conserved from bacteria to humans. Nature 410:490–494

Liu X, Miller MJ, Joshi MS, Thomas DD, Lancaster JR Jr (1998) Accelerated reaction of nitric oxide with O_2 within the hydrophobic interior of biological membranes. Proc Natl Acad Sci USA 95:2175–2179

Martinez-Ruiz A, Lamas S (2004) Detection and proteomic identification of S-nitrosylated proteins in endothelial cells. Arch Biochem Biophys 423:192–199

Martinez-Ruiz A, Lamas S (2004) S-nitrosylation: a potential new paradigm in signal transduction. Cardiovasc Res 62:43–52

Mata CG, Lamattina L (2001) Nitric Oxide Induces Stomatal Closure and Enhances the Adaptive Plant Responses against Drought Stress. Plant Physiol 126:1196–1204

Matsumoto A, Comatas KE, Liu L, Stamler JS (2003) Screening for nitric oxide-dependent protein-protein interactions. Science 301:657–661

Minning DM, Gow AJ, Bonaventura J, Braun R, Dewhirst M, Goldberg DE, Stamler JS (1999) Ascaris haemoglobin is a nitric oxide-activated deoxygenase. Nature 401:497–502

Mohr S, Hallak H, de Boitte A, Lapetina EG, Brune B (1999) Nitric oxide-induced S-glutathionylation and inactivation of glyceraldehyde-3-phosphate dehydrogenase. J Biol Chem 274:9427–9430

Mohr S, Stamler JS, Brune B (1996) Posttranslational modification of glyceraldehyde-3-phosphate dehydrogenase by S-nitrosylation and subsequent NADH attachment. J Biol Chem 271:4209–4214

Motohashi K, Kondoh A, Stumpp MT, Hisabori T (2001) Comprehensive survey of proteins targeted by chloroplast thioredoxin. Proc Natl Acad Sci USA 98:11224–11229

Neill SJ, Desikan R, Clarke A, Hancock JT (2002) Nitric oxide is a novel component of abscisic Acid signaling in stomatal guard cells. Plant Physiol 128:13–16

Nelson EJ, Connolly J, McArthur P (2003) Nitric oxide and S-nitrosylation: excitotoxic and cell signaling mechanism. Biol Cell 95:3–8

Nikitovic D, Holmgren A (1996) S-nitrosoglutathione is cleaved by the thioredoxin system with liberation of glutathione and redox regulating nitric oxide. J Biol Chem 271:19180–19185

Noctor G, Foyer CH (1998) Ascorbate and glutathione: keeping active oxygen under control. Annu Rev Plant Physiol Plant Mol Biol 49:249–279

Perazzolli M, Dominici P, Romero-Puertas MC, Zago E, Zeier J, Sonoda M, Lamb C, Delledonne M (2004) Arabidopsis nonsymbiotic hemoglobin AHb1 modulates nitric oxide bioactivity. Plant Cell 16:2785–2794

Perez-Mato I, Castro C, Ruiz FA, Corrales FJ, Mato JM (1999) Methionine adenosyltransferase S-nitrosylation is regulated by the basic and acidic amino acids surrounding the target thiol. J Biol Chem 274:17075–17079

Planchet E, Jagadis Gupta K, Sonoda M, Kaiser WM (2005) Nitric oxide emission from tobacco leaves and cell suspensions: rate limiting factors and evidence for the involvement of mitochondrial electron transport. Plant J 41:732–743

Rhee KY, Erdjument-Bromage H, Tempst P, Nathan CF (2005) S-nitroso proteome of *Mycobacterium tuberculosis*: Enzymes of intermediary metabolism and antioxidant defense. Proc Natl Acad Sci USA 102:467–472

Rockel P, Strube F, Rockel A, Wildt J, Kaiser WM (2002) Regulation of nitric oxide (NO) production by plant nitrate reductase in vivo and in vitro. J Exp Bot 53:103–110

Ruiz F, Corrales FJ, Miqueo C, Mato JM (1998) Nitric oxide inactivates rat hepatic methionine adenosyltransferase In vivo by S-nitrosylation. Hepatology 28:1051–1057

Russwurm M, Koesling D (2004) NO activation of guanylyl cyclase. Embo J 23:4443–4450

Sakamoto A, Ueda M, Morikawa H (2002) Arabidopsis glutathione-dependent formaldehyde dehydrogenase is an S-nitrosoglutathione reductase. FEBS Lett 515:20–24

Scharfstein JS, Keaney JF Jr, Slivka A, Welch GN, Vita JA, Stamler JS, Loscalzo J (1994) In vivo transfer of nitric oxide between a plasma protein-bound reservoir and low molecular weight thiols. J Clin Invest 94:1432–1439

Shenton D, Grant CM (2003) Protein S-thiolation targets glycolysis and protein synthesis in response to oxidative stress in the yeast Saccharomyces cerevisiae. Biochem J 374:513–519

Smith JN, Dasgupta TP (2000) Kinetics and mechanism of the decomposition of S-nitrosoglutathione by L-ascorbic acid and copper ions in aqueous solution to produce nitric oxide. Nitric Oxide 4:57–66

Sokolovski S, Blatt MR (2004) Nitric oxide block of outward-rectifying K^+ channels indicates direct control by protein nitrosylation in guard cells. Plant Physiol 136:4275–4284

Song JJ, Lee YJ (2003) Differential role of glutaredoxin and thioredoxin in metabolic oxidative stress-induced activation of apoptosis signal-regulating kinase 1. Biochem J 373:845–853

Stamler JS (1994) Redox signaling: nitrosylation and related target interactions of nitric oxide. Cell 78:931–936

Stamler JS, Toone EJ, Lipton SA, Sucher NJ (1997) (S)NO signals: translocation, regulation, and a consensus motif. Neuron 18:691–696

Stöhr C, Strube F, Marx G, Ullrich WR, Rockel P (2001) A plasma membrane-bound enzyme of tobacco roots catalyses the formation of nitric oxide from nitrite. Planta 212:835–841

Taldone FS, Tummala M, Goldstein EJ, Ryzhov V, Ravi K, Black SM (2005) Studying the S-nitrosylation of model peptides and eNOS protein by mass spectrometry. Nitric Oxide 13:176–187

Tedeschi G, Cappelletti G, Negri A, Pagliato L, Maggioni MG, Maci R, Ronchi S (2005) Characterization of nitroproteome in neuron-like PC12 cells differentiated with nerve growth factor: Identification of two nitration sites in alpha-tubulin. Proteomics 5:2422–2432

Trujillo M, Alvarez MN, Peluffo G, Freeman BA, Radi R (1998) Xanthine oxidase-mediated decomposition of S-nitrosothiols. J Biol Chem 273:7828–7834

van der Vliet A, Hoen PA, Wong PS, Bast A, Cross CE (1998) Formation of S-nitrosothiols via direct nucleophilic nitrosation of thiols by peroxynitrite with elimination of hydrogen peroxide. J Biol Chem 273:30255–30262

Viner RI, Williams TD, Schoneich C (1999) Peroxynitrite modification of protein thiols: oxidation, nitrosylation, and S-glutathiolation of functionally important cysteine residue(s) in the sarcoplasmic reticulum Ca-ATPase. Biochemistry 38:12408–12415

Yamasaki H (2000) Nitrite-dependent nitric oxide production pathway: implications for involvement of active nitrogen species in photoinhibition in vivo. Philos Trans R Soc Lond B Biol Sci 355:1477–1488

Yamazaki D, Motohashi K, Kasama T, Hara Y, Hisabori T (2004) Target Proteins of the Cytosolic Thioredoxins in Arabidopsis thaliana. Plant Cell Physiol 45:18–27

Zeidler D, Zähringer U, Gerber I, Dubery I, Hartung T, Bors W, Hutzler P, Durner J (2004) Innate immunity in Arabidopsis thaliana: Lipopolysaccharides activate nitric oxide synthase (NOS) and induce defense genes. Proc Natl Acad Sci USA 101:15811–15816

Zsombok A, Schrofner S, Hermann A, Kerschbaum HH (2005) A cGMP-dependent cascade enhances an L-type-like Ca^{2+} current in identified snail neurons. Brain Res 1032:70–76

Plant Cell Monogr (6)
L. Lamattina · J. C. Polacco: Nitric Oxide in Plant Growth
DOI 10.1007/7089_2006_085/Published online: 11 October 2006

Enzymatic Sources of Nitric Oxide during Seed Germination

Marcela Simontacchi · Sebastián Jasid · Susana Puntarulo (✉)

Physical Chemistry-PRALIB, School of Pharmacy and Biochemistry,
University of Buenos Aires, Junín 956, C1113AAD Buenos Aires, Argentina
susanap@ffyb.uba.ar

Abstract NO is an inorganic free radical gaseous molecule that has been shown to play an unprecedented range of roles in biological systems. Taking into account these numerous functions and the reports indicating that NO can regulate processes related to plant growth and development, endogenous sources of NO need to be clarified as well as the effect of the variations of NO levels upon the plant life cycle. We consider here the main endogenous sources of cellular NO in plant tissues, and the contribution of enzymatic sources upon seed germination. Non-enzymatic generation of NO from NO_2^- under conditions of low pH could be of considerable importance since significant amounts of NO_2^- can be found in plant tissues. However, at least under the reported experimental conditions, enzymatic activity seems to be more relevant to total NO generation, allowing a strict control of NO steady state concentration. The complexity of the overall scenario presented here shows the need for further studies into NO production and consumption pathways in germinating seeds exposed to fluctuating environmental conditions.

1 How Do Plants Come Into Contact with NO?

NO is an inorganic free radical gaseous molecule that has been shown to play an unprecedented range of roles in biological systems. Neutral NO has a single electron in its $2p$–π antibonding orbital and the removal of this electron forms NO^+, while the addition of one more electron to NO forms NO^- (Stamler et al. 1992). The broader chemistry of NO involves a redox array of species with distinctive properties and reactivities: NO^+ (nitrosonium), NO^- (nitroxyl anion), and NO (NO radical) (Fig. 1):

i NO^+ (*nitrosonium*). This species seems to be involved in the formation of a variety of nitroso-compounds that are generated effectively under neutral physiological conditions (Stamler et al. 1992).

ii NO^- (*nitroxyl anion*). *S*-Nitrosothiols are believed to be a minor product of the reaction of NO^- with disulfides (Stamler et al. 1992).

iii NO (*NO radical*). From a biological point of view the important reactions of NO are those with O_2 and its various redox forms and with transition metal ions.

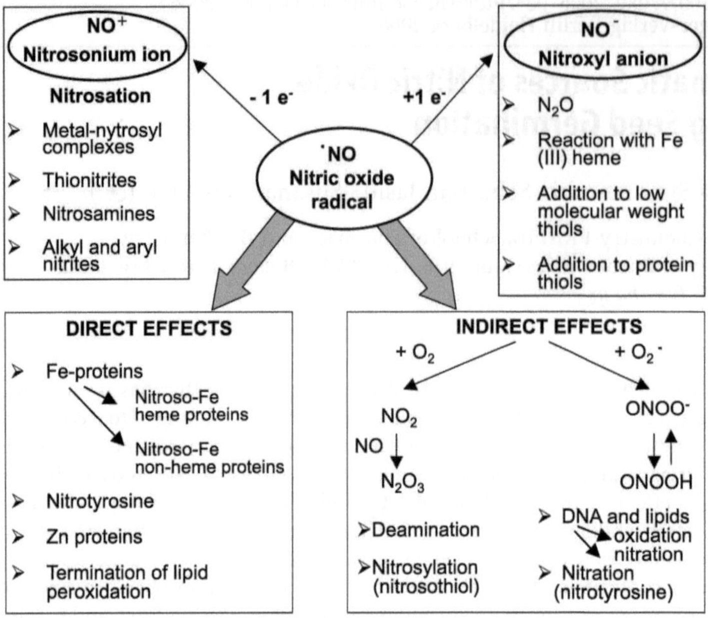

Fig. 1 Nitric oxide ($NO^{\cdot}$) may be oxidized by one electron to give nitrosonium ion (NO^+) or reduced by one electron to form nitroxyl anion (NO^-), which are important intermediates in the chemistry of NO. Nitric oxide ($NO^{\cdot}$) is a free radical and it can rapidly react with an unpaired electron to yield both direct and indirect effects. Oxygen and superoxide anion (O_2^-) react irreversibly with NO to form N_2O_3 and the powerful oxidant peroxynitrite anion ($ONOO^-$), respectively. The main direct reaction of NO is with transition metals

The biological half life of NO, generally assumed to be in the order of seconds, strongly depends upon its initial concentration (Saran et al. 1990). The potential reactions of NO are numerous and depend on many different factors. The site and source of production, as well as the concentration of NO collectively determine whether NO will elicit direct or indirect effects (Fig. 1). A relative balance between oxidative and nitrosative stress should be carefully evaluated. Moreover, to fully understand the complexity of the biological effects of NO the first aspect to be considered is that plant tissues are in contact with both external (atmospheric NO, soil NO) and internally generated NO (endogenous NO) (Fig. 2).

1.1 Atmospheric NO

Substantial NO is emitted from plants into the atmosphere. On the other hand, atmospheric NO is a major greenhouse pollutant produced by combustion of fossil fuels. The emission rate of NO_x (i.e., NO, NO_2) is about

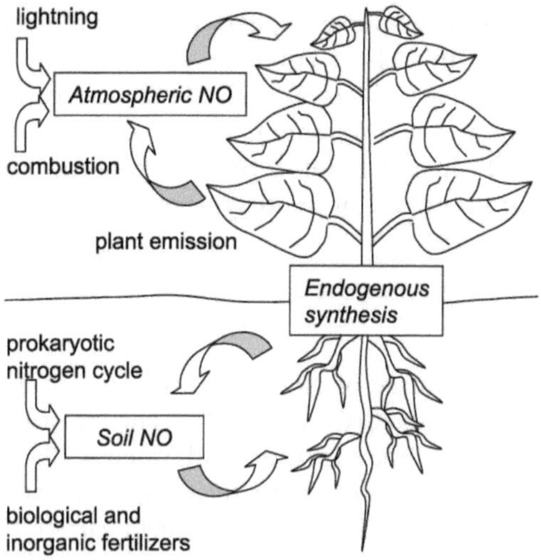

Fig. 2 NO exchange between plants and the environment. NO produced by plant endogenous synthesis could be partially emitted to the surroundings. Plants could also take up NO from the atmosphere and the soil. In the atmosphere, NO can be generated by combustion of fossil fuels, lightning, or biological processes, meanwhile soil NO comes from bacterial nitrifying–denitrifying activities or nitrogen fertilization

260×10^9 kg/year from both anthropogenic (such as the combustion of fossil fuels from both stationary and mobile sources) and non-anthropogenic (such as lightning and biological processes) sources (Elstner and Oßwald 1991). The equilibrium concentration of NO is only 1.1×10^{-10} ppm. The mechanism for formation of NO_x from N_2 and O_2 during combustion starts by the formation of both O and N atoms at very high combustion temperatures (Manahan 1994) or as indicated by reactions 1 and 2, where M is a highly energetic third body that imparts enough energy to the molecular N_2 and O_2 to break their chemical bonds. Once formed, O and N atoms participate in a chain reaction for the formation of NO:

$$O_2 + M \rightarrow O + O + M \quad (1)$$

$$N_2 + M \rightarrow N + N + M\,, \quad (2)$$

Some authors observed in different ecosystems that a post-fire growth and abundance of seedlings was greater in burned areas. However, the effect of the ash generated during fires was not proven under controlled experimental conditions (Thanos and Georghiou 1988). The smoke evolved during wildfires is the most important chemical stimulus for germination of the "fire-type" species (Brown and Van Staden 1997). De Lange and Boucher (1990) were the first to report that plant-derived smoke stimulates seed germination but smoke-stimulated germination has been reported for many fire- and non-fire-dependent species. The smoke appears to be an almost universal signal to seeds. Interestingly, this effect could be mimicked by exogenous application of NO (Leshem 1996; Gouvea et al. 1997), and explained by the fact that NO_x species are active components of smoke (Giba et al. 1999). However, atmospheric NO can affect plants; the exposure to 10 ppm of NO causes a reversible decrease in the rate of photosynthesis.

1.2 NO in the Soil

Soils are an important source of NO and contribute to almost 20% of the global atmospheric NO budget (Conrad 1995). The emission depends on soil levels of NH_4^+ and NO_3^- ions (Tornton and Valente 1996). Microbially derived N_2O and NO are products of denitrification, nitrification, and reduction of NO_3^- to NH_4^+ (Colliver and Stephenson 2000; Zumft 1997). Prokaryotic respiratory NO_2^- reductases (NiRs) are able to catalyze the one-electron reduction of NO_2^- into NO during denitrification (Zumft 1997).

High NO emissions are observed from fertilized soils, but their effects on the physiology of plants are largely unknown (Lamattina et al. 2003). In this regard, a significant increase in NO and N_2O emission from soils has been detected in soils amended with biological and inorganic fertilizers compared with non-fertilized soils (Bremner 1997; Paul et al. 1993). The soil source of NO is similar in magnitude to fossil fuel emissions of NO (Davidson and Kingerlee 1997). Since it has been described that NO promotes a significant increase in chlorophyll content and chloroplast membrane density in maize plants growing with low iron concentration (Graziano et al. 2002), it was proposed that NO derived from nitrogen fertilization could be responsible for the health of nitrogen-fertilized plants (Lamattina et al. 2003).

1.3 Endogenous Synthesis

Evolution of NO_x was first reported from herbicide-treated soybean leaves (Klepper 1978). NO emission from *Pisum sativum* foliage was also described by the pioneering work of Leshem (1996). More recently it was suggested that global NO_x emissions from boreal coniferous forest may be comparable to those produced by worldwide industrial and traffic sources (Hari et al. 2003). NO production in plants is induced by both biotic and abiotic stimuli, such as drought, salt stress, and pathogen infection (Lamattina et al. 2003) and has a central role in coordinating plant growth and development with environmental conditions. There are several potential sources of NO in plants, and their relevance is considered to vary depending on plant organ and on the physiological situation of the organism. In the present report, the main endogenous sources of cellular NO production in plants are considered and the contribution of enzymatic sources upon seed germination are analyzed.

2 Non-enzymatic Sources of NO in Plants

Non-enzymatic generation of NO from NO_2^- under conditions of low pH could be of considerable importance since significant amounts of NO_2^- can be found in the plant tissues (Caro and Puntarulo 1998). The reduction of NO_2^- to NO can be achieved non-enzymatically at acidic pH values in the presence of reductants such as ascorbic acid and glutathione (Bethke et al. 2004), or from protonated NO_2^-, as proposed by Yamasaki et al. (1999) (reaction 3)

$$2HNO_2 \rightarrow NO + NO_2 + H_2O\,. \tag{3}$$

2.1 Apoplastic Reduction of NO_2^-

It has been reported that barley aleurone layers produce NO when NO_2^- is added, probably via a non-enzymatic reduction in the acidic apoplast (Bethke et al. 2004). NO_2^- entering the grain from the soil solution or released by the embryo axis, the scutellum, or the aleurone layer to the apoplast/endosperm cavity would result in an NO production that could be sensed by the other tissues (Bethke et al. 2004). Soils contain NO_2^- in a concentration that can be greater than several hundred micromolar. However, values of 10–50 μM are more common in agricultural soils (Stevens et al. 1998). Adequate conditions for non-enzymatic NO_2^- reduction exist in the apoplast and have been invoked to explain NO_2^- effects on germinating seeds (Giba et al. 1998; Beligni and Lamattina 2000; Bethke et al. 2004).

3 Enzymatic Sources of NO in Plants

3.1 The Arginine-Dependent Pathway

In animals, it is widely accepted that NO production is catalyzed predominantly by the activity of nitric oxide synthases (NOS), which are heme-containing proteins related to the cytochrome P_{450} family. These enzymes catalyze the conversion of L-arginine (Arg) to L-citrulline and NO using NADPH and O_2 as cosubstrates, and employing FAD, FMN, tetrahydrobiopterin (BH_4), and calmodulin (CaM) as cofactors (Stuehr 1999; Bogdan

2001). The existence of one class of NOS (AtNOS1) has been well characterized in *Arabidopsis*, whereas several enzymatic activities, named NOS-like because they are inhibited by arginine homologs, have been described in a variety of plants. Cueto et al. (1996) and Ninnemann and Maier (1996) were the first to show the activity of NOS-like enzymes in legumes. Since then, a significant number of reports have shown the presence of NOS-like enzymes in different plant species. Recently a NOS-like activity was reported in pea stems assessed as N^{ω}–nitro-L-arginine (LNNA)-sensitive NO generation by the hemoglobin assay (Qu et al. 2006). The activities reported were about 15 and 4 nmol min^{-1} mg^{-1} protein for microsomal and cytosolic fractions, respectively.

Immunological analysis suggested that NOS-like proteins were probably localized in the cytoplasm of maize (Ribeiro et al. 1999) and in sunflower hypocotyls (del Río et al. 2004). Electron microscopy and immunogold-labeling with antibodies made against rabbit brain NOS or murine iNOS has confirmed the subcellular localization of a NOS-like protein in the matrix of peroxisomes, in chloroplasts, and in the nucleus of tobacco, maize, and pea (Barroso et al. 1999; Ribeiro et al. 1999). More recently, the presence of a NOS-like enzyme was suggested in peroxisomes by assessing the presence of NO by EPR (Corpas et al. 2004). The presence of NO was also confirmed employing the fluorescence dye DAF-2 DA and by ozone chemiluminiscence (Corpas et al. 2004). This NOS-like activity seems to be immuno-related to animal iNOS and the activity in peroxisomes was about 5 nmol NO min^{-1} mg^{-1} protein. Tobacco cells treated with cryptogein showed early increases in NO-dependent DAF-2 DA fluorescence located in chloroplasts occurring within the first 3 min (Foissner et al. 2000). Elicitation by cryptogein of DAF-2 DA-loaded cells in the presence of the NOS inhibitor N^G-monomethyl-L-arginine (L-NMMA) resulted in reduced fluorescence, suggesting the participation of a plant NOS (Foissner et al. 2000).

3.1.1
Mitochondrial AtNOS1

No protein or gene in plants was identified that had any sequence similarity to the complete animal NOS proteins. Huang et al. (1997) isolated an enzyme of 60 kDa from the neurons of the snail *Helix pomatia*, which implied the existence of a new type of NOS that appeared to have no relationship to mammalian NOS. The snail enzyme had no consensus binding sites for NADPH, FAD, or arginine (Huang et al. 1997). Interestedly, a gene in *Arabidopsis* encodes a protein that is 16% identical to the snail protein (Guo et al. 2003). This protein, which has been termed AtNOS1, was fused to GST and expressed in *E. coli* (Guo et al. 2003), where NO synthesis increased dramatically in crude *E. coli* extracts as compared to controls. The enriched fusion protein showed NOS activity that was dependent on NADPH, Ca^{2+} and CaM, and was independent of added FAD, FMN or BH_4. These features implied that this protein with a V_{max} of 5.0 nmol min^{-1} mg^{-1} protein is distinct from the NOS-like activity described by Corpas et al. (2004).

Guo et al. (2003) showed that the *AtNOS1* mutant plants (Arabidopsis mutants lacking this enzyme) showed a lower NO level than the wild type. Leaves from the mutant had only 20% of the NOS activity of the wild type, assessed by the citrulline assay (Guo et al. 2003), DAF-FM fluorescence, and the use of NO-specific electrodes (He et al. 2004). At present, AtNOS1 activity is the only reported arginine-dependent NOS in plants (Crawford 2006). Since mutations in this gene reduce NO accumulation in vivo but not completely, it seems that other mechanisms are involved in NO synthesis. The evidence to date indicates that *AtNOS1* encodes a novel NOS enzyme that behaves most like the constitutive class of mammalian NOS enzymes (eNOS and nNOS), showing Ca^{2+}-dependent regulation but little transcriptional control (Crawford 2006). In wild-type leaf extracts Guo et al. (2003) reported an activity of 0.1 pmol citrulline min^{-1} mg^{-1} protein.

Computational analysis of the protein sequence revealed that AtNOS1 has a high probability of being targeted to the mitochondria (Guo and Crawford 2005). GFP fusion protein and confocal microscopy confirmed that the *Arabidopsis thaliana* protein NOS1 is localized in mitochondria.

3.2
NO_2^--Dependent Pathways

It has long been known that plants can emit NO from leaves (Klepper 1979; Nishimura et al. 1986) and that this release correlates with tissue NO_2^- levels. NO_2^- is metabolized in large amounts by plant cells. In plant cells NO_2^- is normally reduced to NH_4^+ by the plastidic nitrite reductase (NiR) enzyme. In conditions where nitrate reductase (NR) is highly active (e.g., anoxia or treatment with uncouplers) NO_2^- concentration can increase up to 100-fold over normal values (Rockel et al. 2002). Excessive cellular NO_2^- can be excreted into the surrounding medium (Stoimenova et al. 2003).

NO_2^-, as the anion of a diffusible weak neutral acid, may accumulate in compartments with a pH more alkaline than the cytosol, such as illuminated chloroplasts. Data from Foissner et al. (2000) in elicited cells indicated a higher NO production in chloroplasts and in the vicinity of chloroplasts, suggesting NO_2^- accumulation in the chloroplast as a possible NO source (Klepper 1991). Also, imaging of NO in tobacco leaves challenged by abiotic stressors showed that guard cells were the first cell type to respond, and that NO-dependent fluorescence was initially evident in the plastids (Gould et al. 2003).

NiRs are chloroplast enzymes. Because NO lies between NO_2^- and NH_4^+ in oxidation state, it has been considered as a potential intermediate, at least in non-enzymatic reactions, leading to NO_2^- reduction. The role of plastidic NiR in the synthesis of NO has been discarded by experiments performed in tobacco plants expressing an antisense NiR construct because these plants with very low NiR activity were found to accumulate NO_2^- and produce NO with an emission rate about 100-fold higher than in the wild-type plants (Morot-Gaudry-Talarmain et al. 2002).

3.2.1
Cytosolic NO_3^- Reductase

NO_3^- reductase (NR) was the first enzymatic source of NO to be identified. Assimilatory NR is a large homodimeric enzyme with a molecular mass between 200 and 250 kDa. Each monomer houses three prosthetic groups, FAD, heme, and molybdenum cofactor which transfer electrons from NADH to the enzyme substrate (normally NO_3^-). NR is structurally similar to NOS from animal tissues since it contains a heme and an NAD(P)H/FAD-binding domain. It should be noted that both NR and animal NOS are flavoenzymes and members of the ferredoxin-$NADP^+$ reductase family (Lamattina et al. 2003). NR is a highly regulated enzyme with many levels of control. The expression is induced at the transcriptional level by NO_3^-, light or sugars (Campbell 1999; Kaiser et al. 1999), and it is also regulated posttranscriptionally by phosphorylation. The NR gene is inducible in many plant families, although exceptions, such as the *Leguminosae*, show both constitutive and inducible forms of NR.

It has been shown that this well-known plant protein, apart from its role in NO_3^- reduction and assimilation, can also catalyze the reduction of NO_2^- to NO. It was demonstrated that in vivo NR activity is responsible for releasing NO (Harper 1981; Klepper 1987). These experiments were conducted under anaerobic conditions and resulted in NO_2^- accumulation, and it was established that NR was needed for NO formation (Dean and Harper 1988; Klepper 1990). It was also demonstrated that cytosolic NADH-dependent NR from maize is capable of reducing not only NO_3^- to NO_2^- but also NO_2^- to NO. This opened up the possibility for cytosolic NR to be involved in NO production in plants (Yamasaki et al. 1999; Yamasaki and Sakihama 2000). This line of experiments culminated in the demonstration that purified NR reduces NO_2^- to NO (Yamasaki et al. 1999; Yamasaki 2000; Rockel et al. 2002). However, for maize NR the Km for NO_2^- is between 100 and 300 μM, whereas NO_3^- is a competitive inhibitor with a Ki of 50 μM (Rockel et al. 2002). Thus, under physiological conditions where NO_3^- levels are high (1–5 mM) and NO_2^- content is low (10 μM in illuminated leaves), a significant NO production from NR would not be expected (Crawford 2006). On the other hand, under anaerobic conditions, NO_2^- levels increase and NO release can increase 100-fold (Rockel et al. 2002). In this regard, high NO levels can be generated by using NR-deficient lines that accumulate NO_2^- (Morot-Gaudry-Talarmain et al. 2002; Planchet et al. 2005).

NR produced NO in vitro, at saturating NADH and NO_2^- concentrations, at about 1% of its NO_3^- reduction capacity (Rockel et al. 2002). In spinach extracts the modulation of NR activity, either by activation or inactivation, affected NO production in a similar way in the presence of added NO_2^- and NADH. The maximum rate of NO release was 2 nmol h^{-1} g^{-1} FW in leaves of sunflowers adapted to constant NO_3^- nutrition (Rockel et al. 2002) under light conditions.

3.2.2 Plasma Membrane-Bound Nitrite:NO Reductase Activity

An NO_2^--reducing activity different from the soluble NiR located in plastids has been recently described. This activity catalyzes the reduction of NO_2^- to NO using cytochrome *c* as an electron donor (Stöhr et al. 2001). However, the physiological electron donor has not yet been identified (Stöhr and Stremlau 2006). Nitrite:NO reductase (Ni-NOR), a 310-kDa plasma membrane-bound enzyme from tobacco roots copurified with the root plasma membrane fraction, was found to be different from the plasma membrane-bound NR (PM-NR) (Stöhr et al. 2001). Crude extracts from tobacco root showed a rate of NO formation of 6 nmol h^{-1} mg^{-1} protein, while the NO-producing activity increased to 250 nmol h^{-1} mg^{-1} protein in the PM preparations (Stöhr et al. 2001). This synthesis of NO could be linked to NO_2^- production by PM-NR and might be one of the primary signals that report the presence of exogenous NO_3^- in roots.

3.2.3 Mitochondrial-Dependent NO_2^- Reduction

Plant mitochondria can produce NO under anoxia and sufficient NO_2^- supply; these prerequisites may be faced by plant roots under water logging conditions, or by aqueous plants and algae in eutrophic waters (Planchet et al. 2005; Tischner et al. 2004). Under O_2 concentrations as low as 1%, even root tissues completely lacking NR reduced added NO_2^- to NO, indicating that NR was not the only mechanism for NO_2^--dependent NO formation. Root NO formation was blocked by inhibitors of mitochondrial electron transport (myxothiazol and SHAM) (Gupta et al. 2005). A mitochondrial-dependent NO_2^--reducing activity seems to be part of the NO production by *Arabidopsis thaliana* in response to pathogen attack (Modolo et al. 2005). In this system, the predominant activity appears to be NO_2^- reduction by mitochondria whereas NR activity is apparently more important in providing NO_2^- for NO production. Inhibitors of mitochondrial electron transport inhibit NO synthesis, suggesting that electrons from the mitochondrial electron transport chain drive NO_2^- reduction. This raises the possibility that NO release under hypoxic or anoxic conditions depends on NR for the production of NO_2^-, but it is the mitochondria that produce the bulk of NO from NO_2^- (Crawford 2006).

4 Germination and NO Production in Plants

During evolution, plants developed a special organ, the seed, which ensures their spatial dispersion. Germination, usually defined as the rupture of the seed coat by the radicle and/or coleoptile, will increase the rate of NO_2^- entry from the soil solution. This source of NO_2^- may be augmented by NO_2^- produced enzymatically by NR (Ferrari and Varner 1970). Either NO_2^- or sodium nitroprusside (SNP) improved the germination rate of *Suaeda salsa* under saline conditions (Li et al. 2005). Moreover, organic nitrogenous compounds and inorganic NO_3^- promote germination in many seeds (Grubišić 1992), and it has been shown that NO_3^- significantly increases the light sensitivity and decreases the light requirement for seeds (Toole et al. 1955). Germination of photoblastic lettuce seeds is a phytochrome-dependent process above 25 °C. It was demonstrated that NO donors, such as SNP, are able to promote germination in the dark to the same extent as either gibberellin treatment or a 5-min pulse of white light (Beligni and Lamattina 2000). NO has been assumed to act through a pathway similar to gibberellin, which is involved in phytochrome-dependent reactions (Beligni and Lamattina 2000). However, seeds were also able to germinate in the light, in the presence of the NO scavenger cPTIO, suggesting that light and NO can stimulate germination by different pathways (Beligni and Lamattina 2000). On the other hand, NO donors enhanced the phytochrome A-dependent germination response to red light in photodormant *Arabidopsis* seeds (Batak et al. 2002).

4.1 NO as an Endogenous Metabolite Link to Germination in Embryonic Axes

Since the addition of exogenous NO showed such a clear effect in the promotion of seed germination, endogenous NO generation in an active developing tissue was assessed in our laboratory at the onset of germination of soybean and sorghum seeds, employing the highly specific and sensitive technique EPR (Caro and Puntarulo 1999; Simontacchi et al. 2004). A distinctive and unique EPR signal ($g = 2.03$ and $a_N = 12.5$ G), which enabled a fingerprint-like identification of NO, was detected in the homogenates from axes isolated from either soybean or sorghum seeds imbibed in the presence of NO_3^- in a time-dependent manner (Fig. 3). The NO steady-state concentration in homogenates from sorghum embryonic axes showed a maximum at 24–30 h of imbibition, coincident with the initiation of an active germination in sorghum seeds. NO steady state concentration in homogenates from soybean embryonic axes had a maximum at 15–24 h of imbibition. These data strongly suggest that NO is a physiological product during development of the embryonic axes at a critical stage of the initiation of germination. Moreover, in soybean axes the increase in NO content slightly preceded the initiation

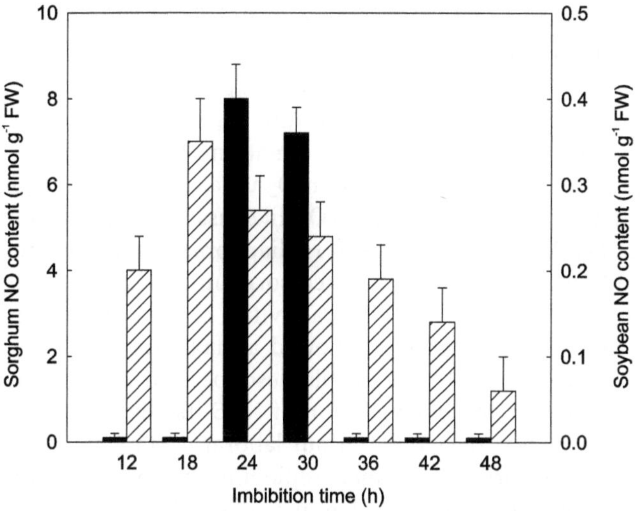

Fig. 3 NO content during germination. Embryonic axes from sorghum (■) and soybean (▨) were excised from seeds imbibed for up to 48 h in the presence of 12 mM NO_3^- (sorghum) or 15 mM NO_3^- (soybean). EPR signals from MGD-Fe-NO adduct were double integrated and compared to a standard solution for NO quantification

of phase III of development and the sharp increase in O_2 consumption (Caro and Puntarulo 1998). NO seems to be one of the possible in vivo metabolites of NO_3^- utilization, since increasing NO_3^- resulted in a significant increase in the EPR spectra of homogenates from soybean embryonic axes imbibed over 24 h (Caro and Puntarulo 1998).

Given that NO is induced by environmental changes, as well as constitutively produced, external and internal cues may converge on the regulation of endogenous NO status (He et al. 2004). Recently, a key role of NO in germination has been confirmed by experiments in the Arabidopsis *AtNOS1* mutant in which seed germination was inhibited (Crawford 2006).

4.2 Enzymatic NO Generation During Germination

In soybean embryonic axes a close association was reported between the activity of NR and the content of NO after 24 h of imbibition, suggesting that this enzymatic activity could be an important physiological source of NO in the axes (Caro and Puntarulo 1998). However, in soybean homogenates, NO generation did not completely depend on NR activity. NOS-like activity, assessed as NADPH-diaphorase, was detected in the homogenates and was inhibited 48 and 28% by the competitive inhibitors of NOS, *N*-nitro-L-arginine (L-NA) and L-NMMA, respectively. In this frame, NO seems to be a physiological metabolite in soybean embryonic axes, generated by the activities of both NR and NOS-like (Table 1). However, non-enzymatic reduction of NO_2^- could not be discarded.

Sorghum embryonic axes follow a biphasic process over the initial 48 h of germination. The early increase in both O_2 uptake and axis weight defines germination phase I, whereas germination phase II is characterized by the assumption of a marked increase in O_2 uptake and growth (Simontacchi et al. 2003). The activities of NR and NOS-like (assessed as a N^G-nitroarginine methyl ester, L-NAME-sensitive NADPH-diaphorase enzymatic activity) were measured in homogenates from isolated sorghum embryonic axes incubated in the presence of 12 mM NO_3^-. The activities of the NADH-dependent NR and NOS-like in the homogenates showed a maximum after 24–30 h of incubation, in agreement with the profile described for NO content (Table 1). In order to determine the participation of NOS-like enzymes in the synthesis of NO, an assay was developed in our laboratory in which the rate of NO generation was measured over time, employing the spin-trapping EPR technique in the presence of arginine, calcium, and NADPH. The measured activity (2.2 ± 0.3 pmol min^{-1} mg^{-1} FW) was inhibited not only by preincubation with 5 mM L-NAME (75%) but also by preincubation with 1 mM L-NMMA (69%), suggesting the participation of a NOS-like activity in the generation of NO during germination (Simontacchi et al. 2004). As previously reported with soybean embryonic axes, a close association between NO content in the homogenates

Table 1 NR and NOS-like activities in sorghum and soybean embryonic axes

Time (h)	Sorghum NR (nmol h^{-1} mg^{-1} prot)	NOS-like	Soybean NR (nmol h^{-1} mg^{-1} prot)	NOS-like
12	nd	nd	8 ± 1	1.7 ± 0.1
18	nd	0.3 ± 0.1*	13 ± 2*	1.6 ± 0.2
24	0.44 ± 0.05*	0.8 ± 0.1*	16 ± 1*	1.5 ± 0.1
30	0.50 ± 0.06*	1.2 ± 0.2*	12 ± 2*	1.3 ± 0.3
36	0.27 ± 0.04*	0.8 ± 0.1*	8 ± 1	1.1 ± 0.2*
42	0.19 ± 0.05*	0.15 ± 0.05*	4 ± 1	0.9 ± 0.3*
48	0.05 ± 0.02	0.07 ± 0.04	1.0 ± 0.5*	0.7 ± 0.2*

Embryonic axes were excised from *Sorghum bicolor* (L.) Moench seeds incubated over Long Ashton solution; the total nitrate concentration was 12 mM. Soybean seeds (*Glycine max* var Hood) were incubated over Steinberg solution containing 15 mM NO_3^-. NR activity was measured by Griess reaction and expressed as nmol NO_2^- h^{-1} mg^{-1} prot. NOS-like activity was assessed as L-NAME sensitive-NADPH diaphorase and expressed as nmol formazan h^{-1} mg^{-1} prot
nd non-detectable
* Significantly different from 12 h values, at $p < 0.05$ (Statview for Windows, SAS Institute, version 5.0)

and NO_3^- supplementation in the incubation medium (up to 30 mM NO_3^-) was determined. These results suggested a role for NR activity in the synthesis of NO during germination, either reducing NO_2^- to NO or providing NO_2^-, which in turn could be converted into NO in mitochondria (Planchet et al. 2005).

5 Conclusions and Perspectives

Taking into account the numerous functions that over the last decade have been assigned to NO, and the reports indicating that NO can regulate processes related to plant growth and development, to further analyze the role of NO in cell signaling one aspect in need of study is the endogenous sources of generation and the variations of NO levels over the plant cycle. The data summarized here indicate that NO is clearly an endogenous metabolite in plants and that its content is strictly controlled upon germination. The cellular mechanisms for NO detoxification could involve not only release to the extracellular medium (allowed by NO permeability through lipid membranes), but also reaction with O_2^- to generate peroxynitrite, as well as other yet-to-be-identified reactions. Still unclear is the physiological relevance of NO/O_2^- cross-talk in the bulk of cellular components, in the sense that peroxynitrite could represent a new messenger or that NO/O_2^- represents the annihilation of the message borne by either species.

A critical issue to be further explored is the physiological sources of NO in an actively developing tissue at the onset of germination. Figure 4 summarizes the reported possible sources of NO in the embryonic axis. The main pathways proposed to be responsible for endogenous generation of NO are: (i) non-enzymatic reactions, (ii) NR activity, and (iii) NOS-like enzymatic activity. Regarding the enzymatic sources, it should be pointed out that both NR and NOS-like activities could be modulated in vivo by physiological conditions and that the relative contribution of each of these enzymes to the overall generation of NO could be altered by effectors. The contribution of NR activity to the physiological production of NO in soybean axes was suggested either in soybean, (where NR is a constitutive enzyme) or in sorghum (where NR is inducible). Standard assays for assessing NR and NOS-like activities indicated that both were of the same order of magnitude during early germination (Simontacchi et al. 2004). Also, there is a significant increase in NR and NOS-like activities before the beginning of active growth. Parallel increase in these activities and the NO content in the axes does not prove unequivocally that NO production depends only on NR and NOS-like enzymatic activities. However, the experimental evidence justifies their further analyses as NO sources.

There are still important aspects related to NO consumption and other alternative sources of NO that require characterization. Even though both NR

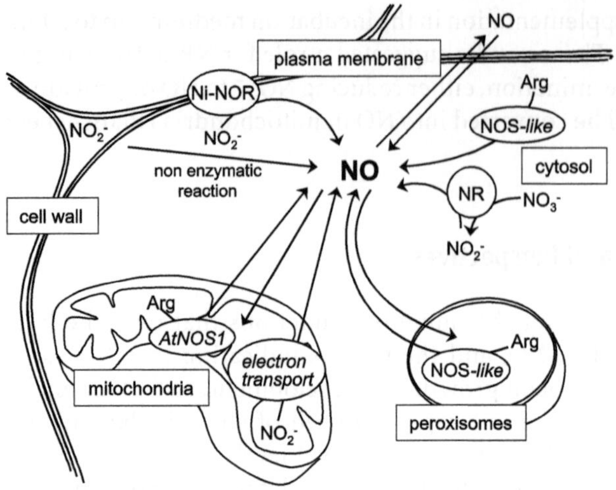

Fig. 4 Possible sources of NO during germination. NO can be generated from NO_2^- in the apoplastic space (chemical reduction of NO_2^-), in plasma membrane (Ni-NOR), in the cytosol (NR), and inside the mitochondria (the electron carriers can reduce NO_2^- to NO). Arginine-dependent NO synthesis was postulated to take place in mitochondria (AtNOS1), in cytosol, and in peroxisomes (NOS-like)

and NOS-like activities are higher in soybean than in sorghum, NO steady state concentration at the initial stages of germination is almost ten times higher in sorghum than in soybean. The complexity of the overall scenario presented here shows the need for further studies into NO production and consumption pathways, mainly in germinating seeds exposed to fluctuating environmental conditions. Thus, this information opens the possibility of designing protocols to enhance successful plant development.

Acknowledgements This study was supported by grants from the University of Buenos Aires (B017, B012), Agencia Nacional de Promoción Científica y Tecnológica (ANPCyT) (PICT 11187), and CONICET (PIP 5706). S.P. and M.S. are career investigators and SJ is a fellow from CONICET.

References

Barroso JB, Corpas FJ, Carreras A, Sandalio LM, Valderrama R, Palma JM, Lupiáñez JA, del Río LA (1999) Localization of nitric oxide synthase in plant peroxisomes. J Biol Chem 274:36729–36733

Batak I, Dević M, Giba Z, Grubišić D, Poff K, Konjević R (2002) The effects of potassium nitrate and NO-donors on phytochrome A- and phytochrome B-specific induced germination of *Arabidopsis thaliana* seeds. Seed Sci Res 12:253–259

Beligni MV, Lamattina L (2000) Nitric oxide stimulates seed germination and de-etiolation, and inhibits hypocotyl elongation, three light-inducible responses in plants. Planta 210:215–221

Bethke PC, Badger MR, Jones RL (2004) Apoplastic synthesis of nitric oxide by plant tissues. Plant Cell 16:332–341

Bogdan C (2001) Nitric oxide and the regulation of gene expression. Trends Cell Biol 11:66–75

Bremner JM (1997) Sources of nitrous oxide in soils. Nutr Cycl Agroecosyst 49:7–16

Brown NAC, Van Staden J (1997) Smoke as a germination cue: a review. Plant Growth Regul 22:115–124

Campbell WH (1999) Nitrate reductase structure, function and regulation: bridging the gap between biochemistry and physiology. Annu Rev Plant Physiol Plant Mol Biol 50:277–303

Caro A, Puntarulo S (1998) Nitric oxide decreases superoxide anion generation by microsomes from soybean embryonic axes. Physiol Plant 104:357–364

Caro A, Puntarulo S (1999) Nitric oxide generation by soybean embryonic axes. Possible effect on mitochondrial function. Free Radic Res 31:205–212

Colliver BB, Stephenson T (2000) Production of nitrogen oxide and dinitrogen oxide by autotrophic nitrifiers. Biotechnol Adv 18:219–232

Conrad R (1995) Soil microbial processes involved in production and consumption of atmospheric trace gases. Adv Microb Ecol 14:207–250

Corpas FJ, Barroso JB, Carreras A, Quiros M, Leon AM, Romero-Puertas MC, Esteban FJ, Valderrama R, Palma JM, Sandalio LM, Gomez M, del Rio LA (2004) Cellular and subcellular localization of endogenous nitric oxide in young and senescent pea plants. Plant Physiol 136:2722–2733

Crawford NM (2006) Mechanisms for nitric oxide synthesis in plants. J Exp Bot 57:471–478

Cueto M, Hernández-Perea O, Martín R, Ventura ML, Rodrigo J, Lamas S, Golvano MP (1996) Presence of nitric oxide synthase activity in roots and nodule of *Lupinus albus*. FEBS Lett 398:159–164

Davidson EA, Kingerlee W (1997) A global inventory of nitric oxide emissions from soils. Nutr Cycl Agroecosyst 48:37–50

Dean JV, Harper JE (1988) The conversion of nitrite to nitrogen oxide(s) by the constitutive NAD(P)H-nitrate reductase enzyme from soybean. Plant Physiol 88:389–395

De Lange JH, Boucher C (1990) Autoecological studies on *Audouinia capitata* (Bruniaceae). I. Plant-derived smoke as seed germination cue. S Afr J Bot 56:585–588

del Río LA, Corpas FJ, Barroso JB (2004) Nitric oxide and nitric oxide synthase activity in plants. Phytochemistry 65:783–792

Elstner EF, Oßwald W (1991) Air pollution: involvement of oxygen radicals (a mini review) Free Radic Res Commun 12–13:795–807

Ferrari TE, Varner JE (1970) Control of nitrate reductase activity in barley aleurone layers. Proc Natl Acad Sci USA 65:729–736

Foissner I, Wendehenne D, Langebartels C, Durner J (2000) In vivo imaging of an elicitor-induced nitric oxide burst in tobacco. Plant J 23:817–824

Giba Z, Grubišić D, Todorović S, Sajc L, Stojakocić D, Konjević R (1998) Effect of nitric oxide-releasing compounds on phytochrome-controlled germination of Empress tree seeds. Plant Growth Regul 26:175–181

Giba Z, Grubišić D, Konjević R (1999) How the seeds sense environmental changes? The role of nitrogenous compounds. Arch Biol Sci 51:121–129

Gould KS, Lamotte O, Klinguer A, Pugin A, Wendehenne D (2003) Nitric oxide production in tobacco leaf cells: a generalized stress response? Plant Cell Environ 26:1851–1862

Gouvêa CMCP, Souza JF, Magalhães ACN, Martins IS (1997) NO-releasing substances that induce growth elongation in maize root segments. Plant Growth Regul 21:183–187

Graziano M, Beligni MV, Lamattina L (2002) Nitric oxide improves internal iron availability in plants. Plant Physiol 130:1852–1859

Grubišić D, Giba Z, Konjević R (1992) The effect of organic nitrates in phytochrome-controlled germination of *Paulownia tomentosa* seeds. Photochem Photobiol 56:629–632

Guo FQ, Crawford NM (2005) Arabidopsis nitric oxide synthase1 is targeted to mitochondria and protects against oxidative damage and dark-induced senescence. Plant Cell 17:3436–3450

Guo FQ, Okamoto M, Crawford NM (2003) Identification of a plant nitric oxide synthase gene involved in hormonal signaling. Science 302:100–103

Gupta KJ, Stoimenova M, Kaiser WM (2005) In higher plants, only root mitochondria, but not leaf mitochondria reduce nitrite to NO, in vitro and in situ. J Exp Bot 56:2601–2609

Hari P, Raivonen M, Vesala T, Munger J, Pilegaard K, Kulmala M (2003) Atmospheric science: ultraviolet light and leaf emission of NO_x. Nature 422:134

Harper JE (1981) Evolution of nitrogen oxide(s) during in vivo nitrate reductase assay of soybean leaves. Plant Physiol 68:1488–1493

He Y, Tang RH, Hao Y, Stevens R, Cook C, Ahn SM, Yang Z, Chen L, Guo FQ, Fiorani F, Jackson R, Crawford N, Pei ZM (2004) Nitric oxide represses the *Arabidopsis* floral transition. Science 305:1968–1971

Huang S, Kershbaum HH, Engel E, Hermann A (1997) Biochemical characterization and histochemical localization of nitric oxide synthase in the nervous system of the snail, *Helix pomatia*. J Neurochem 69:2516–2528

Kaiser WM, Weiner H, Huber SC (1999) Nitrate reductase in higher plants: a case study for transduction of environmental stimuli into control of catalytic activity. Physiol Plant 105:385–390

Klepper LA (1978) Nitric oxide (NO) evolution from herbicide-treated soybean plants. Plant Physiol 61:S65

Klepper LA (1979) Nitric oxide (NO) and nitrogen dioxide (NO_2) emissions from herbicide-treated soybean plants. Atmos Environ 13:537–542

Klepper LA (1987) Nitric oxide emissions from soybean leaves during in vivo nitrate reductase assays. Plant Physiol 85:96–99

Klepper LA (1990) Comparison between NO_x evolution mechanism of wild-type and *nr1* mutant soybean leaves. Plant Physiol 93:26–32

Klepper L (1991) NO_x evolution by soybean leaves treated with salicylic acid and selected derivatives. Pestic Biochem Physiol 39:43–48

Lamattina L, García-Mata C, Graziano M, Pagnussat G (2003) Nitric oxide: The versatility of an extensive signal molecule. Annu Rev Plant Biol 54:109–136

Leshem YY (1996) Nitric oxide in biological systems. Plant Growth Regul 18:155–159

Li W, Liu X, Khan MA, Yamaguchi S (2005) The effect of plant growth regulators, nitric oxide, nitrate, nitrite and light on the germination of dimorphic seeds of *Suaeda salsa* under saline conditions. J Plant Res 118:207–214

Manahan SE (1994) Gaseous inorganic air pollutants. In: Environmental chemistry, 6th edn. Lewis, Boca Raton, FL, 11:338–345

Modolo LV, Augusto O, Almeida IMG, Magalhães JR, Salgado I (2005) Nitrite as the major source of nitric oxide production by *Arabidopsis thaliana* in response to *Pseudomonas syringae*. FEBS Lett 579:3814–3820

Morot-Gaudry-Talarmain Y, Rockel P, Moureaux T, Quillere I, Leydecker MT, Kaiser WM, Morot-Gaudry JF (2002) Nitrite accumulation and nitric oxide emission in relation to cellular signaling in nitrite reductase antisense tobacco. Planta 215:708–715

Ninnemann H, Maier J (1996) Indications for the occurrence of nitric oxide synthases in fungi and plants and the involvement in photoconidiation of *Neurospora crassa*. Photochem Photobiol 64:393–398

Nishimura H, Hayamizu T, Yanagisawa Y (1986) Reduction of NO_2 to NO by rush and other plants. Environ Sci Technol 20:413–416

Paul JW, Beauchamp EG, Zhang X (1993) Nitrous and nitric oxide emission during nitrification and denitrification from manure-amended soil in the laboratory. Can J Soil Sci 73:539–553

Planchet E, Gupta KJ, Sonoda M, Kaiser WM (2005) Nitric oxide emission from tobacco leaves and cell suspensions: rate limiting factors and evidence for the involvement of mitochondrial electron transport. Plant J 41:732–743

Qu Y, Feng H, Wang Y, Zhang M, Cheng J, Wang X, An L (2006) Nitric oxide functions as a signal in ultraviolet-B induced inhibition of pea stems elongation. Plant Sci 170:994–1000

Ribeiro EA, Cunha FQ, Tamashiro WMSC, Martins IS (1999) Growth phase-dependent subcellular localization of nitric oxide synthase in maize cells. FEBS Lett 445:283–286

Rockel P, Strube F, Rockel A, Wildt J, Kaiser WM (2002) Regulation of nitric oxide (NO) production by plant nitrate reductase in vivo and in vitro. J Exp Bot 53:103–110

Saran M, Michel C, Bors W (1990) Reaction of NO with O_2^-. Implications for the action of endothelium-derived relaxing factor (EDRF). Free Radic Res Commun 10:221–226

Simontacchi M, Sadovsky L, Puntarulo S (2003) Profile of antioxidant content upon developing of *Sorghum bicolor* seeds. Plant Sci 164:709–715

Simontacchi M, Jasid S, Puntarulo S (2004) Nitric oxide generation during early germination of sorghum seeds. Plant Sci 167:839–847

Stamler JS, Singel DJ, Loscalzo J (1992) Biochemistry of nitric oxide and its redox-activated forms. Science 258:1898–1902

Stevens RJ, Laughlin RJ, Malone JP (1998) Soil pH affects the processes reducing nitrate to nitrous oxide and di-nitrogen. Soil Biol Biochem 30:1119–1126

Stöhr C, Stremlau S (2006) Formation and possible roles of nitric oxide in plant roots. J Exp Bot 57:463–470

Stöhr C, Strube F, Marx G, Ullrich WR, Rockel P (2001) A plasma membrane - bound enzyme from tobacco roots catalyses the formation of nitric oxide from nitrite. Planta 212:835–841

Stoimenova M, Libourel IGL, Ratcliff RG, Kaiser WM (2003) The role of nitrate reduction in the anoxic metabolism of roots. II. Anoxic metabolism of tobacco roots with or without nitrate reductase activity. Plant Soil 253:155–167

Stuehr DJ (1999) Mammalian nitric oxide synthases. Biochim Biophys Acta 1411:217–230

Tischner R, Planchet E, Kaiser WM (2004) Mitochondrial electron transport as a source for nitric oxide in the unicellular green alga *Chlorella sorokiniana*. FEBS Lett 576:151–155

Thanos CA, Georghiou K (1988) Ecophysiology of fire-stimulated seed germination in *Cistus incanus* ssp *creticus* (L.) Heywood and *C. salvifolius* L. Plant Cell Environ 11:841–849

Toole EH, Toole VK, Bortwick HA, Hendricks SB (1955) Photocontrol of Lepidium seed germination. Plant Physiol 30:15–21

Tornton FC, Valente RJ (1996) Soil emissions of nitric oxide and nitrous oxide from no-till corn. Soil Sci Soc Am J 60:1127–1133

Yamasaki H (2000) Nitrite-dependent nitric oxide production pathway: implication for involvement of active nitrogen species in photoinhibition in vivo. Philos Trans R Soc Lond B Biol Sci 355:1477–1488

Yamasaki H, Sakihama Y (2000) Simultaneous production of nitric oxide and peroxynitrite by plant nitrate reductase: in vitro evidence for the NR-dependent formation of active nitrogen species. FEBS Lett 468:89–92

Yamasaki H, Sakihama Y, Takahashi S (1999) An alternative pathway for nitric oxide production in plants: new features of an old enzyme. Trends Plant Sci 4:128–129

Zumft WG (1997) Cell biology and molecular basis of denitrification. Microbiol Mol Biol Rev 61:533–616

Plant Cell Monogr (6)
L. Lamattina · J. C. Polacco: Nitric Oxide in Plant Growth
DOI 10.1007/7089_2006_086/Published online: 25 October 2006

Seeking the Role of NO in Breaking Seed Dormancy

Zlatko Giba[1] · Dragoljub Grubišić[2] (✉) · Radomir Konjević[1]

[1]Institute of Botany, Faculty of Biology, University of Belgrade, Takoska 43, 11000 Belgrade, Serbia and Montenegro

[2]"Siniša Stanković" Institute for Biological Research, Bulevar despota Stefana 142, 11060 Belgrade, Serbia and Montenegro
frc@ibiss.bg.ac.yu

Abstract In the course of evolution, higher plants developed a special reproductive organ, the seed, which ensures their spatio-temporal distribution and perpetuation of the species. The fate of the future plant is almost completely determined when the seed "decides" to germinate. A number of dormancy mechanisms serve to detect surrounding conditions and define the appropriate point in time for germination. To ensure survival of the future seedling, environmental conditions have to be detected, integrated, and translated through different signaling molecules at the seed level, even before germination starts. One of the ten smallest molecular species known, nitric oxide, is now recognized as an endogenous mediator of seed germination, external dormancy-breaking agent, and outer information carrier that provides the seeds with integral information on the factors most important for plant growth and development.

1 Introduction

The process of seed germination consists of imbibition (rapid initial water uptake), a plateau phase (high metabolic activity, but little change in water content), and subsequent radicle protrusion and growth (coincident with increase of water content). In terms of regulation of germination, the plateau phase is of primary interest. Generally, the length of this germination phase is extended by dormancy mechanisms, while factors that promote germination do so by shortening this phase.

In comparison with other plant organs, the remarkable property of seeds is that they are resting organs generally having a low moisture content ranging from 7 to 15%. In this state, seeds are able to survive for long time periods, often for many years. Some higher plants, such as the Balkan endemic species *Ramonda serbica* and *R. nathaliae*, can attain much lower water contents in their vegetative tissues (as low as 2–3%) without any repercussion for future life after rehydration (Quartacci et al. 2002). In this state (called anabiosis), almost no metabolic activity can be detected, whereas quiescent seeds show very low, but measurable, metabolic activity.

To germinate, quiescent seeds in the presence of suitable temperature and aeration need only to be hydrated. However, seeds of many species do not germinate even under optimum external conditions. These seeds are dormant and must be exposed to certain environmental stimuli, i.e., must experience some special situations, before they start to germinate. There is no unambiguous definition of this phenomenon, probably due to the fact that dormancy is manifested and broken in various ways in different species (Bewley and Black 1982; Nikolaeva et al. 1985). Dispersion of the seeds, mediated by animals or wind, can be considered an adaptive trait that ensures "distribution in space" of seed germination. Similarly, dormancy can be interpreted as an adaptive trait that ensures and optimizes the "distribution in time" of seed germination. By means of these two mechanisms, higher plants expand their range.

1.1 Action of Light

Plants can detect environmental light conditions at the seed level. As an environmental stimulus, light usually acts as a dormancy-breaking agent, but it can also inhibit germination. This is one of the most delicately regulated types of dormancy. Positively photoblastic seeds are not equipped with large amounts of food reserves (Dedonder et al. 1988). Light conditions for photosynthesis in the surroundings must be detected early at the seed level and these seeds are therefore equipped with appropriate photoreceptors.

Among plant photoreceptors only phytochromes have been shown to directly mediate induction of germination (Bortwick et al. 1954). Phytochrome B and phytochrome A regulate red light-induced germination, whereas induction of germination by far-red light, when it occurs, is mediated only by phytochrome A (Shinomura et al. 1998). Hennig et al. (2002) showed that phytochrome E is also involved in light-induced seed germination. When triggered, and before translocation into nuclei, these photoreceptors initiate a signal transduction cascade in the cytoplasm; this probably involves G-proteins, cGMP, and the Ca^{2+}/Cam complex (Bowler et al. 1994; Mustilli and Bowler 1997). Finally, it is believed that changes in seed sensitivity to stored gibberellins as well as de novo gibberellin synthesis lead to completion of germination (Toyomasu et al. 1998; Schwachtje and Baldwin 2004).

2 Nitrates and Nitrites Stimulate Seed Germination

2.1 Nitrates

Inorganic nitrates significantly increase the light sensitivity and decrease the light requirement of seeds (Toole et al. 1955). Although this effect of nitrates and nitrites has been known for a long time (Lehmann 1909) and several hypotheses as to their interaction with phytochrome have been formulated (Hilton 1985; Grubišić and Konjević 1990), almost nothing is known about the mechanism of action. The proposed connection between nitrogenous compounds, the pentose-phosphate pathway, and seed dormancy (Roberts 1973) has never been experimentally confirmed (Adkins et al. 1984; Cohn 1989). To elucidate this phenomenon, different authors have used different experimental model systems, such as germination of *Sysimbrium officinale* seeds (Hilhorst and Karssen 1989), the phytochrome-controlled germination of *Spirodela polyrhiza* turions (Appenroth et al. 1992), or fern spore germination (Haas and Scheuerlein 1991). For example, seeds of *Sisymbrium officinale* are not even photoblastic if they do not germinate in the presence of potassium nitrate or if endogenous concentrations of nitrates in the seeds are too low (Karssen and Hilhorst 1992).

2.2 Nitrites

Similarly to nitrates, inorganic nitrites stimulate seed germination of different species (Bewley and Black 1982). In the majority of cases, the stimulative effect is pH-dependent. Maximal activity of inorganic nitrites is obtained at around pH 3 (Cohn et al. 1983). It was shown that a stimulative effect of nitrite is obtained even without direct contact with the nitrite solution. Stimulation of germination can be achieved if grains are placed in a closed container in the presence of a vessel filled with an acidified nitrite solution. Assuming dormancy-breaking activity of some gaseous compound evolved from the acidified nitrite solution, the same authors showed that nitrogen dioxide breaks the dormancy of *Oryza sativa* when applied to imbibed or unimbibed caryopses (Cohn and Caste 1984). These results were largely overlooked until it was shown that NO donors such as organic nitrates can stimulate seed germination (Grubišić et al. 1991, 1992).

2.3 Organic Nitrates

Although used as anti-angina drugs in human medicine for more than 100 years (Ignarro 2002), until the beginning of the last decade of the 20th century, there were no data on the physiological effects of organic nitrates (e.g., nitroglycerin) in plants. This effect of organic nitrates was shown for the first time in stimulation of *Stellaria media* seed germination (Grubišić et al. 1991) and was further examined using light-induced *Paulownia tomentosa* seed germination as a model system (Grubišić et al. 1992). In both cases, nitroglycerin produced the same stimulative effect as potassium nitrate, but at approximately 100 times lower concentrations. It has been demonstrated that plant tissues can easily metabolize nitroglycerin into dinitro and mononitro derivatives (Goel et al. 1996). Appearance of nitric oxide after administration of nitroglycerin to *P. tomentosa* seeds was detected in EPR measurements (Giba et al. 1998). Nitroglycerin conversion to NO is mediated by the activity of aldehyde dehydrogenase in animal systems (Zhiqiang et al. 2002) but involvement of this enzyme in bioactivation of nitroglycerin is not yet confirmed in plants.

3 Nitric Oxide as Endogenous Mediator of Seed Germination

Possible effectiveness of NO as a dormancy-breaking agent in seed germination was for the first time anticipated by Hendricks and Taylorson (1974). Using different types of respiratory inhibitors such as sodium azide and potassium cyanide, they tentatively concluded that NO might be physiologically active, acting as a dormancy-breaking agent in seed germination. Studying effects of azide in animal systems, Murad came to a similar conclusion more than a decade later, additionally linking NO effects with cGMP formation. Several years after this, Ignarro (1987) established that the postulated EDRF (endothelium-derived relaxing factor) is NO. Because of the biological significance of the newly discovered physiologically active molecule, they were awarded the Nobel Prize in Physiology and Medicine together with Robert Furchgot for the year 1998.

In the plant kingdom, the first experimentally confirmed phenomenon promoted by NO was the stimulation of seed germination (Grubišić and Konjević 1990; Grubišić et al. 1991, 1992; Giba et al. 1994).

3.1 A Pharmacological Approach

Most of the current information about the function of NO in plants comes from pharmacological studies using NO donors, NO scavengers, and specific inhibitors. Induction/stimulation of a germination response after application of NO to seeds does not by itself establish that NO mediates the physiological process under study, and especially does not prove that NO is an endogenous mediator of the process. Nitric oxide is applied to seeds by using NO donors, substances that generate NO spontaneously, or after metabolic activation. Very rarely, the application of NO gas is used (Keeley and Fotheringham 1997). The applied substances might be effective per se, and during experimentation it is necessary to use more than one donor and corresponding chemical or structural analogs unable to release NO. Examining the physiological activity of nitroglycerin in *P. tomenosa* seed germination, Grubišić et al. (1992) used different types of organic nitrates (isosorbide mono- and dinitrate, pentaerythritol tetranitrate etc.). All of them showed a similar maximum of activity at a concentration around 10^{-4} M. Chemical analogs of nitroglycerin, such as triacetin (glycerol triacetate) or glycerol, alone were completely ineffective in *P. tomentosa* seed germination, suggesting the importance of the presence of nitro groups in the molecule for its physiological activity. Subsequent in vivo detection of NO after administration of nitroglycerin to seeds (Giba et al. 1998) and correlation with germination dynamics and dose dependency clearly indicate that this effect is NO-mediated. These results were additionally confirmed by using other, very different compounds with NO-releasing properties. *S*-Nitroso acetylpenicillamine (SNAP) and 3-morpholinosydnonimine (SIN-1) also potentiated germination of these seeds. Again, appropriate inactive compounds, chemical and structural analogs unable to release NO (acetylpenicillamine and molsidomine), expressed either a negligible effect or none at all (Giba et al. 1997, 1998).

Subsequently, experiments confirmed NO activity in seed germination and agreed well with parallel EPR measurements (Giba et al. 1998). For example, the presence of potassium nitrite in the incubation medium stimulates light-induced *P. tomentosa* seed germination. Its maximal effect is achieved at concentrations similar to those of inorganic nitrates, around 10^{-2} M. Acidification of the nitrite solution down to pH 3.5 does not contribute greatly to its effectiveness. Although the extent of germination was slightly higher, the maximum of nitrite activity remained at the same concentration. However, at a pH equivalent to its pK (3.3) or lower, i.e., under conditions when nitrite releases NO spontaneously and behaves as an NO donor, its effectiveness dramatically increases. Under these conditions, nitrite stimulates germination at almost two orders of magnitude lower concentrations. This stimulation is similar to activity of the other mentioned NO donors tested in the same experimental system. A slightly smaller shift in effectiveness can be obtained if seeds are treated with a mixture of potassium nitrite and a reductant, which produces NO at a pH close to neutral (Giba et al. 1998).

3.2
Sodium Nitroprusside

One of the most frequently used NO donors in this type of experiment is sodium nitroprusside [$Na_2Fe(CN)_5NO$]. In research on the effects of different substances with electron-accepting properties in *P. tomentosa* seed germination, it was found that only high concentrations (around 10^{-2} M) of compounds with standard redox potentials equal to or higher than $E_0' = +360$ mV (inorganic nitrates, hexacyanoferrate (III), or hexachloroirridate (IV)) potentiate the germination of *P. tomentosa* seeds (Giba et al. 1994). Similar results were obtained with seeds of other species (Plummer et al. 2001). Some authors reported the critical redox potential for dormancy-breaking activity to be at slightly lower levels of about $E_0' = +300$ mV (Estabroock and Yoder 1998). A notable exception to all of these findings was the stimulative ef-

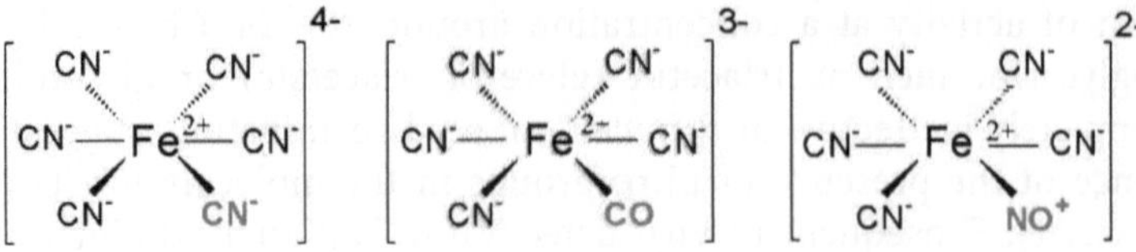

Fig. 1 Structure of different ferrous complexes used in germination assays. The anionic parts of three analogous complex ferrous compounds. From *left* to *right*: potassium hexacyanoferrate (II), potassium carbonyl prussiate, and sodium nitroprusside

fectiveness of sodium nitroprusside in seed germination (Giba et al. 1994). In light-induced *P. tomentosa* seed germination, this very low potential electron acceptor was effective in concentrations similar to those of NO-releasing compounds, but with an even broader range of promotive concentrations (Giba et al. 1997, 1998). These results were later confirmed by Beligni and Lamattina (2000) who used *Lactuca sativa* as the experimental system, and in *A. thaliana* (Batak et al. 2002). The effectiveness of SNP in *P. tomentosa* seed germination was tested by application of two of its analogs, hexacyanoferrate (II) and carbonyl prussiate ($K_4[Fe(CN)_6]$ and $K_3[Fe(CN)_5CO]$) (Giba et al. 1998). In the inner ligand sphere, these molecules contain, respectively, cyanide anion (CN^-) or carbon monoxide (CO) instead of nitrosonium ion (NO^+) as the sixth ligand. Both substances at 10^{-2} M had a minor effect on seed germination. The three isoelectronic ligands NO^+, CO, and CN^- provide corresponding ferrous complexes with practically the same electron configuration (Fig. 1).

However, only sodium nitroprusside is able to release nitric oxide into the solution. This NO release from sodium nitroprusside occurs only after nitroprusside has undergone reduction and released some or all of the bound cyanides. Apart from NO, possible byproducts of nitroprusside decomposition, such as ferrous iron or cyanide ions, do not enhance *P. tomentosa* seed germination (Živković et al. 2005). In some other experimental systems, cyanide by itself was capable of stimulating germination, rendering questionable the mechanism of nitroprusside action (Sarath et al. 2005). Although this puzzling stimulatory effect of cyanides has been known for a long time (Bewley and Black 1982), exogenously applied KCN in concentrations of up to 10^{-2} M has no effect on germination of *P. tomentosa* seeds, except if applied together with sodium azide, which by itself causes total inhibition of germination. Surprisingly, in this case the cyanide ions allow germination in the presence of inhibitory concentrations of azide (Živković et al. 2005).

3.3 Gibberellic Acid Nitrite

One trend in the development of NO donors is the synthesis of different "chimeric" molecules. These so-called "hybrid NO donors" are substances that combine the known physiological activity of one molecule with the NO-donating function. In animal systems this combination not only generates a new drug with mixed functions, but in some cases also shows novel activities (Hou et al. 1999; Granik and Grigorev 2002).

Trying to combine the physiological effectiveness of NO donors in seed germination and gibberellic activity in one molecule, we synthesized gibberellic acid nitrite (nGA_3). This molecule, containing two – O – NO groups instead of hydroxyl groups on the gibban skeleton, was tested in germination of seeds of *Stellaria media* and *P. tomentosa*. These seeds differ with respect to their light requirement: *P. tomentosa* seeds have an absolute light requirement for germination, while *S. media* seeds, although light-sensitive, germinate to a certain extent in darkness (depending on temperature). Additionally, GA_3 can replace the light requirement of *P. tomentosa* seeds, enabling germination to occur in darkness, while exogenously applied GA_3 in concentrations of up to 10^{-3} M is ineffective in germination of *S. media* seeds.

The light-requiring seeds of these two species (*P. tomentosa* and *S. media*) were used to test the possibility that the nGA_3 could be a stimulator of seed germination that combines the activity of gibberellic acid by itself with the effectiveness of organic nitrates and/or organic nitrites in seed germination. Here we employed the rarely used technique of filling dry seeds with active matter by immersing seeds in methylene chloride solutions of the substances to be tested. It was shown that gibberellic activity of nGA_3 does not exist after nitrosation, i.e., after incorporation of – O – NO groups instead of hydroxyl groups into the GA_3 molecule. Although the gibberellic activity of nGA_3 is lost, the experimental substance keeps NO-releasing properties and acts as an NO donor in both experimental systems (Jovanović et al. 2005).

Some other similar compounds were synthesized and tested, such as nitrosalicylic acid, where the NO-releasing moiety is directly attached to the C atom via a C – N bond. In general, it was concluded that this approach is also promising as a means of achieving targeted delivery of NO in cells and tissues.

4
Endogenous Sources of NO in Seeds

Possible sources of NO in the plant kingdom were discussed for the first time at the European Symposium on Plant Photomorphogenesis held at Sitges (Spain) in 1995. There was general agreement that if there is any enzymatic endogenous NO production in plants, it probably would not be mediated by animal-like NO-synthases (Giba et al. 1995). Rather, apart from non-enzymatic routes, the enzymatic apparatus of plants present opportunities to produce NO in several different ways using available nitrates and nitrites already present in the soil.

In plants, especially under stress conditions, nitric oxide can be produced non-enzymatically through light-mediated conversion of NO_2 by carotenoids (Cooney et al. 1994) and enzymatically by nitrate and nitrite reductases (Dean and Harper 1988; Yamasaki et al. 1999; Stöhr et al. 2001), probably via activity of class 2 hemoglobins and possibly by means of other nitric oxide synthase-independent mechanisms (Kozlov et al. 1999; Harrison 2002). It was found by in vitro measurements that the NO-generating capacity of nitrate reductase can account only for approximately 1% of the total activity (Rockel et al. 2002). Based on effective concentrations of NO donors of around 10^{-4} M, it can be assumed that the most frequently used concentrations of nitrates in seed germination (around 10^{-2} M) would be sufficient for physiological activity via the NO pathway. Examining stimulatory effects of NO donors in barley seed germination, Bethke et al. (2003) showed that aleurone cells have powerful proton pumps allowing external pH to be around 3. In addition, with the presence of reduced ascorbate in the apoplastic space the microenvironment around the barley aleurone layer is ideally suited for non-enzymatic production of NO from nitrite (Bethke et al. 2003).

Enzymes that transform L-arginine into citrulline and NO (Palmer et al. 1988) are found in practically all divisions of the living world (for a review, see Torreilles 2001). Even though some authors suggested such a possibility (Leshem 1996; Cueto et al. 1996; Corpas et al. 2001), neither a gene nor any protein with a higher sequence similar to known animal-like NOS has been cloned or isolated in plants. Even in the sequenced *Arabidopsis* genome, no gene coding for protein with a sequence similar to those of large animal NOS proteins has been found. These facts suggest that the NOS activity in plants comes from different types of enzymes.

Endogenous sources of NO are often classified as enzymatic and non-enzymatic, but obviously in plants these sources can also be classified as nitrate-mediated and arginine-mediated. After retraction of papers by members of Klessig's group concerning NOS activity of a protein of the glycine decarboxylase complex (named variant P), it becomes clear that the only candidate for this type of NO synthesis in plants is AtNOS1, a 60 kDa protein encoded by a gene with a sequence similar to one from snail (*Helix pomatia*), which is implicated in NO synthesis (Guo et al. 2003). This is the only plant gene encoding NOS that has been identified to date.

4.1 Nitric Oxide Action in Seed Germination

Seed germination is a crucial event in the life of higher plants. Dormant seeds do not germinate unless specific environmental signals or events occur. It is not surprising that nitrate and/or NO activity in seed germination is coupled with phytochrome, a basic pigment system by which seeds detect their surroundings. Based on the elegant experimentation of Shinomura et al. (1996)

it was shown that exogenously applied nitrates and NO donors predominantly affect phyA-induced germination, while phyB-induced germination is affected to a far lesser extent (Batak et al. 2002). Constitutively expressed phyB serves other purposes. Light perceived mainly by phyB in the seeds generates a signal that preconditions the future seedling to its most likely environment, i.e., seedling development is predetermined by the light environment experienced by the seeds (Mazzaella et al. 2005).

In germination of light-requiring seeds, NO activity can be attributed to activation of soluble guanylate cyclase. It has been hypothesized that cGMP is a member of the phytochrome transduction chain. On the other hand, though a large number of species show linkage between the photoblastic response and NO-stimulated germination, a number of species do not. This implies more than one mechanism of NO action. The proposed cGMP-dependent mechanism is still questionable because the only plant guanylate cyclase enzyme identified to date is not NO-dependent (Ludidi and Gehring 2003). Nitric oxide action in seed germination is not exclusively connected with photoblastism. In non-photoblastic seeds, activity is evident in NO-mediated activation of β-amylase (Zhang et al. 2005). However, the molecular mechanisms by which NO operates are still largely unknown, and major effort will be needed to identify direct targets of NO in seed germination.

The role postulated to date for nitrogenous oxides as endogenous regulatory molecules in various physiological processes is perhaps overestimated. These extremely reactive molecules can interact with a large number of different molecules, most of them members of signal transduction pathways of a wide variety of plant hormones or other signaling compounds. It is hard to discriminate experimentally between a normal regulatory physiological role of nitrogenous oxides and a simple side effect of their reactivity. Even in a simple aqueous solution, nitric oxide is transformed into several radical and non-radical species (Fig. 2). These compounds exist in equilibrium. Almost

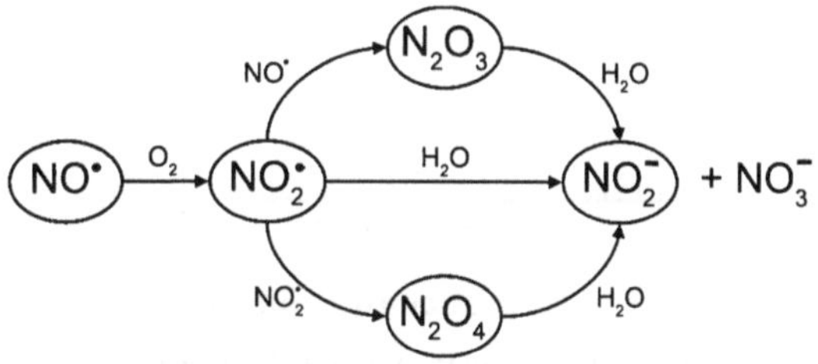

Fig. 2 The chemistry of nitrogen oxides is simple in aqueous solution. Nitric oxide reacts with molecular oxygen to form NO_2 radical as the first reaction product. NO_2 exists as the dimer N_2O_4 or can be transformed to N_2O_3 after reaction with NO radical. All three nitrogenous oxides in reaction with water produce nitrite anions, but in the presence of additional oxidizing species can be transformed into the final oxidation product, nitrate. The chemistry of nitrogen oxides in gaseous and aqueous phases differs in minor details

all of them can be physiologically active and individually capable of breaking dormancy. This complicated situation is similar to that in animal systems, where the NO_2 radical mimics typical NO-mediated effects (Davidson et al. 1996), and the nitrite anion can be physiologically active by itself (Bryan et al. 2005).

Another question is the stability of NO in living systems. The reaction of nitric oxide with atmospheric triplet oxygen is the main reason why this molecule is termed "unstable". Chemical oxidation of NO to NO_2 is a third-order reaction, and its rate depends on the square of the NO concentration: $k_3\,[NO]^2[O_2]$. Hence, oxidation of NO is significant only at a relatively high NO concentration. At low concentrations, NO can be considered "stable". The half-life of NO is of the order of hours in micromolar concentrations, and several orders of magnitude higher in nanomolar concentrations (Conrad 1996). Because of the presence of nitrogen oxides in the soil and/or lower troposphere, plants have developed specific mechanisms to protect against their elevated concentrations. Some authors have speculated that the presence of alternative oxidase (AOX) in plants additionally serves such purposes (Millar and Day 1996). AOX lowers the probability of superoxide formation and thereby protects cells from the NO/O_2^- reaction product, peroxynitrite. Huang et al. (2002) demonstrated transcriptional activation of the AOX gene family as one of the earliest events after treatment with NO in *Arabidopsis*

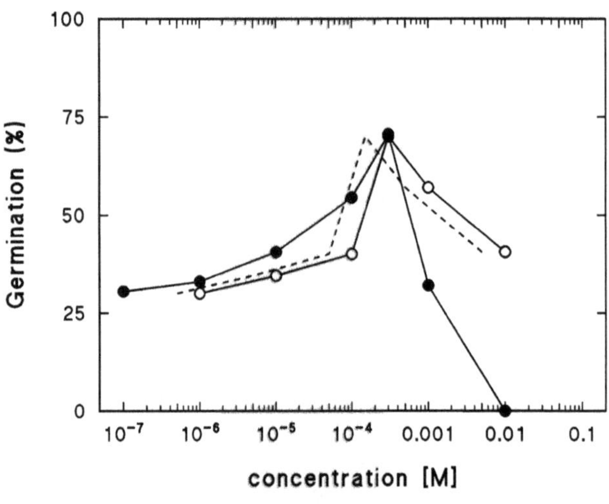

Fig. 3 Effect of Angeli's salt and DEAE/NONOate on light-induced *P. tomentosa* seed germination. Seeds were imbibed in different concentrations of Angeli's salt (*open circles*) or DEAE/NONOate (*solid circles*) for 3 days, irradiated for 5 min with red light, returned to darkness, and left to germinate for an additional 7 days. The germination percentage was determined 10 days after the start of imbibition. The *dotted line* represents germination at calculated concentrations of nitroxyl liberated from Angeli's salt. Under these conditions, the nitroxyl anion exists only in its protonated form, HNO

cell suspension cultures. Plant non-symbiotic hemoglobins are also known to detoxify NO in a NAD(P)H-dependent manner (for a review see Delledonne 2005).

Much attention has recently been given to redox siblings of NO (Bartberger et al. 2002). NO can exist as a free radical, nitrosonium cation, and nitroxyl anion ($NO^{\cdot}$, NO^{+}, NO^{-}). These are distinct moieties that are chemically related. However, it seems that in *P. tomentosa* seed germination the above mentioned NO redox forms are equally active (Fig. 3). The maximal stimulative effect of Angeli's salt (NO^{-} donor) and diethylamine NONOate ($NO^{\cdot}$ donor) on light-induced *P. tomentosa* seed germination is obtained at concentrations of around 10^{-4} M. This is similar to the effectiveness of SNP, a presumed NO^{+} donor (Hou et al. 1999). However, future investigations will probably yield new data on this interesting redox system.

5
NO as Exogenous Mediator of Seed Germination

A possible ecological role has been suggested for the nitrate requirement in seed germination. The nitrate requirement may provide seeds with information on the nitrogen status of the soil and the proximity of already established plants (Pons 1989). If the nitrate dependence of dormant seeds allows them to sense the nitrogen status of the ecosystem before germination starts, this capacity can be considered part of the survival strategy of plants (Goudey et al. 1988; Hsiao and Quick 1996; Bell et al. 1999). Additionally, Thanos and Rundel (1995) suggested that increased soil nitrate content after wildfires (Polglase et al. 1986) can stimulate germination from the soil seed bank, thus triggering repair of fire-damaged ecosystems.

After discovery of NO action in seed germination, the theory of Pons was generalized. Nitrogen oxides that are present in the soil (as "soil trace gases"), or in the lower troposphere, may be carriers of information about soil nitrate content and microbial activity, thereby indicating soil quality (Giba et al. 1999). Regular annual variation of soil trace gas fluxes (Nielsen et al. 1996) can provide seeds with information about seasonal and climate changes in their surroundings. Nitrogenous oxides are thus outer information carriers providing seeds with integral data about many important factors required for successful germination and seedling growth (Giba et al. 2003). It must be stressed that the proposed delicate regulation of germination processes in ecosystems can be easily disturbed by pollutants containing NO and related compounds. Pollutants can trigger seed germination when external conditions are inappropriate for seedling growth. This would have irreversible direct consequences for the future population and community structure.

Nitrogenous oxides are present in the smoke generated by wildfires, whose stimulative effect can be considered a special case, i.e., a part of the general mechanism whereby ecosystems regulate the timing and sequence of seed germination from soil seed banks (Giba et al. 2005).

5.1 Smoke and Seed Germination

Smoke evolved during wildfires is one of the most important chemical stimuli of the germination of fire-type species. De Lange and Boucher (1990) were the first to report that plant-derived smoke is a potent stimulator of seed germination. After that discovery, germination stimulated by smoke and its extracts was demonstrated for many fire-dependent and non-fire-dependent species. There are several excellent reviews on this topic (Brown and van Staden 1997; van Staden et al. 2000; Minorsky 2002).

Slow combustion of dry or fresh plant material produces compounds that are water-soluble (liquid smoke) and that stimulate the germination of a wide variety of seeds. Although the physiological activity of smoke or smoke extracts in seed germination is well documented so far, progress in identifying the active component of smoke has been slow and limited. Baldwin and coworkers (1994) tested, alone or in combination, more than 200 compounds found in smoke fractions. It was later concluded that similar types of compounds are present in smoke extracts derived from different plant materials (van Staden et al. 1995). Aqueous extract of plant-derived smoke is a complex mixture of thousands of components (Adriansz et al. 2000). Several compounds found in smoke have been singled out as being responsible for its stimulating activity. A possible role was suggested for the gaseous plant hormone ethylene and for short chain fatty acids such as octanoic acid (Jäger et al. 1996). Adriansz et al. (2000) identified 1,8 cineol as an active germination stimulator in smoke. Attempts to elucidate the physiological mechanism of the well-known stimulating effect of inorganic nitrates in seed germination (Lehmann 1909; Grubišić and Konjević 1990; Thanos and Rundel 1995) led to the conclusion that gaseous nitrogenous compounds such as NO are potent stimulaters of seed germination (Grubišić et al. 1992; Giba et al. 1994, 1997). NO exhibited maximal activity in almost 100-fold lower concentration than inorganic nitrates (Giba et al. 1998). These findings and the fact that nitrogenous oxides are inevitable byproducts of wildfires supported the conclusion that these oxides are germination cues, i.e., active components of smoke (Keeley and Fotheringham 1997; for a review see Giba et al. 2003). However, after discovery that the combustion of pure cellulose has a stimulatory effect on germination (Baldwin et al. 1994) some authors speculated that the germination signal from smoke is a non-nitrogenous compound that likely contains only C, H and O (Preston et al. 2004). Very recently, Flemmatti et al. (2004) found that the active factor from smoke that stimulates seed germination may be butenolide, a compound that is partly similar in molecular structure to strigol.

5.2
More than One Physiological Activity of Smoke

Although much work has been done and several hypotheses have been put forward, there is still a considerable amount of uncertainty related to this topic. None of the substances mentioned above has physiological characteristics that could ensure all the physiological activities of smoke or liquid smoke extracts.

If simultaneously applied with liquid smoke, maximal effectiveness of different gibberellins in *P. tomentosa* seed germination can be obtained at lower concentrations (down to $< 0.1\times$) than without liquid smoke (Todorović et al. 2005). In *A. thaliana* seed germination, liquid smoke enhances the stimulative effect of gibberellic acid if applied in combination with suboptimal concentrations of GA_3 (Fig. 4). Nitrogenous compounds (nitrates and/or NO donors) do not enhance gibberellin-induced germination (Todorović et al. 2005). The observed changes in sensitivity to gibberellins link the activity of smoke with cytokynins (Tomas and Van Staden 1995) and the so-called "permissive" action of cytokynins on gibberellin activity in seed germination (Khan 1977). This implies direct changes caused by liquid smoke in the level of GA-receptors and/or GA-binding proteins, such as have already been proposed by some authors (Schwachtje and Baldwin 2004). During chemical processes in flame, it can be supposed that very low molecular weight compounds are

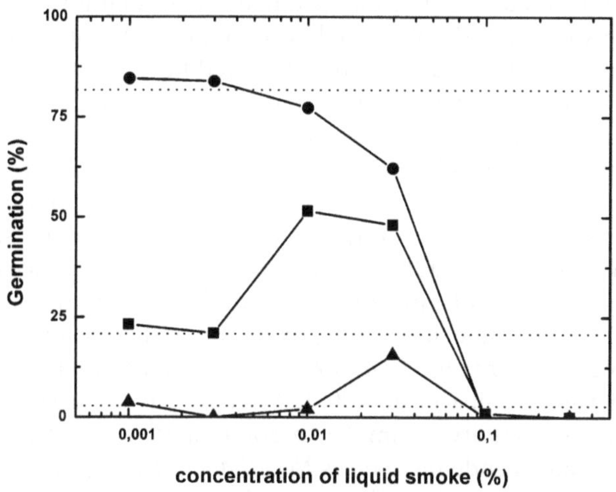

Fig. 4 Effect of liquid smoke on *A. thaliana* seed germination in the the presence of suboptimal concentrations of gibberellic acid. Seeds were imbibed, in darkness, in a mixture of different concentrations of liquid smoke (v/v) and 10^{-5} M (*triangles*), 10^{-4} M (*squares*) or 3×10^{-4} M (*circles*) concentrations of GA_3. The final germination percentage was determined after 5 days. The *dotted lines* represent germination induced only by the corresponding GA_3 solutions

formed that can act as molecular chaperones (Bennion and Daggett 2003). Although there is no inherent gibberellic activity of smoke, its activity can still be linked with gibberellins. In contrast to gibberellins, liquid smoke cannot replace the light requirement of *P. tomentosa* seeds, but it can overcome inhibition of light-induced germination caused by exogenous application of growth retardants, i.e., inhibitors of gibberellin biosynthesis (Todorovic et al. 2005).

On the basis of these findings it can be concluded that the active component of smoke must share properties of gibberellins or their analogs, cytokinins and/or low molecular weight chaperones, or nitrogenous compounds such as active nitrogen species, i.e., NO radicals and related compounds. It is hard to believe that one hypothetical molecule from smoke could possess the activity of all the mentioned physiologically active substances. For example, even the simplest "combination", gibberellic acid nitrite (nGA_3) (the chimeric molecule of GA_3 and organic nitrites) has no gibberellic activity, but is active in seed germination only as an NO donor (Jovanović et al. 2005). It has already been proposed by some authors (e.g., Doherty and Cohn 2000) that there may be multiple active components of liquid smoke, which can act together to break dormancy.

5.3 Are Nitrogenous Oxides Active Components of Smoke?

It has been shown that vapors from acidified nitrite solutions or illuminated nitroprusside solutions stimulate seed germination without direct contact of seeds with solution (Cohn and Castle 1984; Bethke et al. 2004). Similarly, liquid smoke vapor stimulates germination of *P. tomentosa* seeds (Todorovic et al. 2005). Although Keeley and Fotheringham postulated it in 1998, some authors excluded the possibility that nitrogenous oxides are active components of smoke. This opinion is mainly based on the observation of Baldwin and coworkers (Preston et al. 2004) that pure cellulose-derived smoke is active in seed germination. Additionally, there is a notable absence of evidence indicating the presence of nitrite ions in liquid smoke extracts. Nitrites are products of NO "decomposition" (i.e., oxidation and subsequent hydration), but can also be a source for non-enzymatic NO formation by dismutation at acidic pH values (Giba et al. 1998). Moreover, it is possible to demonstrate stimulatory activity of smoke in some species of nitrate-insensitive seeds, such as Balkan endemic species *Nepetha rtanjensis* (unpublished data). Similarly, stimulatory activity of liquid smoke is sustained if applied in combination with the NO scavenger cPTIO (Light and van Staden 2003).

However, when considering the relevance of the presence of nitrogenous oxides in smoke for its physiological activity, the situation can be much more complicated. Because of the chemical composition of air (99% molecular nitrogen and triplet oxygen), nitrogenous oxides are inevitable by-products of flames (Hayhurst and McLean 1974). Nitrogenous oxides can be transformed, stabilized, and deposited on the seed coat or soil particles in the form of a wide variety of organic or inorganic nitrogenous compounds, most of them stable and possibly physiologically active per se or as NO donors. (Cohn 1996; Morikawa et al. 2002). Yamasaki (2005) estimated the number of possible generated nitrogen-containing signaling agents to be more than 10^4. Charred wood, for example, is a potent stimulator of seed germination. Interesting demonstrations of the influence exerted by atmospheric NO on several crucial physiological processes can be found in the literature. Pharmacologists at the Free University of Berlin obtained a number of unexpected results in experiments with guanylate cyclase. When enzyme activity was measured in close proximity to the city's highway loop, they found a strong correlation with atmospheric NO concentration (Friebe et al. 1996). Additionally, forensic studies have firmly established that a clear indicator of smoke poisoning in victims is the level of nitrite/nitrate in whole blood, arising after conversion of the NO/NO_2 mixture from smoke (Kage et al. 2000). Another problem is presented by experiments on cPTIO and smoke action, whose results are inconclusive. An inevitable by-product of the NO-scavenging action of cPTIO is nitrite anion, a well known stimulator of seed germination. It is believed that nitrite is active by itself, even in animal systems (Bryan et al. 2005). Some authors found that the "specific" NO scavenger cPTIO inhibits the stimulatory action of cyanide-releasing compounds in seed germination of species that are sensitive to cyanides. The explanation that cyanide induces endogenous NO production in seeds seems unconvincing (Sarath et al. 2005).

5.4
"Cold Plasma" Stimulates Seed Germination

Finally, mention must be made here of the remarkable stimulatory effect of non-equilibrium (low temperature) air plasma pretreatment on seed germination of different species. "Cold plasma" can be considered a kind of cold flame obtainable at low gas pressure and low temperatures, even ones equal to room temperature (Hippler et al. 2001). Several minutes of exposure to air-derived cold plasma, in a plasma reactor, stimulate light-induced germination of *P. tomentosa* seeds while argon-derived cold plasma, for example, is almost completely ineffective (Živković et al. 2004). Considering this phenomenon in various plant species, different authors suggested that the pretreatment process results in covalent or non-covalent attachment of plasma-produced molecular fragments on the seed envelope (Volin et al. 2000; Dubinov et al. 2000). During cold plasma chemical processes in a mixture of gases such as air, so-called "active nitrogen species" are generated that are physiologically active in *P. tomentosa* seed germination. It must be stressed that the products generated during cold plasma chemical processes are similar to those generated in flames (Hayhurst and McLean 1974). Interestingly, free nitrite ions were not detected in the imbibing medium of air plasma-pretreated *P. tomentosa* seeds (Živković et al. 2004), but FTIR analysis reveals the presence of NO moieties attached on the seed coat of these seeds (unpublished data).

Some authors speculate that active components of smoke may act by directly oxidizing cellulose or other components in the seed coat, producing a product that subsequently elicits germination (Keeley and Fotheringham 1998; see also Preston et al. 2004). This can be explained by assuming that there is no active component of smoke, but that one arises during interaction between smoke components and the seed coat. However, a mixture of active compounds (including nitrogenous oxides) is likely to be the active principle of smoke.

6 Conclusion

It is well established that NO is physiologically active in seed germination. Besides elucidation of direct molecular targets of NO as an endogenous factor for germination, major effort will be needed to place the available experimental data in a chemical ecological context including the effectiveness of NO as a pollutant. One of the most profound capabilities of plants at the seed level is their ability to escape in time and select an environment by sensing ecologically informative external signals. The data obtained so far provide a strong conceptual and experimental foundation for asking the right questions and planning future work.

References

Adkins SW, Simpson GM, Naylor JM (1984) The physiological basis of seed dormancy in *Avena fatua*. IV. Alternative respiration and nitrogenous compounds. Physiol Plant 60:234–238

Adriansz TD, Rummey JM, Bennett IJ (2000) Solid phase extraction and subsequent identification by gas chromatography-mass spectrometry of a germination cue present in smoky water. Anal Lett 33:2793–2804

Appenroth KJ, Augsten H, Mohr H (1992) Photophysiology of turion germination in *Spirodela polyrhiza* (L.) Schleiden. X. Role of nitrate in the phytochrome-mediated response. Plant Cell Environ 15:743–748

Baldwin LT, Staszak-Kozinski L, Davidson R (1994) Up in smoke: I. Smoke-derived germination cues for postfire annual *Nicotiana attenuata* Torr ex Watson. J Chem Ecol 20:2345–2371

Bartberger MD, Liu W, Ford E, Miranda KM, Switzer C, Fukuto JM, Farmer P, Wink DA, Houk K (2002) The reduction potential of nitric oxide (NO) and its importance to NO biochemistry. Proc Natl Acad Sci USA 99:10958–10963

Batak I, Dević M, Giba Z, Grubišić D, Poff KL, Konjević R (2002) The effects of potassium nitrate and NO donors on phytochrome A- and phytochrome B-specific induced germination of *Arabidopsis thaliana* seeds. Seed Sci Res 12:253–259

Bell DT, King LA, Plummer JA (1999) Ecophysiological effects of light quality and nitrate on seed germination in species from Western Australia. Aust J Ecol 24:2–10

Beligni MV, Lamattina L (2000) Nitric oxide stimulates seed germination and de-etiolation, and inhibits hypocotyl elongation, three light-inducible responses in plants. Planta 210:215–221

Bennion BJ, Dagett V (2004) Counteraction of urea-induced protein denaturation by trimethylamine *N*-oxide: A chemical chaperone at atomic resolution. Proc Natl Acad Sci USA 101:6433–6438

Bethke P, Beligni V, Badger M, Jacobsen J, Jones R (2003) Nitric oxide as an endogenous regulator of seed physiology. Plant Biol 25–30 July, Honolulu, Hawai, USA, Abs. 25004

Bethke P, Gubler F, Jacobsen JV, Jones R (2004) Dormancy of *Arabidopsis* seeds and barley grains can be broken by nitric oxide. Planta 219:847–855

Bewley JD, Black M (1982) Physiology and biochemistry of seeds in relation to germination. II. Viability, dormancy and environmental control. Springer, Berlin Heidelberg New York

Bortwick HA, Hendricks SB, Toole EH, Toole VK (1954) Action of light on lettuce-seed germination. Bot Gazette 115:205–225

Bowler C, Neuhaus G, Yamagata H, Chua NH (1994) Cyclic GMP and calcium mediate phytochrome phototransduction. Cell 77:73–81

Brown NAC, van Staden J (1997) Smoke as a germination cue: a review. Plant Growth Reg 22:115–124

Bryan NS, Fernandez B, Bauer S, Garcia-Saura M, Milsom A, Rassaf T, Maloney R, Bharti A, Rodriguez J, Feelish M (2005) Nitrite is a signaling molecule and regulator of gene expression in mammalian tissues. Nature 1:290–297

Cohn MA, Butera DL, Hughes LA (1983) Seed dormancy in red rice. III. Response to nitrite, nitrate and ammonium ions. Plant Physiol 73:381–384

Cohn MA, Castle L (1984) Dormancy in red rice. IV. Response of unimbibed and imbibing seeds to nitrogen dioxide. Physiol Plant 60:552–556

Cohn MA (1989) Factors influencing the efficacy of dormancy-breaking chemicals. In: Taylorson RB (ed) Recent advances in the development and germination of seeds. Plenum, New York, pp 261–267

Conrad R (1996) Soil microorganisms as controllers of atmospheric trace gases. (H_2, CO, CH_4, OCS, N_2O, and NO). Microbiol Rev 60:609–640

Cooney RV, Harwood PJ, Custer LJ, Franke AA (1994) Light-mediated conversion of nitrogen dioxide to nitric oxide by carotenoids. Environ Health Persp 102:460–462

Corpas F, Barosso J, Del Rio A (2001) Peroxisomes as a source of reactive oxygen species and nitric oxide signal molecules in plant cells. Trends Plant Sci 6:145–150

Cueto M, Hernandez-Perera O, Martin R, Bentura ML, Rodrigo J, Lamas S, Golvano MP (1996) Presence of nitric oxide synthase activity in roots and nodules of *Lupinus albus*. FEBS Lett 398:159–164

Davidson CA, Kaminski PM, Migdan W, Wollin MS (1996) Nitrogen dioxide causes pulmonary arterial relaxation via thiol nitrosation and NO formation. Am J Physiol 270:H1038-H1043

Dean JV, Harper JE (1988) The conversion of nitrite to nitrogen oxide(s) by the constitutive NAD(P)H-nitrate reductase enzyme from soybean. Plant Physiol 88:389–395

Dedoner A, Rethy R, De Petter H, Fredericq H, De Greef J (1988) Preliminary screening experiments on the effects of light and GA_3 in the germination of different seed species. In: De Greef J (ed) Photoreceptors and plant development. University Press, Antwerp, pp 431–435

Delledeonne M (2005) NO news is good news for plants. Curr Opin Plant Biol 8:1–7

Doherty LC, Cohn MA (2000) Seed dormancy in red rice (*Oryza sativa*). XI. Commercial liquid smoke elicits germination. Seed Sci Res 10:415–421

Dubinov A, Lazarenko E, Selemir V (2000) Effect of glow discharge air plasma on grain crops seed. IEEE Trans Plasma Sci 20:180–183

Estabroock EM, Yoder JI (1998) Plant-plant communications: rhizosphere signaling between parasitic angiosperms and their hosts. Plant Physiol 116:1–7

Flemmatti G, Ghisalberti E, Dixon K, Trengove R (2004) A compound from smoke that promotes seed germination. Science 305:977–977

Friebe A, Malkewitz J, Schultz G, Koesling D (1996) Positive effects of pollution. Nature 382:120–120

Furchgott RF (1988) Studies on relaxation of rabbit aorta by sodium nitrite: the basis for the proposal that the acid-activable inhibitory factor from bovine retractor penis is inorganic nitrite and the endothelium-derived relaxing factor is nitric oxide. In: Vanhoutte PM (ed) Vasodilatation: vascular smooth muscle, peptides, autonomic nerves and endothelium. Raven, New York, pp 401–414

Giba Z, Grubišić D, Konjević R (1992) Sodium nitroprusside – stimulated germination of common chick weed (*Stellaria media* L.) seeds. Arch Biol Sci 44:17P–18P

Giba Z, Grubišić D, Konjević R (1994) The effect of electron acceptors on the phytochrome-controlled germination of *Paulownia tomentosa* seeds. Physiol Plant 91:290–294

Giba Z, Grubišić D, Konjević R (1995) The effect of nitric oxide-releasing compounds on the germination of *Paulownia tomentosa* seeds. European symposium on photomorphogenesis in plants, 9–13 July 1995, Sitges, Barcelona, Spain, p 23

Giba Z, Grubišić D, Sajc L, Stojaković Đ, Konjević R (1997) Effect of some nitric oxide donors and methylene blue on light-induced germination of *Paulownia tomentosa* seeds. Arch Biol Sci 49:15P–16P

Giba Z, Grubišić D, Todorović S, Sajc L, Stojaković Đ, Konjević R (1998) Effect of nitric oxide – releasing compounds on phytochrome – controlled germination of Empress tree seeds. Plant Growth Reg 26:175–181

Giba Z, Grubišić D, Konjević R (1999) How the seeds sense environmental changes? The role of nitrogenous compounds. Arch Biol Sci 51:121–129

Giba Z, Grubišić D, Konjević R (2003) Nitrogen oxides as environmental sensors for seeds. Seed Sci Res 13:187–196

Giba Z, Grubišić D, Konjević R (2005) Nitric oxide and seed germination. In: Magalhaes JR, Singh RP, Passos LP (ed) Nitric oxide signaling in higher plants. Studium, Houston, pp 239–275

Goel A, Kumar G, Payne GF, Dube SK (1997) Plant cell biodegradation of a xenobiotic nitrate ester, nitroglycerin. Nat Biotechnol 15:174–177

Goudey JS, Saini HS, Spencer MS (1998) Role of nitrate in regulating germination of *Synapsis arvensis* L (wild mustard). Plant Cell Environ 11:9–12

Granik VG, Grigorev NB (2002) Exogenous donors of nitric oxide (a chemical aspect). Russian Chem Bull 51:1375–1422

Grubišić D, Konjević R (1990) Light and nitrate interaction in phytochrome-controlled germination of *Paulownia tomentosa* seeds. Planta 181:239–243

Grubišić D, Giba Z, Konjević R (1991) Organic nitrates stimulated germination of common chick weed (*Stellaria media* L) seeds. Arch Biol Sci 43:7P–8P

Grubišić D, Giba Z, Konjević R (1992) The effect of organic nitrates in phytochrome-controlled germination of *Paulownia tomentosa* seeds. Photochem Photobiol 56:629–632

Guo F-Q, Okamoto M, Crawford NM (2003) Identification of a plant nitric oxide synthase gene involved in hormonal signaling. Science 302:100–103

Haas CJ, Scheuerlein R (1991) Nitrate effect on phytochrome-mediated germination in fern spores: Investigation on the mechanism of nitrate action. J Plant Physiol 138:350–357

Harrison R (2002) Structure and function of xanthine oxidoreductase: Where are we now? Free Radic Biol Med 33:774–797

Hendricks SB, Taylorson RB (1974) Promotion of seed germination by nitrate, nitrite, hydroxylamine and ammonium salts. Plant Physiol 54:304–309

Henning L, Stoddart WM, Dieterle M, Whitelam GC, Schäfer E (2002) Phytochrome E controls light-induced germination of Arabidopsis. Plant Physiol 128:194–200

Hayhurst AN, McLean AG (1974) Mechanism for producing NO from nitrogen in flames. Nature 251:303–305

Hilhorst HWM, Karssen CM (1989) Nitrate reductase independent stimulation of seed germination of *Sysimbrium officinale* L. (hedge mustard). Ann Bot 63:131–137

Hilton JR (1985) The influence of light and potassium nitrate on the dormancy and germination of *Avena fatua* L. seed stored buried and under natural conditions. J Exp Bot 36:974–979

Hippler R, Pfau S, Schmidt M, Shoenbach KH (2001) Low temperature plasma physics. Wiley, Berlin

Hou YC, Janczuk A, Wang PG (1999) Current trends in the development of nitric oxide donors. Curr Pharm Des 5:417–447

Hsiao AI, Quick WA (1996) The roles of inorganic nitrogen salts in maintaining phytocrome- and gibberellin A_3-mediated germination control in skotodormant lettuce seeds. J Plant Growth Reg 15:159–165

Huang X, von Rad U, Durner J (2002) Nitric oxide induces transcriptional activation of the nitric oxide – tolerant alternative oxidase in *Arabidopsis thaliana* suspension cells. Planta 215:914–923

Ignarro LJ, Buga GM, Wood KS, Byrns RE, Chaudhury G (1987) Endothelium-derived relaxing factor produced and released from artery and vein is nitric oxide. Proc Natl Acad Sci USA 84:9265–9269

Ignarro LJ (2002) After 130 years the molecular mechanism of action of nitroglycerin is revealed. Proc Natl Acad Sci USA 99:7816–7817

Jovanović V, Giba Z, Đoković D, Milosavljević S, Grubišić D, Konjević R (2005) Gibberellic acid nitrite stimulates germination of two species of light-requiring seeds via the nitric oxide pathway. Ann NY Acad Sci 48:476–481

Jäger AK, Strydom A, van Staden J (1996) The effect of ethylene, octanoic acid, and plant-derived smoke extracts on the germination of light-sensitive lettuce seeds. Plant Growth Reg 19:197–210

Kage S, Kudo K, Ikeda N (2000) Determination of nitrate in blood by gas chromatography and gas chromatography-mass spectrometry. J Chromatogr B 742:363–368

Karssen CM, Hilhorst HWM (1992) The ecology of regeneration in plant communities. In: Fenner M (ed) Seeds. CAB, Wallingford, UK, pp 327–348

Keeley JE, Fotheringham CJ (1997) Trace gas emissions and smoke-induced seed germination. Science 276:1248–1250

Keeley JE, Fotheringham CJ (1998) Mechanism of smoke-induced seed germination in a post-fire chaparral annual. J Ecol 86:27–36

Khan AA (ed) (1977) The physiology and biochemistry of seed dormancy and germination. North Holland, Amsterdam

Kozlov AV, Staniek K, Nohl H (1999) Nitrite reductase activity is a novel function of mammalian mitochondria. FEBS Lett 454:127–130

Lehmann E (1909) Zur Keimungphysiologie und -biologie von *Ranunculus sclereatus* L. und einigen anderen Samen. Berichte der Deutschen Botanischen Gesellschaft 27:476–494

Leshem YY (1996) Nitric oxide in biological systems. Plant Growth Reg 18:155–159

Light ME, van Staden J (2003) The nitric oxide specific scavenger carboxy-PTIO does not inhibit smoke stimulated germination of Grand Rapids lettuce seeds. South Afr J Bot 69:217–220

Ludidi N, Gehring C (2003) Identification of novel protein with guanylyl cyclase activity in *Arabidopsis thaliana*. J Biol Chem 278:6490–6494

Mazzaella MA, Arana MV, Staneloni RJ, Perelman S, Batiller MJR, Muschietti J, Cerdian PD, Chen K, Sanchez RA, Zhu T, Chory J, Casal J (2005) Phytochrome control of the *Arabidopsis* transcriptome anticipates seedling exposure to light. Plant Cell 17:2507–2516

Millar AH, Day DA (1996) Nitric oxide inhibits the cytochrome oxidase but not alternative oxidase of plant mitochondria. FEBS Lett 398:155–158

Minorsky P (2002) Smoke-induced germination. Plant Physiol 128:1167–1168

Morikawa H, Fukunaga K, Takahashi M, Kawamura Y, Sakamoto A (2002) Formation of unidentified nitrogen (UN) in plants – UN from nitrates. Plant Cell Physiol 43:S128

Mustilli AC, Bowler C (1997) Tuning into the signals controlling photoregulated gene expression in plants. EMBO J 16:5801–5806

Nikolaeva MG, Rasumova MV, Gladkova VN (1985) In: Danilova F (ed) Reference book of dormant seed germination. Nauka, Leningrad (in Russian)

Nielsen T, Pilegaard K, Egelov AH, Granby K, Hummelshoj P, Jensen NO, Skov H (1996) Atmospheric nitrogen compounds: occurrence, composition and deposition. Sci Tot Environ 189/190:459–465

Palmer RMJ, Ashton DS, Moncada AS (1988) Vascular endothelial cells synthesize nitric oxide from L-arginine. Nature 333:664–666

Polglase PJ, Attiwill PM, Adams MA (1986) Immobilization of soil nitrogen following wildfire in two eucalypt forests of south-eastern Australia. Acta Oecol/Oecol Plant 7:261–271

Preston CA, Becker R, Baldwin IT (2004) Is NO news good news? Nitrogen oxides are not components of smoke that elicits germination in two smoke-stimulated species, *Nicotiana attenuata* and *Emmenanthe penduliflora*. Seed Sci Res 14:73–79

Plummer JA, Rogers AD, Turner DW, Bell DT (2001) Light, nitrogenous compounds, smoke and GA_3 break dormancy and enhance germination in the Australian everlasting daisy *Shoenia filifolia* subsp. *subuilifolia*. Seed Sci Tech 29:321–330

Pons TL (1989) Breaking of seed dormancy by nitrate gap detection mechanism. Ann Bot 63:139–143

Quartacci MF, Glišić O, Stevanović B, Navari-Izzo F (2002) Plasma membrane lipids in the resurrection plant *Ramonda serbica* following dehydration and rehydration. J Exp Bot 53:2159–2166

Roberts EH (1973) Oxidative processes and the control of seed germination. In: Heydecker W (ed) Seed ecology. Butterworths, London, pp 189–218

Rockel P, Strube F, Rockel A, Wildt J, Kaiser WM (2002) Regulation of nitric oxide (NO) production by plant nitrate reductase in vivo and in vitro. J Exp Bot 53:103–110

Sarath G, Bethke PC, Jones R, Baird LM, Hou G, Mitchell RB (2006) Nitric oxide accelerates seed germination in warm-season grasses. Planta 223:1154–1164

Stöhr C, Strube F, Marx G, Ulrich W, Rockel P (2001) A plasma-membrane bound enzyme of tobacco roots catalyses the formation of nitric oxide from nitrite. Planta 212:835–841

Schwachtje J, Baldwin IT (2004) Smoke exposure alters endogenous gibberellin and abscisic acid pools and gibberellin sensitivity while eliciting germination in the post-fire annual, *Nicotiana attenuata*. Seed Sci Res 14:51–60

Shinomura T, Nagatani A, Hanzawa H, Kubota M, Watanabe M, Furuya M (1996) Action spectra for phytochrome A- and B-specific photoinduction of seed germination in *Arabidopsis thaliana*. Proc Natl Acad Sci USA 93:8129–8133

Thanos CA, Rundel PW (1995) Fire-followers in chaparral: nitrogenous compounds trigger seed germination. J Ecol 83:207–216

Todorović S, Giba Z, Živković S, Grubišić D, Konjević R (2005) Stimulation of empress tree germination by liquid smoke. Plant Growth Reg 47:141–148

Thomas TH, van Staden J (1995) Dormancy break of celery (*Apium graveolens* L.) seeds by plant-derived smoke extract. Plant Growth Reg 17:195–198

Toole EH, Toole VK, Bortwick HA, Hendricks SB (1955) Photocontrol of *Lepidium* seed germination. Plant Physiol 30:15–21

Torreilles J (2001) Nitric oxide: one of the more conserved and widespread signaling molecules. Frontiers Biosci 6:1161–1172

Toyomasu T, Kawaide H, Mitsuhashi W, Inoue Y, Kamiya Y (1998) Phytochrome regulates gibberellin biosynthesis during germination of photoblastic lettuce seeds. Plant Phys 118:1517–1523

Van Staden J, Drewes FE, Brown NAC (1995) Some chromatographic characteristics of germination stimulants in plant-derived smoke extracts. Plant Growth Reg 17:241–249

Van Staden J, Brown NAC, Jäger A, Johnson TA (2000) Smoke as a germination cue. Plant Species Biol 15:167–178

Volin J, Denes F, Young R, Park SM (2000) Modification of seed germination performance through cold plasma chemistry technology. Crop Sci 40:1706–1718

Yamasaki H, Sakihama Y, Takahashi S (1999) An alternative pathway for nitric oxide production in plants: new features of an old enzyme. Trends Plant Sci 4:128–129

Yamasaki H (2005) Nitric oxide research in plant biology: its past and future. In: Magalhaes JR, Singh RP, Passos LP (ed) Nitric oxide signaling in higher plants. Studium, Houston, pp 1–23

Zhang H, Shen WH, Zhang W, Xu L (2005) A rapid response of β-amylase to nitric oxide but not gibberellin in wheat seeds during the early stage of germination. Planta 220:708–716

Zhiqiang C, Zhang J, Stamler JS (2002) Identification of the enzymatic mechanism of nitroglycerin bioactivation. Proc Natl Acad Sci USA 99:8306–8311

Živković S, Puač N, Giba Z, Grubišić D, Petrović ZL (2004) The stimulatory effect of non-equilibrium (low temperature) air plasma pretreatment on light-induced germination of Paulownia tomentosa seeds. Seed Sci Tech 32:693–701

Živković S, Giba Z, Grubišić D, Konjević R (2005) The counteracting effect of potassium cyanide in sodium azide-inhibited germination of Paulownia tomentosa Steud. seeds. Arch Biol Sci 57:29–34

Sarath G, Bethke PC, Jones R, Baird LM, Hou G, Mitchell RB (2006) Nitric oxide accelerates seed germination in warm season grasses. Planta 223:1154–1164
Stöhr C, Strube F, Marx G, Ullrich W, Rockel P (2001) A plasma membrane-bound enzyme of tobacco roots catalyses the formation of nitric oxide from nitrite. Planta 212:835–841
Schwachtje J, Baldwin IT (2004) Smoke exposure alters endogenous gibberellin and abscisic acid pools and gibberellin sensitivity while eliciting germination in the post-fire annual, Nicotiana attenuata. Seed Sci Res 14:51–60
Shinomura T, Nagatani A, Hanzawa H, Kubota M, Watanabe M, Furuya M (1996) Action spectra for phytochrome A- and B-specific photoinduction of seed germination in Arabidopsis thaliana. Proc Natl Acad Sci USA 93:8129–8133
[illegible]
[illegible]
[illegible]
[illegible]
[illegible]

Plant Cell Monogr (6)
L. Lamattina · J. C. Polacco: Nitric Oxide in Plant Growth
DOI 10.1007/7089_2006_087/Published online: 25 October 2006

Nitric Oxide Functions as Intermediate in Auxin, Abscisic Acid, and Lipid Signaling Pathways

Natalia Correa-Aragunde · María Luciana Lanteri · Carlos García-Mata · Arjen ten Have · Ana María Laxalt · Magdalena Graziano · Lorenzo Lamattina (✉)

Instituto de Investigaciones Biológicas, Facultad de Ciencias Exactas y Naturales, Universidad Nacional de Mar del Plata, CC 1245, 7600 Mar del Plata, Argentina
lolama@mdp.edu.ar

Abstract Nitric oxide (NO) is a chemical messenger that actively operates in the plant kingdom. In recent years, NO has been shown to be involved in many and diverse growth, developmental, and physiological processes in plants. It has been shown that NO takes part in different hormone signaling pathways and also acts in concert with well-characterized second messengers. In this chapter, we discuss findings that contribute to the understanding of the role/s of NO during root organogenesis and stomatal movement, focusing on the interrelations between NO and the phytohormones auxin and abscisic acid. We emphasize the requirement of calcium as an essential intermediate present in the NO-mediated responses. Finally, novel data concerning the cross-talk between NO and phosphatidic acid signaling pathways in response to (a)biotic stresses are presented and discussed.

1 Introduction

The last decade has witnessed milestones in nitric oxide (NO) research in plant biology. An impressive number of publications on the functions of NO in the plant kingdom show that NO is involved in controlling growth, developmental, and patho/physiological processes. As a result, NO is now considered a second messenger implicated in many plant cell signaling events, frequently under the control of hormonal stimuli. Table 1 summarizes publications that reported experimental evidence supporting a role for NO in mediating phytohormone-regulated processes. Most of those results will be discussed throughout this chapter and in other contributions in the present volume.

Here, we will focus on the cross-talk between auxin and NO as well as between abscisic acid (ABA) and NO in two well-described hormone-regulated physiological processes, namely root development and stomatal movement. Finally, we will discuss the implications of new data from our laboratory involving a NO-mediated activation of phospholipid signaling in tomato cell cultures. Convergent results have appeared in the last few years

Table 1 Phytohormone-regulated processes mediated by NO

Process	Plant species	Organ/tissue	Refs.
Auxin			
Root tip elongation	*Zea mays*	Root	Gouvea et al. 1997
Adventitious root formation	*Cucumis sativus*	Hypocotyl	Pagnussat et al. 2002, 2003, 2004; Lanteri et al. 2006
Lateral root formation	*Solanum lycopersicum*	Root	Correa-Aragunde et al. 2004
Primary root growth inhibition	*Solanum lycopersicum*	Root	Correa-Aragunde et al. 2004
Gravitropic response	*Arabidopsis thaliana*	Root	Hu et al. 2005
Cell cycle activation	*Arabidopsis thaliana*	Cell suspension	Otvos et al. 2005
	Solanum lycopersicum	Root	Correa Aragunde et al. 2006
Root hair development	*Lactuca sativa*, *Arabidopsis thaliana*	Root	Lombardo et al. 2006
ABA			
Stomatal closure	*Vicia faba*	Leaf epidermis	Garcia-Mata and Lamattina 2001, 2002
	Arabidopsis thaliana	Leaf epidermis	Desikan et al. 2002; Neill et al. 2002
Induction of antioxidant enzymes	*Stylosanthes guianensis*	Leaf	Zhou et al. 2005
Citokinin			
Betalaine accumulation	*Amaranthus caudatus*	Leaf	Scherer and Holk 2002

that firmly suggest that NO and some phospholipid signals are operative in common regulated signaling pathways during plant responses to (a)biotic stresses.

2 Auxin as a Major Root Growth Regulator

Plant growth and morphogenesis are under the control of both environmental stimuli and endogenous signals such as plant hormones. Auxin, the first hormone discovered in plants, is one of the most important morphogenic compounds that shape the whole plant body (Friml 2003; Sachs 2004). Auxin regulates, in concert with other plant growth regulators, embryo and fruit development, vascular tissue differentiation, control of lateral branching and root growth, and tropic responses such as apical dominance, gravitropism, and phototropism (Abel and Theologis 1996). Roots are an attractive system to study auxin action because of their morphological simplicity and well-characterized responses. The response triggered by auxin depends on the cell type, developmental stage, and physiological condition as well as on the auxin concentration within the cell.

Auxin is mainly synthesized in the shoot apical meristem and transported, cell by cell, to the root by auxin-efflux and -influx transporters. In roots, transport is more complex, with two distinct polarities. Auxin moves acropetally (toward the root apex) through the central cylinder and basipetally (from the root apex toward the base) through the outer layer of cells (Jones 1998). It is well known that auxin translocation is essential for auxin-mediated processes in roots. Mutants deficient in auxin transport often have altered gravitropic response and display few adventitious roots (ARs) and lateral roots (LRs) (Marchant et al. 1999, 2002; Xu et al. 2005). In addition, auxin perception and signaling pathways are also essential to complete auxin's biological function. Genetic and molecular approaches in Arabidopsis have resulted in the identification of many components of the auxin signaling pathway. Several lines of evidence indicate that auxin induces gene expression by targeting members of the Aux/IAA family of transcriptional repressor proteins for degradation via the 26S proteasome. Auxin-regulated ubiquinitation of the Aux/IAA proteins is mediated by the F-box protein TIR1 (Leyser 2002). Recent results have shown that TIR1 is an auxin receptor (Dharmasiri et al. 2005; Kepinski and Leyser 2005). Additionally, microRNAs and signal molecules such as NO are also involved in auxin-mediated responses in roots (Guo et al. 2005; Mallory et al. 2005; Lanteri et al. 2006a). After more than a century of research, the auxin-mediated signaling pathways still remain to be completely elucidated.

2.1 NO Functions in Auxin-Induced Signaling Pathways Driving Root Growth and Developmental Processes

NO plays a central role in determining the morphology and developmental pattern of roots (Fig. 1). Gouvea et al. (1997) provided the first evidence for the participation of NO in an auxin-induced process in roots. NO accumulation in response to auxin treatment was shown by Pagnussat et al. (2002) in cucumber explants during AR formation. Subsequently, it was demonstrated that auxin application to roots resulted in localized NO production during LR and root hair (RH) formation, as well as asymmetric NO accumulation in the root tip during the gravitropic response (Correa-Aragunde et al. 2004; Lombardo et al. 2006; Hu et al. 2005). Auxin-induced AR, LR, and RH formation as well as root gravitropic response were prevented by the application of the specific NO scavenger cPTIO, suggesting a key role for endogenous NO in mediating those processes. Moreover, application of NO was able to reverse the inhibitory effect of the basipetal auxin transport inhibitor

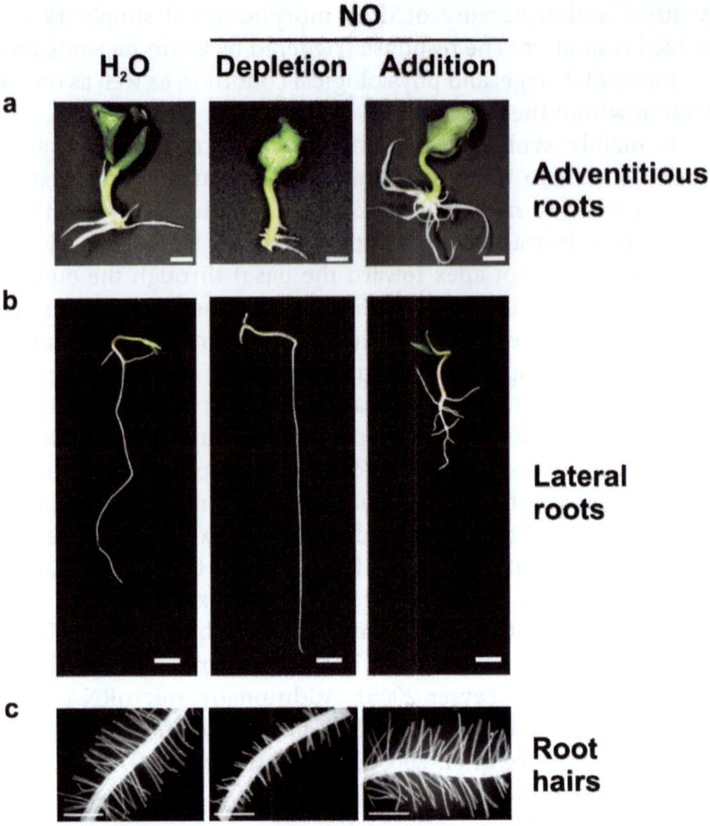

Fig. 1 Effect of NO addition or depletion on adventitious root, lateral root, and root hair development. **a** Primary roots were removed from hypocotyls of 10-day-old cucumber seedlings. Photographs represent explants incubated with H_2O, 200 μM cPTIO (NO depletion), or 10 μM SNP (NO addition) for 5 days. The *scale bar* represents 1 cm. **b** Germinated seeds of tomato were treated with H_2O, 1 mM cPTIO (NO depleted), or 200 μM SNP (NO addition). Representative photographs of seedlings after 5 days of treatment. The *scale bar* represents 1 cm. **c** Arabidopsis (Col-0) seeds were germinated in ATS medium with 1% sucrose and 0.7% agar for 3 days and then transferred to fresh medium containing H_2O, 500 μM cPTIO (NO depletion), or 1 μM SNP (NO addition). Representative photographs of root hairs were taken after 4 days of treatment. The *scale bar* represents 0.5 mm

1-naphtylphthalamic acid (NPA; Pagnussat et al. 2003; Correa-Aragunde et al. 2004; Hu et al. 2005). Taken together these results suggest that NO is an important molecule operating downstream of auxin through a linear signaling pathway during root growth and development.

There are several sources of NO in plants and most of them have been described in roots. Enzymatic sources include a nitrate reductase (NR), an NO synthase (NOS) and a root-specific plasma membrane-bound nitrite reductase activity (Rockel et al. 2002; Guo et al. 2003; Stöhr et al. 2001; see also other chapters in this volume). In addition, from the activity of microorganisms (among them plant-growth-promoting rhizobacterium like *Azospirillum*), it was suggested than the rhizosphere is an important source of NO (Creus et al. 2005).

To study the contribution of the NO sources in root morphological responses, pharmacological approaches using both NR and NOS inhibitors have been employed. In particular, application of NR or NOS inhibitors resulted in reduced NO accumulation and gravitropic bending. Since these inhibitors did not completely prevent the response, apoplastic non-enzymatic NO production was suggested (Hu et al. 2005). Genetic approaches were followed using the Arabidopsis *nos1*, *nia1*, *nia2*, and *nia1/nia2* mutants in order to demonstrate the requirement for the enzymes NOS and NR in NO-mediated responses (Desikan et al. 2002; Guo et al. 2003; Lombardo et al. 2006). In general, these mutants are defective in NO production and display altered NO-mediated phenotypes. Our recent results support the participation of both NOS- and NR-catalyzed NO synthesis during RH formation (Lombardo et al. 2006).

2.2 NO Regulates Cyclic GMP- and MAPK-Dependent Signaling Pathways During Adventitious Root Formation

Cyclic GMP is an intracellular signaling molecule whose concentration is transiently increased in response to NO via regulation of guanylate cyclase activity (GC; Neill et al. 2003). It was shown that the GC inhibitor, 6-anilino-5, 8-quinilinedione (LY83583), was able to reduce AR formation in both auxin- and NO-treated explants. This inhibition was reversed by treatment with the membrane-permeable cGMP analog 8-Br-cGMP. This evidence strongly supports NO operating downstream of auxin in inducing AR formation through the GC-catalyzed synthesis of cGMP (Pagnussat et al. 2003). cGMP can act via cADPR, which was shown to regulate cytosolic Ca^{2+} concentration in various plant species (Sanders et al. 1999). Results obtained with compounds that block cADPR/ryanodine-sensitive Ca^{2+} channels or cADPR synthesis indicate that cADPR is involved in AR formation (Lanteri et al. 2006b). Calcium efflux from intracellular stores can also be mediated by the ligand inositol 1,4,5-triphosphate (IP_3; Allen et al. 1995). Blockers of IP_3-regulated Ca^{2+} channels significantly suppressed AR formation induced by auxin or by NO. It was also demonstrated that Ca^{2+}-dependent protein kinases (CDPKs) are downstream messengers in the Ca^{2+}-regulated signaling pathways triggered by auxin and NO (Lanteri et al. 2006b).

Auxin activates a mitogen-activated protein kinase (MAPK) signaling cascade during AR formation (Pagnussat et al. 2004). Additional results using the specific NO scavenger cPTIO support an NO-dependence for this activation.

This MAPK cascade seems to be cGMP-independent, since the NO-induced MAPK activity was not affected by the GC inhibitor LY83583 (Pagnussat et al. 2004). Overall, the available data indicate a scenario in which basipetal transport of auxin induces an NO production where ARs develop, through an as-yet unknown mechanism. Then, NO triggers a bifurcated signaling pathway that includes:

1. An increase in the levels of cGMP, cADPR, and possibly other second messengers, which results in elevation of both cytosolic Ca^{2+} concentration and CDPK activity.
2. Induction of a cGMP-independent MAPK cascade. The activation of all the pathways triggered by NO seems to be required for AR formation since the stimulatory effect of auxin and NO is abolished when one of these pathways is compromised. The use of mutants with modified NR or NOS expression will certainly help to determine the NO source/s during AR formation.

2.3 NO Modulates the Expression of Cell Cycle Regulatory Genes During Lateral Root Formation

LR formation initiates in the pericycle. There, cells adjacent to the xylem poles of the vascular cylinder divide to form a LR primordium. The LR primordium grows through the overlying cell layers of the parent root and eventually breaks through the epidermis and emerges (Laskowski et al. 1995; Malamy and Benfey 1997; Dubrovsky et al. 2001; Casimiro et al. 2003). It was shown that NO induces LR formation (Fig. 1b) as part of the auxin signaling cascade (Correa-Aragunde et al. 2004). The timing of experimental depletion of endogenous NO after auxin treatment and the subsequent blockage of LR formation suggest that NO is required for the first steps of LR development, i.e., cell cycle progression and LR primordia establishment (Correa-Aragunde et al. 2006).

Auxin has been reported extensively to be involved in regulation of cell division in different plant organs and cell cultures (Ferreira et al. 1994; Burssens et al. 2000; Himanen et al. 2002, 2004; Roudier et al. 2003). Auxin exerts its effect on LR formation by cell cycle stimulation at the G1-to-S transition (Himanen et al. 2002). Auxin application induces the expression of A-, B-, and D-type cyclins as well as cyclin-dependent kinases (CDKs). Additionally, auxin represses the expression of the kip-related proteins (KRPs) gene family involved in the arrest of cells in the G1 phase (De Veylder et al. 2001; Himanen et al. 2002). The analysis of cell cycle regulatory genes showed that NO induces the expression of the cyclins *CYCA2;1*, *CYCD3;1* and the cyclin-dependent kinase *CDKA1* during LR formation in tomato (Correa-Aragunde et al. 2006). Interestingly, the auxin-induced expression of cell cycle regulatory genes was prevented or delayed when NO was scavenged with cPTIO, suggesting that endogenous NO is required for auxin action. The NO effect was more evident for the gene involved in the G1-to-S phase transition *CYCD3;1* (Correa-Aragunde et al. 2006). Low levels of *CYCD3;1* and high levels of the CDK inhibitor *KRP2* mRNA could be observed in NO-depleted roots, conditions where the expression of *CDKA1* seems to be constitutive. Subsequent addition of auxin or NO to the NO-depleted roots results in a dramatic decrease of *KRP2* transcript levels, which is accompanied by the accumulation of *CYCD3;1* (Correa-Aragunde et al. 2006). Moreover, treatment with the auxin transport inhibitor NPA induces cell cycle arrest in the G1 phase (Himanen et al. 2002). NO is able to rapidly activate *CYCD3;1* mRNA accumulation in the presence of NPA, suggesting that NO induces *CYCD3;1* in auxin-depleted roots (Correa-Aragunde et al. 2006).

In another recent report, NO was shown to stimulate the activation of cell division and embryogenic cell formation in leaf protoplast-derived cells of alfalfa (Otvos et al. 2005). Several NO donors stimulated BrdU incorporation into DNA while NOS inhibitors acted in the opposite way. In alfalfa cell suspension, NO transiently induces *CYCA2;1*, and *CYCD3;1* mRNA level, and CDKA;1,2 and CDKB;1,2 activities (Otvos et al. 2005). The NO effect on cell division in alfalfa leaf cell suspensions occurs only in the presence of auxin, while the effect of NO on LR formation occurs even in the presence of NPA, suggesting that cell suspension responses to NO do not necessarily mimic those in the whole plant. Altogether, the data indicate that NO is able to modulate the expression of the A- and D-type cyclin genes as well as the activity of CDKs taking part in the signaling cascade downstream of auxin. No studies regarding the regulation of B-type cyclins by NO have yet been reported. Since an effect of NO on cell cycle regulatory genes has been shown in LR development and embryogenic cell formation, it would be interesting to know whether NO is required for other growth processes. Also of interest is the potential relationship between NO and other hormones or molecules involved in cell division (cytokinins, ABA, sugars, brassinosteroids).

Evidence presented in this review suggests that auxin triggers a common signaling pathway in all the root developmental process studied so far. The participation of specific NO sources and second messengers such as cGMP, Ca^{+2}, IP_3 or MAPKs were reported in many of those auxin-induced processes. However, the order of events and the interaction between these messengers still remains to be further elucidated. Future studies will contribute to deciphering the complete scenario of auxin action, including the sources of NO and specific targets of NO during root growth and organogenesis. In addition, how auxin causes an increase in NO production remains to be addressed.

3 NO is Downstream of ABA in Controlling both Induction of Stomatal Closure and Inhibition of Stomatal Aperture

Stomata are epidermal pores of the aerial parts of higher plants. The regulation of the pore size is crucial for plants to optimize CO_2 uptake and to minimize transpirational H_2O loss. This process is accurately regulated by a pair of contiguous specialized cells called guard cells. Stomatal movement is regulated by osmotically driven changes of guard cell volume. Stomatal opening occurs when massive uptake of osmotically active solutes (mainly K^+ and Cl^-), induces H_2O uptake, causing a volume increase of guard cells and thus increasing the aperture of the stomatal pore. Conversely, a closing stimulus can trigger a subsequent net loss of osmotically active solutes, the loss of H_2O, the reduction of guard cell volume and, finally, a reduction of stomatal pore size (Assmann and Shimazaki 1999; Blatt 2000).

In order to control gas exchange, plants must sense and integrate multiple external (environmental) and internal (chemical) stimuli. Among them, ABA is probably the most studied and best characterized stimulus. ABA is rapidly produced in roots and leaves in response to drought (Assmann and Shimazaki 1999). Once in the guard cells, ABA induces stomatal closure and inhibits stomatal opening via a complex signaling network that involves a myriad of components, among them Ca^{2+}, K^+, IP_3, MAPK, and H_2O_2 (Fan et al. 2004). Even though guard cell signaling has been thoroughly studied for the last three decades, there are still knowledge gaps in the network and new components are still being discovered. NO has been reported to be among the novel components.

One of the first observations linking NO and stomatal movements showed that NO induces stomatal closure, reduces transpiration rate, and enhances plant adaptive responses to drought stress (Garcia-Mata and Lamattina 2001). It was shown that exogenous application of NO induced stomatal closure, in a Ca^{2+}-dependent manner, in monocot and dicotyledoneous species (Garcia-Mata and Lamattina 2001). Since then, many reports have enlarged the knowledge of NO biology in guard cells (Neill et al. 2002; Desikan et al. 2002; Garcia-Mata et al. 2003). As a result, we now know that in guard cells, endogenous NO is enzymatically synthesized by either nitrate reductase (NR) (Desikan et al. 2002) or/and nitric oxide synthase (NOS) (Guo et al. 2003), depending on the plant species and on the physiological conditions. It was also demonstrated that NO is an active component of ABA- (Garcia-Mata and Lamattina 2002; Neill et al. 2002), H_2O_2-, and UV-B-induced stomatal closure (He et al. 2005; Bright et al. 2006). UV-B light has been reported to induce both H_2O_2 and NO in guard cells (He et al. 2005). Interestingly, in the same work authors showed that NO is also required for endogenous synthesis of H_2O_2 in response to UV-B. With respect to ABA signaling, endogenous NO production (both via NR and NOS) was reported to be required by ABA for the induction of stomatal closure (Garcia-Mata and Lamattina 2002; Neill et al. 2002; Guo et al. 2003). Bright and colleagues (2006) have recently shown that endogenous H_2O_2 production induced by ABA is required for NO synthesis in Arabidopsis guard cells. NO, in turn, induces Ca^{2+} liberation from intracellular Ca^{2+} stores, leading to a rise in cytoplasmic calcium concentration ($[Ca^{2+}]_{cyt}$) and a down-regulation of K^+ concentration through the inhibition of inward rectifying K^+ channels (Garcia-Mata et al. 2003, and discussed in the chapter by Sokolovski and Blatt in this volume). In addition, Sokolovski and colleagues (2005) showed that NO also contributes to stomatal closure by the activation of K^+ efflux through outward rectifying K^+ channels.

Regardless of the advances in the understanding of aspects of NO participation in the induction of stomatal closure, the role of NO in stomatal opening processes is still poorly understood. The first reports on this subject showed that the NR null double mutant *nia1/nia2* did not exhibit altered ABA-dependent inhibition of light-induced stomatal opening (Desikan et al. 2002). However, ABA was not able to inhibit stomatal opening in the Arabidopsis NOS1-deficient mutant, *Atnos1* (Guo et al. 2003). These two reports point towards a NOS-dependent NO generation involved in inhibition of stomatal opening.

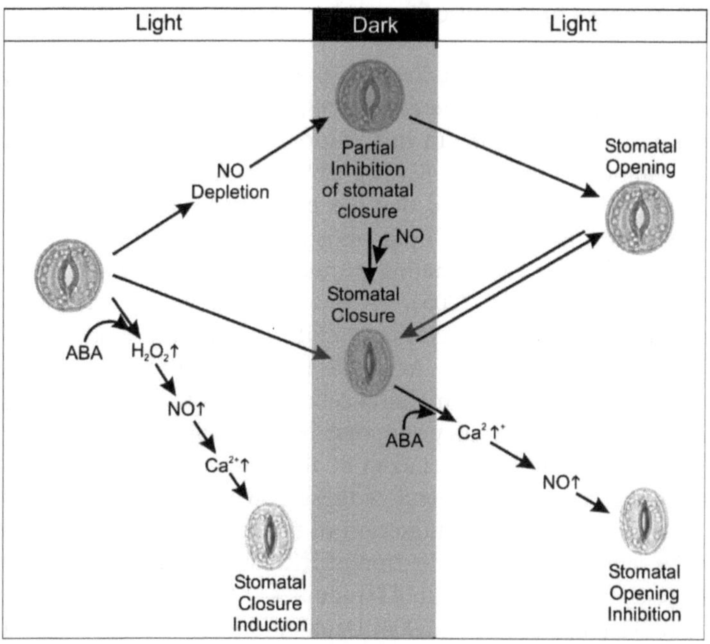

Fig. 2 Simplified model showing the involvement of NO during the regulation of stomatal movement. NO participates both in the induction of stomatal closure and the inhibition of light-induced stomatal opening

These data were recently confirmed in *Vicia faba* epidermal strips (Garcia-Mata and Lamattina; submitted). Pharmacological approaches indicate that NOS-dependent NO production is required by ABA for the inhibition of light-induced stomatal opening. Interestingly, and in contrast to stomatal closure processes, Ca^{2+} seems to be acting upstream of NO by increasing NO production via a Ca^{2+}-dependent NOS activity. The bifurcated pathway in ABA-mediated activation of either stomatal closure or inhibition of stomatal opening is not new for guard cell signaling. Several second messengers (e.g., Ca^{2+}, pH, and protein phosphatases) were reported to be part of the pathways leading either to the aperture or to the closure of the stomatal pore through different mechanisms (Irving et al. 1992; Cousson and Vavasseur 1998; Allen et al. 1999).

This new evidence firmly establishes that NO acts downstream of ABA in the physiological events controlling stomatal transpiration (Fig. 2). We now have a rather accurate idea of the position of NO in the intricate signaling network that regulates stomatal movement. Further studies will strengthen the current models, mainly by identifying the NO targets and the molecular mechanisms through which NO regulates them.

4
Cross-Talk between NO and Phosphatidic Acid Signaling Pathways in Plant Responses to (A)Biotic Stresses

NO has been shown to be involved in the development of the hypersensitive response (HR) (Delledonne et al. 1998, 2001; see also the chapters by De Stefano et al. and by Shapiro in this volume) as well as the development of disease resistance (Durner and Klessig 1999; Klessig et al. 2000; Wendehenne et al. 2004). Another compound that has been shown to be involved in plant disease resistance is the polar lipid phosphatidic acid (PA). PA is a second messenger involved in the signaling of many (a)biotic stresses (for a review see Testerink and Munnik 2005). PA has been shown to accumulate upon treatments with the elicitors xylanase, chitotetraose and flagellin in tomato cells (van der Luit et al. 2000), Avr4 (Cf-4 expressing tobacco cells; de Jong et al. 2004), and Nod factors in alfalfa cells (den Hartog et al. 2003). Very recent evidence in our laboratory supports NO participation in PA and related signaling phospholipid events (Laxalt et al. submitted). Interestingly, PA has many downstream targets. Several of these downstream targets are involved in the regulation or even the induction of the oxidative burst, a process that seems to be required for the induction of HR. First, it has been shown that PA binds the protein kinase PDK 1 in Arabidopsis and that it activates the protein kinase AGC2-1 (Anthony et al. 2004; Deak et al. 1999). AGC2-1 is implicated in oxidative burst-mediated signaling in Arabidopsis (Rentel et al. 2004). Second, PA activates NADPH oxidase (Palicz et al. 2001) which, by means of genetics, has been shown to be required for pathogen-induced ROS formation (Torres et al. 2002; Yoshioka et al. 2003). Third, exogenously applied PA induces an oxidative burst in tobacco cells (de Jong et al. 2004) as well as in tomato cells (Laxalt, unpublished).

PA can principally be synthesized via two signaling pathways (for review see Meijer and Munnik 2003): (i) Phospholipase D (PLD) hydrolyzes mainly structural phospholipids such as phospatidylcholine, resulting directly in PA formation. (ii) Phosphatidyl inositol-phospholipase C (PI-PLC) hydrolyzes the phosphatidylinositols PI, PIP, or PIP_2, resulting in diacylglycerol (DAG) and soluble inositolphosphate (IP, IP_2, or IP_3, respectively). DAG can be phosporylated by DAG kinase (DGK) resulting in PA. The IPs can enter a kinase cascade resulting in IP_{3-6}, which are involved in the induction of Ca^{2+} release from internal stores. Ca^{2+} is renowned as a second messenger and has been shown to induce or stimulate NADPH oxidase, but also PI-PLCs and PLDs. Interestingly, different PLD and PI-PLC isoforms can have different requirements for Ca^{2+} (Munnik et al. 1998).

Xylanase induces NO and PA formation in tomato cell cultures. PA was accumulated via an NO-dependent induction of PI-PLC/DGK in xylanase-elicited tomato cells (Laxalt et al. submitted). Pretreatment of cells with NO scavenger cPTIO or the PI-PLC specific inhibitor U73122 reduced PA formation. Furthermore, it was shown that PA formation, induced by either xylanase or the NO donor SNAP, depends on DGK, further implying the involvement of a PI-PLC rather than a PLD activation. PLD activity is stimulated by xylanase but apparently independently of NO (Laxalt et al. unpublished). Additionally, application of either cPTIO or U73122 inhibits the oxidative burst. Thus, all evidence supports NO activation of the PI-PLC signaling pathway, promoting increases in PA levels. More recent experiments have shown that yet other elicitors, chitosan and AVR4 from *Cladosporium fulvum,* induce NO-dependent PI-PLC/DGK formation of PA (Laxalt, unpublished). Another line of evidence supporting the involvement of NO in PA-mediated signaling pathways is the regulation of stomatal closure. Stomatal closure might be induced as a strategy to prevent penetration by the pathogen. Recent evidence indicates that PA is involved in stomatal closure processes, that this pathway also depends on NO, and that NO affects both PI-PLC/DGK and PLD activities (Distéfano et al., unpublished).

Other elicitors, but not all, induce PA, for example AVR4 (but not AVR9) from *Cladosporium fulvum* does induce PA (de Jong et al. 2004). The elicitor cryptogein has been shown to induce NO, Ca^{2+}, and cell death in tobacco cells (Lamotte et al. 2004). PA was not studied in this system but the Ca^{2+} increase is consistent with a PI-PLC activation. Based on all these results two statements can be made:

1. NO is involved in many plant defense signaling pathways but apparently not in all
2. Events downstream of NO can vary

This is nicely demonstrated in tobacco where NO is induced by both the virus TMV and the oomycete elicitor cryptogein. While TMV-NO induces PAL (Durner et al. 1998), cryptogein-NO does not. An attractive hypothesis is that phospholipid signaling might be a source of diversifying and fine-tuning regulation of downstream NO-mediated responses.

We have as yet no exact knowledge regarding the molecular events that result in NO-mediated PA accumulation. In the case of xylanase, either a PI-PLC or a DGK or both is/are activated, resulting in an increase of PA. If DGK is induced, PA increase is not accompanied by IP_3 accumulation, whereas if a PI-PLC is induced, then PA is accompanied by IP_3 accumulation. Whether IP_3 accumulation results in IP_6 and hence Ca^{2+} release depends on the activity of a cascade of IP kinases. It will be interesting to explore whether the variety of enzymes that are involved might provide the plant with a high flexibility regarding NO-triggered PA accumulation. NO induced by chitosan might activate a set of enzymes different from those activated by NO derived from xylanase. It should be mentioned that due to the nature of the substrate, phospholipid signaling is located at or near the membrane and there may be localized areas, which might contain different phospholipases and/or different phospholipids. This might explain at least some of the discrepancies reported in different systems, especially when phospholipid signaling is combined with other flexible second messengers such as Ca^{2+}.

Future analyses will have to be directed at the characterization of the molecular events that link NO and phospholipid signaling. A great number of genes encoding enzymes involved in phospholipid signaling have been identified. Five tomato PLD genes have been cloned from tomato (Laxalt et al. 2001), the *A. thaliana* genome encodes nine PI-PLCs, and six PI-PLCs genes have been identified in tomato (ten Have and Vossen unpublished). Nitration and nitrosylation events could be alternative regulatory mechanisms connecting NO and phospholipid signaling. Silencing of specific PI-PLCs and DGKs genes could also represent a valuable tool for identifying which genes are regulated in each particular plant response that involves NO production and subsequent PA accumulation.

5 Concluding Remarks

NO has been recognized as a new plant signal molecule in the last decade. However, a number of controversial observations, the complexity of NO production in plants, and special features associated with NO chemistry make this molecule exciting but still mysterious. Some particularities associated with NO as a signal molecule are:

1. NO is constitutively generated in plants through a number of enzymatic and non-enzymatic sources.
2. NO is emitted and absorbed by plants from both leaves and roots.
3. NO is an uncharged free radical that can freely diffuse through biological membranes.
4. NO is one of the agents that contributes to the regulation of redox status of the cell as well as to cellular Ca^{2+} homeostasis. This makes NO a potential effector of any cellular adjustment in which the redox chemistry and changes in Ca^{2+} concentrations are involved

Communication between and within cells is vital for the growth and development of any organism. Cells communicate with each other through chemical signals, including hormones, which are sensed by the cell to generate a secondary wave of messages. The classical plant hormones (auxins, ethylene, gibberellins, cytokinins, and abscisic acid) were discovered more than 50 years ago. However, receptors for the hormone action have been only identified in the last few years, concurrently with the most important discoveries of NO actions in plant biology. Currently we know that auxin/receptor interaction induces targeted degradation of transcriptional repressors (Kepinski and Leyser 2005; Dharmasiri et al. 2005), and that ABA/receptor promotes the accumulation of transcriptional repressor through an effect on RNA processing (Razem et al. 2006). How the interactions of auxin and ABA with their respective receptors could be affected by NO is an intriguing question that will surely be elucidated in the near future. Moreover, it is equally important to know if the interaction between hormones and receptors is a prerequisite for the observed downstream enhancement of NO production.

Even if the results support a clear NO-dependence of physiological responses under the control of auxins and ABA in root development and stomatal movement, how so ubiquitous a molecule as NO accomplishes such diverse and apparently unrelated effects remains elusive. To discover how NO could be simultaneously and specifically involved in a myriad of cellular actions triggered by hormonal stimuli and phospholipid signaling is a major challenge. Evolution is a resultant of forces that increases the complexity of: (i) molecular networks within and between cells, and (ii) interrelations between individuals and populations. In making these parallels, a current challenge in our understanding is to appreciate the relevance of NO in keeping synchronized the complexity of plant cell homeostasis. Is NO to plant cell homeostasis as oxygen is to respiration?

Acknowledgements We acknowledge M.C. Luciano for her help in searching bibliography. This work was supported by Agencia Nacional de Promoción Científica y Tecnológica (ANPCyT, PICTs 1-9767/00 and 1-14457/03 to L.L. and PICT 20058 to A.M.L.), Consejo Nacional de Investigaciones Científicas y Técnicas (CONICET, PIP 0898/98 and PIP R1908/05 to L.L.), Fundación Antorchas, and institutional grants from Universidad Nacional de Mar del Plata (UNMdP), Argentina. C.G.M., A.M.L. and L.L. are members of the Permanent Research Staff and N.C-A., M.L.L. and M.G. are Postgraduate Fellows from CONICET, Argentina. L.L. is a fellow from the J.S. Guggenheim Foundation.

References

Abel S, Theologis A (1996) Early genes and auxin action. Plant Physiol 111:9–17

Allen GJ, Muir SR, Sanders D (1995) Release of Ca^{2+} from individual plant vacuoles by both InsP3 and cyclic ADP-ribose. Science 268:735–737

Allen GJ, Kuchitsu K, Chu SP, Murata Y, Schroeder JI (1999) Arabidopsis abi1–1 and abi2–1 phosphatase mutations reduce abscisic acid-induced cytoplasmic calcium rises in guard cells. Plant Cell 11:1785–1798

Anthony RG, Henriques R, Helfer A, Meszaros T, Rios G, Testerink C, Munnik T, Deak M, Koncz C, Bogre L (2004) A protein kinase target of a PDK1 signaling pathway is involved in root hair growth in Arabidopsis. EMBO J 23:572–581

Assmann SM, Shimazaki K (1999) The multisensory guard cell. Stomatal responses to blue light and abscisic acid. Plant Physiol 119:809–816

Blatt M (2000) Ca^{2+} signaling and control of guard-cell volume in stomatal movements. Curr Opin Plant Biol 3:196–204

Bright J, Desikan R, Hancock JT, Weir IS, Neill SJ (2006) ABA-induced NO generation and stomatal closure in Arabidopsis are dependent on H_2O_2 synthesis. Plant J 45:113–122

Burssens S, de Almeida Engler J, Beeckman T, Richard C, Shaul O, Ferreira P, Van Montagu M, Inze D (2000) Developmental expression of the Arabidopsis thaliana CycA2;1 gene. Planta 211:623–631

Casimiro I, Beeckman T, Graham N, Bhalerao R, Zhang H, Casero P, Sandberg G, Bennett MJ (2003) Dissecting Arabidopsis lateral root development. Trends Plant Sci 8:165–171

Correa-Aragunde N, Graziano M, Chevalier C, Lamattina L (2006) Nitric oxide modulates the expression of cell cycle regulatory genes during lateral root formation in tomato. J Exp Bot 57:581–588

Correa-Aragunde N, Graziano M, Lamattina L (2004) Nitric oxide plays a central role in determining lateral root development in tomato. Planta 218:900–905

Cousson A, Vavasseur A (1998) Putative involvement of cytosolic Ca^{2+} and GTP-binding proteins in cyclic-GMP-mediated induction of stomatal opening by auxin in *Commelina communis*. Planta 206:308–314

Creus C, Graziano M, Casanovas E, Pereyra M, Simontacchi M, Puntarulo S, Barassi C, Lamattina L (2005) Nitric oxide is involved in the Azospirillum brasilense-induced lateral root formation in tomato. Planta 221:297–303

Deak M, Casamayor A, Currie RA, Downes CP, Alessi DR (1999) Characterisation of a plant 3-phosphoinositide-dependent protein kinase-1 homologue which contains a pleckstrin homology domain. FEBS Lett 451:220–226

de Jong CF, Laxalt AM, Bargmann BO, de Wit PJ, Joosten MH, Munnik T (2004) Phosphatidic acid accumulation is an early response in the Cf-4/Avr4 interaction. Plant J 39:1–12

Delledonne M, Xia Y, Dixon RA, Lamb C (1998) Nitric oxide functions as a signal in plant disease resistance. Nature 394:585–588

Delledonne M, Zeier J, Marocco A, Lamb C (2001) Signal interactions between nitric oxide and reactive oxygen intermediates in the plant hypersensitive disease resistance response. Proc Natl Acad Sci USA 98:13454–13459

Desikan R, Graffiths R, Hancock J, Neill S (2002) A new role for an old enzyme: nitrate reductase-mediated nitric oxide generation is required for abscisic acid-induced stomatal closure in Arabidopsis thaliana. Proc Natl Acad Sci USA 99:16314–16318

De Stefano M, Vandelle E, Polverari A, Ferrarini A, Delledonne M (2007) Nitric oxide-mediated signaling functions during the plant hypersensitive response. In: Lamattina L, Polacco JC (eds) Nitric oxide in plant growth. Plant Cell Monographs, vol 6. Springer, Berlin Heidelberg New York (in press)

De Veylder L, Beeckman T, Beemster GT, Krols L, Terras F, Landrieu I, van der Schueren E, Maes S, Naudts M, Inze D (2001) Functional analysis of cyclin-dependent kinase inhibitors of Arabidopsis. Plant Cell 13:1653–1668

Dharmasiri N, Dharmasiri S, Estelle M (2005) The F-box protein TIR1 is an auxin receptor. Nature 435:441–445

Dubrovsky JG, Rost TL, Colon-Carmona A, Doerner P (2001) Early primordium morphogenesis during lateral root initiation in Arabidopsis thaliana. Planta 214:30–36

Durner J, Klessig DF (1999) Nitric oxide as a signal in plants. Curr Opin Plant Biol 2:369–374

Durner J, Wendehenne D, Klessig DF (1998) Defense gene induction in tobacco by nitric oxide, cyclic GMP, and cyclic ADP-ribose. Proc Natl Acad Sci USA 95:10328–10333

Fan LM, Zhao Z, Assmann SM (2004) Guard cells: a dynamic signaling model. Curr Opin Plant Biol 7:537–546

Ferreira PC, Hemerly AS, Engler JD, van Montagu M, Engler G, Inze D (1994) Developmental expression of the Arabidopsis cyclin gene cyc1At. Plant Cell 6:1763–1774

Friml J (2003) Auxin transport – shaping the plant. Curr Opin Plant Biol 6:7–12

Garcia-Mata C, Gay R, Sokolovski S, Hills A, Lamattina L, Blatt MR (2003) Nitric oxide regulates K^+ and Cl^- channels in guard cells through a subset of abscisic acid-evoked signaling pathways. Proc Natl Acad Sci USA 100:11116–11121

Garcia-Mata C, Lamattina L (2001) Nitric oxide induces stomatal closure and enhances the adaptive plant responses against drought stress. Plant Physiol 126:1196–1204

Garcia-Mata C, Lamattina L (2002) Nitric oxide and abscisic acid cross talk in guard cells. Plant Physiol 128:790–792

Gouvea CMCP, Souza JF, Magalhaes ACN, Martins IS (1997) NO releasing substances that induce growth elongation in maize root segments. Plant Growth Regul 21:183–187

Guo FQ, Okamoto M, Crawford NM (2003) Identification of a plant nitric oxide synthase gene involved in hormonal signaling. Science 302:100–103

Guo HS, Xie Q, Fei JF, Chua NH (2005) MicroRNA directs mRNA cleavage of the transcription factor NAC1 to downregulate auxin signals for arabidopsis lateral root development. Plant Cell 17:1376–1386

He J, Xu H, She X, Song X, Zhao W (2005) The role and the interrelationship of hydrogen peroxide and nitric oxide in the UV-B-induced stomatal closure in broad bean. Funct Plant Biol 32:237–247

Himanen K, Boucheron E, Vanneste S, de Almeida Engler J, Inze D, Beeckman T (2002) Auxin-mediated cell cycle activation during early lateral root initiation. Plant Cell 14:2339–2351

Himanen K, Vuylsteke M, Vanneste S, Vercruysse S, Boucheron E, Alard P, Chriqui D, Van Montagu M, Inze D, Beeckman T (2004) Transcript profiling of early lateral root initiation. Proc Natl Acad Sci USA 101:5146–5151

Hu X, Neill S, Tang Z, Cai W (2005) Nitric oxide mediates gravitropic bending in soybean roots. Plant Physiol 137:663–670

Irving HR, Gehring CA, Parish RW (1992) Changes in cytosolic pH and calcium of guard cells precede stomatal movements. Proc Natl Acad Sci USA 89:1790–1794

Jones AM (1998) Auxin transport: down and out and up again. Science 282:2201–2203
Kepinski S, Leyser O (2005) The Arabidopsis F-box protein TIR1 is an auxin receptor. Nature 435:446–451
Klessig DF, Durner J, Noad R, Navarre DA, Wendehenne D, Kumar D, Zhou JM, Shah J, Zhang S, Kachroo P, Trifa Y, Pontier D, Lam E, Silva H (2000) Nitric oxide and salicylic acid signaling in plant defense. Proc Natl Acad Sci USA 97:8849–8855
Lamotte O, Gould K, Lecourieux D, Sequeira-Legrand A, Lebrun-Garcia A, Durner J, Pugin A, Wendehenne D (2004) Analysis of nitric oxide signaling functions in tobacco cells challenged by the elicitor cryptogein. Plant Physiol 135:516–529
Lanteri ML, Graziano M, Correa-Aragunde N, Lamattina L (2006a) From cell division to organ shape: nitric oxide is involved in auxin-mediated root development. In: Balušca F, Manusco S, Volkmann D (eds) Communication in plants: neuronal aspects of plant life. Springer, Berlin Heidelberg New York, pp 123–136
Lanteri ML, Pagnussat GC, Lamattina L (2006b) Calcium and calcium-dependent protein kinases are involved in nitric oxide- and auxin-induced adventitious root formation in cucumber. J Exp Bot 57:1341–1351
Laskowski MJ, Williams ME, Nusbaum HC, Sussex IM (1995) Formation of lateral root meristems is a two-stage process. Development 121:3303–3310
Laxalt AM, ter Riet B, Verdonk JC, Parigi L, Tameling WI, Vossen J, Haring M, Musgrave A, Munnik T (2001) Characterization of five tomato phospholipase D cDNAs: rapid and specific expression of LePLD beta 1 on elicitation with xylanase. Plant J 26:237–247
Leyser O (2002) Molecular genetics of auxin signaling. Annu Rev Plant Biol 53:377–398
Lombardo MC, Graziano M, Polacco J, Lamattina L (2006) Nitric oxide functions as a positive regulator of root hair development. Plant Signal Behav 1:28–33
Malamy JE, Benfey PN (1997) Down and out in Arabidopsis: the formation of lateral roots. Trends Plant Sci 2:390–396
Mallory AC, Bartel DP, Bartel B (2005) MicroRNA-directed regulation of Arabidopsis AUXIN RESPONSE FACTOR17 is essential for proper development and modulates expression of early auxin response genes. Plant Cell 17:1360–1375
Marchant A, Bhalerao R, Casimiro I, Eklof J, Casero PJ, Bennett M, Sandberg G (2002) AUX1 promotes lateral root formation by facilitating indole-3-acetic acid distribution between sink and source tissues in the Arabidopsis seedling. Plant Cell 14:589–597
Marchant A, Kargul J, May ST, Muller P, Delbarre A, Perrot-Rechenmann C, Bennett MJ (1999) AUX1 regulates root gravitropism in Arabidopsis by facilitating auxin uptake within root apical tissues. EMBO J 18:2066–2073
Meijer HJ, Munnik T (2003) Phospholipid-based signaling in plants. Annu Rev Plant Biol 54:265–306
Neill SJ, Desikan R, Clarke A, Hancock JT (2002) Nitric oxide is a novel component of abscisic acid signaling in stomatal guard cells. Plant Physiol 128:13–16
Neill SJ, Desikan R, Hancock JT (2003) Nitric oxide signaling in plants. New Phytol 159:11–35
Otvos K, Pasternak TP, Miskolczi P, Domoki M, Dorjgotov D, Szucs A, Bottka S, Dudits D, Feher A (2005) Nitric oxide is required for, and promotes auxin-mediated activation of, cell division and embryogenic cell formation but does not influence cell cycle progression in alfalfa cell cultures. Plant J 43:849–860
Pagnussat G, Simontacchi M, Puntarulo S, Lamattina L (2002) Nitric oxide is required for root organogenesis. Plant Physiol 129:954–956
Pagnussat GC, Lanteri ML, Lamattina L (2003) Nitric oxide and cyclic GMP are messengers in the indole acetic acid-induced adventitious rooting process. Plant Physiol 132:1241–1248

Pagnussat GC, Lanteri ML, Lombardo MC, Lamattina L (2004) Nitric oxide mediates the indole acetic acid induction activation of a mitogen-activated protein kinase cascade involved in adventitious root development. Plant Physiol 135:279–286

Palicz A, Foubert TR, Jesaitis AJ, Marodi L, McPhail LC (2001) Phosphatidic acid and diacylglycerol directly activate NADPH oxidase by interacting with enzyme components. J Biol Chem 276:3090–3097

Razem F, El-kereamy A, Abrams S, Hill R (2006) The RNA binding protein, FCA, is an abscisic acid receptor. Nature 439:290–294

Rentel MC, Lecourieux D, Ouaked F, Usher SL, Petersen L, Okamoto H, Knight H, Peck SC, Grierson CS, Hirt H, Knight MR (2004) OXI1 kinase is necessary for oxidative burst-mediated signaling in Arabidopsis. Nature 427:858–861

Rockel P, Strube F, Rockel A, Wildt J, Kaiser W (2002) Regulation of nitric oxide (NO) production by plant nitrate reductase in vivo and in vitro. J Exp Bot 53:103–110

Roudier F, Fedorova E, Lebris M, Lecomte P, Gyorgyey J, Vaubert D, Horvath G, Abad P, Kondorosi A, Kondorosi E (2003) The Medicago species A2-type cyclin is auxin regulated and involved in meristem formation but dispensable for endoreduplication-associated developmental programs. Plant Physiol 131:1091–1103

Sachs T (2004) Self-organization of tree form: a model for complex social systems. J Theor Biol 230:197–202

Sanders D, Brownlee C, Harper JF (1999) Communicating with calcium. Plant Cell 11:691–706

Scherer GFE, Holk A (2002) No donors mimic and No inhibitors inhibit cytokinin action in betalaine accumulation in *Amaranthus caudatus*. Plant Growth Regul 32:345–350

Shapiro AD (2007) Nitric oxide in cell-to-cell communication coordinating the plant hypersensitive response. In: Lamattina L, Polacco JC (eds) Nitric oxide in plant growth. Plant Cell Monographs, vol 6. Springer, Berlin Heidelberg New York (in press)

Sokolovski S, Blatt MR (2007) Nitric oxide and plant ion channel control. In: Lamattina L, Polacco JC (eds) Nitric oxide in plant growth. Plant Cell Monographs, vol 6. Springer, Berlin Heidelberg New York (in press)

Sokolovski S, Hills A, Gay R, García-Mata C, Lamattina L, Blatt MR (2005) Protein phosphorylation is a prerequisite for intracelluar Ca^{2+} release and ion channel control by nitric oxide and abscisic acid in guard cells. Plant J 43:520–529

Stöhr C, Strube F, Marx G, Ullrich WR, Rockel P (2001) A plasma membrane-bound enzyme of tobacco roots catalyses the formation of nitric oxide from nitrite. Planta 212:835–841

Testerink C, Munnik T (2005) Phosphatidic acid: a multifunctional stress signaling lipid in plants. Trends Plant Sci 10:368–375

Torres MA, Dangl JL, Jones JD (2002) Arabidopsis gp91phox homologues AtrbohD and AtrbohF are required for accumulation of reactive oxygen intermediates in the plant defense response. Proc Natl Acad Sci USA 99:517–522

van der Luit AH, Piatti T, van Doorn A, Musgrave A, Felix G, Boller T, Munnik T (2000) Elicitation of suspension-cultured tomato cells triggers the formation of phosphatidic acid and diacylglycerol pyrophosphate. Plant Physiol 123:1507–1516

Wendehenne D, Durner J, Klessig DF (2004) Nitric oxide: a new player in plant signaling and defence responses. Curr Opin Plant Biol 7:449–455

Xu M, Zhu L, Shou H, Wu P (2005) A PIN1 family gene, OsPIN1, involved in auxin-dependent adventitious root emergence and tillering in rice. Plant Cell Physiol 46:1674–1681

Yoshioka H, Numata N, Nakajima K, Katou S, Kawakita K, Rowland O, Jones JD, Doke N (2003) Nicotiana benthamiana gp91phox homologs NbrbohA and NbrbohB participate in H_2O_2 accumulation and resistance to *Phytophthora infestans*. Plant Cell 15:706–718

Zhou B, Guo Z, Xing J, Huang B (2005) Nitric oxide is involved in abscisic acid-induced antioxidant activities in *Stylosanthes guianensis*. J Exp Bot 56:3223–3228

Plant Cell Monogr (6)
L. Lamattina · J. C. Polacco: Nitric Oxide in Plant Growth
DOI 10.1007/7089_2006_088/Published online: 25 October 2006

Nitric Oxide in Cytokinin and Polyamine Signaling: Similarities and Potential Crosstalk

Günther F. E. Scherer

Universität Hannover, Institut für Zierpflanzenbau und Gehölzforschung,
Abt. Molekulare Ertragsphysiologie, Herrenhäuser Str. 2, 30419 Hannover, Germany
scherer@zier.uni-hannover.de

Abstract New roles are emerging for nitric oxide (NO) besides senescence and defense. While NO itself is probably used as a chemical weapon in defense, in several examples the biosynthesis of NO is up-regulated so rapidly that a function of NO as a second messenger in signal transduction seems likely. We investigated this postulate with the signal substances cytokinin and polyamines. After polyamine addition to *Arabidopsis* seedlings, net NO biosynthesis increased with no apparent lag phase, and after zeatin addition, NO increased within 3 min. Thus, the up-regulation of NO levels is faster than activation of gene expression, consistent with our hypothesis of NO being a second messenger. A role for NO in signal transduction constitutes a new finding for both classes of signaling substances – polyamines and cytokinins. Cytokinin–polyamine crosstalk has already been indicated by known overlaps in the physiology of both signals, and the results reported here strengthen the case for this crosstalk. In addition, known multiple functions of polyamines in the physiology of pathogen defense, abiotic stress, hormones, and embryogenesis are already known to involve NO as a mediator. We discuss this new "input" into polyamine physiology.

1 Introduction

Nitric oxide (NO) seems to "invade" plant physiology more and more. After only 10 years since its (re)discovery in plants, NO physiology covers a lot more fields than the two major initial areas, senescence (Leshem 1996) and defence (Delledonne et al. 1998; Durner et al. 1998). It seems that plants employ NO in two roles, as a second messenger in signal transduction and as a chemical weapon in defence. To support the hypothesis that NO is a second messenger it is necessary to prove that NO biosynthesis is rapidly initiated by signals and that the signals are sensed by receptors. In this chapter, we will discuss our initial findings concerning the cytokinin-stimulated NO biosynthesis related to new results that indicate the presence of a polyamine-stimulated NO biosynthesis in plants. The similarities between certain aspects of cytokinin- and polyamine-stimulated NO biosynthesis suggest that the previously noted functional overlaps between the cytokinin- and polyamine-associated physiological responses may use NO as a common messenger.

2 Cytokinin and Polyamine Signaling Employ Induction of NO Biosynthesis

2.1 NO Biosynthetic Enzymes and NO Quantification Methods

2.1.1 NO Biosynthetic Enzymes in Plants

Plants possess two known enzymes for NO biosynthesis, nitrate reductase (NR) and an arginine- and FADH-dependent NO synthase (*At*NOS1). Both were clearly tested as purified enzymes (Harper 1981; Guo et al. 2003) and have functions assigned to them. The production of NO by NR was discovered by Harper (1981) and gained fresh attraction recently (Desikan et al. 2002; Rockel et al. 2002; Morot-Gaudry-Talarmain et al. 2002). Probably, NO production by NR is specifically used during anoxia since nitrite, the substrate for NO synthesis by NR, does not usually accumulate but does under anoxia (Morot-Gaudry-Talarmain et al. 2002). NR-made NO, however, participates in the long-term response to cytokinin, indicating a role of NR in NO homoeostasis (Tun et al. submitted). In addition to its role as an NO-generating enzyme, a role in ABA signaling was assigned to NR (Desikan et al. 2002).

The *At*NOS1 was genetically and enzymologically identified (Guo et al. 2003, 2005). A function in pathogen defence is obvious as *Atnos1* plants are very sensitive to pathogens (Zeidler et al. 2004; Guo et al. 2005, and our own experience in the green house). As the *AtNOS1* gene family consists of six genes, other functions may also be associated with the other five genes, like cytokinin- and/or polyamine-induced NO biosynthesis. Other enzymes for NO biosynthesis are a membrane-bound NR with unknown function (Stöhr et al. 2003), and a mitochondrial enzyme (Tischner et al. 2004). Xanthine dehydrogenase was often mentioned as a hypothetical source of NO (Neill et al. 2003, and citations therein) but this was questioned by experimental results (Tischner et al. 2004). Non-enzymatic sources of NO in plants have been described but the functional significance is not clear (Neill et al. 2003; Bethke et al. 2004). We found that polyamines instantaneously stimulate NO biosynthesis, which may indicate a new enzyme (class) (Tun et al. 2006; Santa-Catarina et al. submitted; Silveira et al. 2006).

2.1.2 Methods of NO Quantification

Investigation of rapid time courses of NO biosynthesis needs methods specific for NO detection and quantification. Methods of NO quantification in plant research are: binding of NO to hemoglobin (Delledone et al. 1998), NO microelectrodes (Yamasaki et al. 1999; Sakihama et al. 2002), chemiluminescence under anoxia in vivo (e.g., Rockel et al. 2002; Tischner et al. 2004), binding to an ESPR spin probe and quantification (Huang et al. 2004; Zeidler et al. 2004; Corpas et al. 2006), the photoacoustic method (Leshem and Pinchasov 2000; Mur et al. 2005), a mass spectrometry method (Conrath et al. 2004), and the use of fluorescent dyes such as DAF2 (4,5-diaminofluoresceine) or derivatives of DAR (DAR-4M AM: diaminorhodamine-4M acetoxymethyl ester; DAR-4M: diaminorhodamine 4M) (Wendehenne et al. 2001; Neill et al. 2003; Lamattina et al. 2003, for reviews). NO released into the water will also generate nitrite, and then, after conversion of nitrite to NO, this NO can be measured by chemiluminescence quantification. This is suitable for enzymatic assays of NO synthase but not for direct in vivo measurements (Corpas et al. 2004, 2006).

Some methods have disadvantages, like the low sensitivity of the hemoglobin method or the restriction to anoxic conditions as in the in vivo chemoluminescence method (Rockel et al. 2002). NO microelectrodes measure NO in the surrounding medium and do not allow tissue analysis, ESPR spin probes require ground tissue, and the photo acoustic method calls for special instrumentation. With the advent of DAF-2 and DAR-4M AM, fluorescent probes became immediately popular. Small samples can be used for quantification of NO in the medium at high sensitivity, and visualization of tissue distribution is quite possible (e g., Foissner et al. 2000; Tun et al. 2001; Desikan et al. 2002; Guo et al. 2003, 2005; Gould et al. 2003; Gabaldon et al. 2004; Gerber et al. 2004, Zeidler et al. 2004; Corpas et al. 2006). However, no ratio imaging is possible with DAF or DAR dyes, which would be desirable for better quantification, and other critical interferences must be taken into account (see the chapter by Arita et al. in this volume).

We started with DAF-2 to measure NO released into the medium from cell cultures (Tun et al. 2001). NO-induced fluorescence increase of DAR derivatives is independent of the pH above 4 in both the binding reaction and the fluorescence emission (Kojima et al. 2000, 2001). DAR derivatives are not light-activated whereas DAF derivatives are light-activated (Broillet et al. 2001; Gould et al. 2003). Therefore, in the follow-up work we used DAR-4M (cell impermeable) for quantification of NO diffusing into media and DAR-4M AM (cell permeable) for in situ fluorescence microscopy. Although fluorescent dyes have been questioned in their specificity because they also react with peroxynitrite (Roychowdhury et al. 2002) it is clear that this peroxynitrite must originate by the reaction of O_2^- and NO. Thus, potentially present peroxynitrite always also indicates NO in tissues and, in the outside medium, DAR-4M clearly quantifies NO, since peroxynitrite is not assumed to diffuse into the medium. This is why in our work for all key experiments both dyes were used to allow quantification of NO release from plant material and description of NO tissue distribution.

The role of nitrite in NO-dependent fluorescence emission was investigated and we had to convince ourselves that nitrite generates a very strong increase of DAR-4M fluorescence in the cuvette *without any plant* (Fig. 1). This strongly discourages the use of nitrite (or induction of nitrite by added nitrate) in conjunction with fluorescence methods and casts doubt on some previous publications. Whenever NO is quantified by methods not using fluorescent compounds (e.g., chemiluminescence), that doubt does not then apply to nitrite/NO physiology in general.

Hydrogen peroxide, as a product of polyamine oxidation (Binda et al. 2002), might also induce side reactions with DAF or DAR compounds. However, with hydrogen peroxide no fluorescence increase with DAR-4M in a purely chemical reaction was observed (Tun et al. 2006). As a rule for experimentation, any uncertainty that peroxynitrite or nitrite may be generated by the presence of nitrate is prevented by good oxygen-rich media, which prevents nitrite accumulation in tissue. Controls for the presence of reactive oxygen species (ROS) by H_2DCF-DA fluorescence and by DAF-2 were seldom made but prove that NO can be reliably quantified by DAF or DAR derivatives (Roychowdhury et al. 2002; Gerber et al. 2004).

The advantage of fluorescence microscopy is the tissue resolution. NO has a half life of 5 s and a diffusion distance of probably no more than 500 μm (Yamasaki 2005). Because of the short half life it is clear that from a source of NO in the tissue a steep gradient will be formed around the source so that tissue

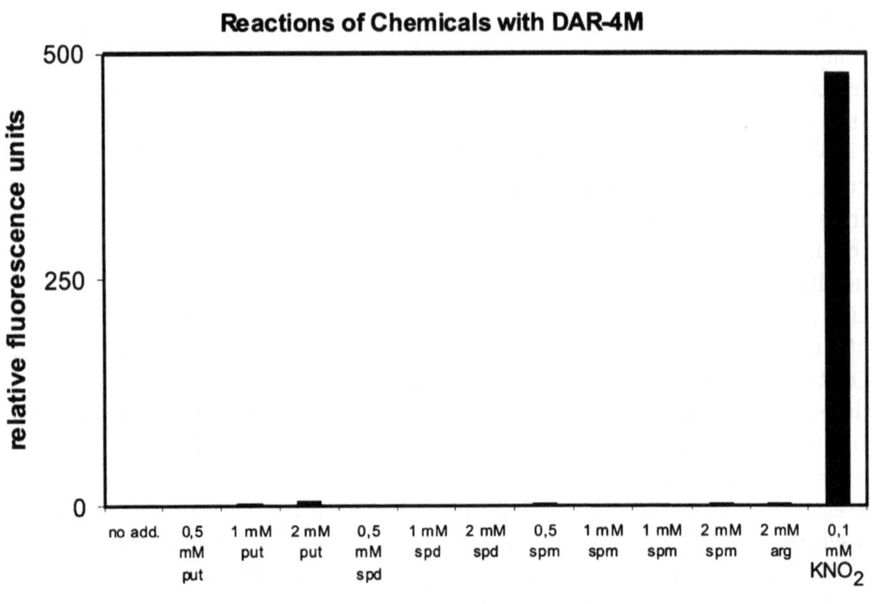

Fig. 1 Reaction of chemicals with DAR-4M. Treatment time was 4 h. Whereas all polyamines changed fluorescence readouts very little with time, KNO_2 instantaneously generated a large readout (shown at 4 h). *put* putrescine, *spd* spermidine, *spm* spermine, *arg* arginine)

imaging will still predominantly show such sources. Our experience shows that cell-to-cell differences are reproducibly apparent as well as NO emission over short or longer time spans (see Tun et al. 2006) and are supported by similar findings on NO tissue distribution by others (Corpas et al. 2004, 2006).

2.1.3
Inhibitors of NO Biosynthesis in Plants

For the plant NOS1 only one (!) inhibitor has been tested with isolated and purified enzyme, L-NAME (N^G-nitro-L-arginine methyl ester) (Guo et al. 2003). AET (2-(2-aminoethyl)2-thiopseudourea) was a reliable inhibitor of cytokinin-induced and polyamine-induced NO biosynthesis (Tun et al. 2001, 2006). It is a guanino group analog (as present in arginine), an arginine analog, and a monoamine analog and also inhibits polyamine-induced NO biosynthesis (Tun et al. 2005b). In our hands, L-NAME and L-NMMA (N^G-monoethyl-L-arginine) were not efficient even though many reports on inhibition of NO biosynthesis by these compounds exist. Inhibitor concentrations in published experiments ranged from 0.1 mM to 1 mM or even 5 mM sometimes. The question must be asked if this is still specific inhibition. Especially, in treatments over several hours, using a toxic derivative of an amino acid at high concentration, side effects on protein biosynthesis seem inevitable. This in turn, may affect NO biosynthesis in an indirect manner. A new plant NO biosynthesis inhibitor reported is aminoguanine (Carimi et al. 2005; Corpas et al. 2006). Regarding our results, the sensitivity of both polyamine- and cytokinin-induced NO to a single inhibitor AET points out to a single (group of) enzyme(s) as NO source, but this is an issue to be investigated.

2.2
Signals Inducing Rapid NO Biosynthesis

In the field of pathogen defence, rapid responses to elicitors were observed and NO biosynthesis was stimulated after 2–3 min (Foissner et al. 2000; Zeidler et al. 2004) and, 10 min after wounding and JA addition (Huang et al. 2004). The other examples of rapid regulation of NO biosynthesis by a signal substance are cytokinin (3 min) and polyamines (no recognizable lag phase, see below), under anoxia to rid the cell from nitrite (Rockel et al. 2002; Morot-Gaudry-Talarmain et al. 2002), and ABA within 30 min (Desikan et al. 2002; Guo et al. 2003). The importance of NO signaling in pathogen defence is widely recognized and often reviewed so that this field is not presented in any detail here (Durner et al. 1999; Wendehenne et al. 2001, 2004; Lamotte et al. 2005; Delledonne 2005). In pathogen defense, *At*NOS1 was identified as the enzyme for rapid NO biosynthesis regulation (Zeidler et al. 2004), while NR is responsible for NO production under anoxia (Rockel et al. 2002; Morot-

Gaudry-Talarmain et al. 2002). Both NR and *At*NOS1 appear to be responsible for NO signaling by ABA (Desikan et al. 2002; Guo et al. 2003).

2.3 Cytokinin Stimulation of NO Biosynthesis and Role in Signal Transduction

Plant cell cultures respond to cytokinin by increased NO biosynthesis within 3 min (Tun et al. 2001). We also quantified NO release from *Arabidopsis* seedlings with DAR-4M by a fluorometric method and observed NO biosynthesis as fluorescence increase of the cell-permeable dye DAR-4M AM by microscopy (Fig. 2). Most of these data are contained in a recent publication (Tun et al. submitted) and methods can be found in a previously published report (Tun et al. 2006). In *Arabidopsis* seedlings, we systematically investigated both signals, cytokinin and polyamines, in induction of NO release into the medium and in NO accumulation in all tissues in response to these signals.

The time course with seedlings was equally as rapid as the published one with cell cultures and in both examples we observed a lag phase of only 2–3 min after which the curves for endogenous and hormone-stimulated NO biosynthesis split. Such a rapid time course cannot be explained by transcription but only by posttranslational activation of an NO-generating enzyme. It should be stressed that the induction of NO biosynthesis is cytokinin-specific. The analog adenine and other hormones tested (IAA, ABA, GA, ACC) did

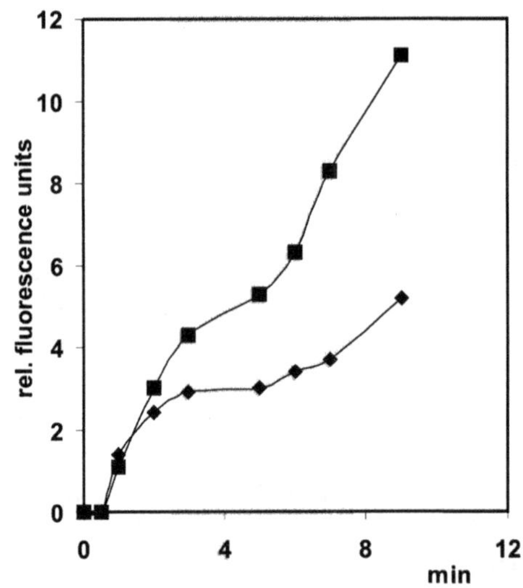

Fig. 2 Rapid induction of NO biosynthesis by zeatin in wild-type *Arabidopsis* seedlings. Fluorescence from DAR-4M was recorded, which increases by binding NO covalently. *Diamonds* endogenous NO release from seedlings into the medium. *Squares* release of NO in the presence of 20 μM zeatin. For method details see Tun et al. (2006)

not stimulate (Tun et al. 2001). A clear hormone specificity is, therefore, indicated. Claims that auxin induces NO are not verified by our experimental methods and materials (cell cultures) but long-term involvement of NO in the auxin–cytokinin balance cannot be excluded and may explain involvement of NO in stimulation of side root formation, an auxin-specific response (Correa-Aragunde et al. 2004).

In wild-type *Arabidopsis* seedlings, weak fluorescence was found in leaves, but more so in trichomes and bundles. In the root-shoot transition zones the NO-induced fluorescence was high, but lesser so in the hypocotyls. Roots accumulated NO to a weaker extent. Cotyledons produced high amounts of NO. In all these organs the bundles are strongly inducing of NO-dependent fluorescence. Meristems of roots produced more NO, and very young leaves and the shoot meristem were high NO accumulators. Zeatin, a cytokinin, induced higher NO biosynthesis in all tissues and this response was strongest in leaves and roots, less so in cotyledons. At higher magnification it was clear that vacuoles as acidic compartments accumulated the cationic dye but probably also certain lipophilic substances, like cuticles of guard cells and trichomes, whereas the guard cells themselves remained dark (Fig. 3). Both the endogenous and the zeatin-induced NO release were sensitive to AET, an enzyme inhibitor, and PTIO, a chemical NO scavenger. The tissue distribution of strong zeatin induction of NO was very similar to polyamine induction of NO tissue accumulation (see below). The strong NO-induced fluorescence accumulation in and around the bundles agrees with reports in the literature (Corpas et al. 2004, 2006). However, subcellular localization of NO-generating enzymes remains an open question. Localization in the peroxisomes was evidenced by an antibody against the animal protein iNOS. iNOS is not present in plant tissue and, moreover, this antibody was contradicted by proteomics methods: it did not recognize any NO biology-related protein (Butt et al. 2003). Hence, the tissue distribution of zeatin-induced NO in our experiments, so far, does not provide a clue to the NO-generating enzyme or the subcellular localization of the relevant enzyme.

In *Arabidopsis*, mutants are available for both enzymes known to generate NO, the NOS1 (Guo et al. 2003, 2005) and NR (Wilkinson and Crawford 1993). We tested whether the absence of these two enzymes affected the hormone response. The responses in *Atnos1* knockout seedlings to zeatin in all other tissues (root tip, root-shoot transition zone, cotyledon, hypocotyls, shoot tips) was indistinguishable from wild-type plants. Furthermore, the morphological responses to cytokinin were indistinguishable in wild-type and *Atnos1* plants. Hence, we conclude that this particular enzyme is not regulated by zeatin.

We next examined the *nia1,2* seedlings, which lack NR (Wilkinson and Crawford 1993). The general cellular and subcellular distribution of zeatin-induced fluorescence was quite similar to the wild-type pattern without exogenous zeatin. However, hypocotyls and all parts of the root of *nia1,2*

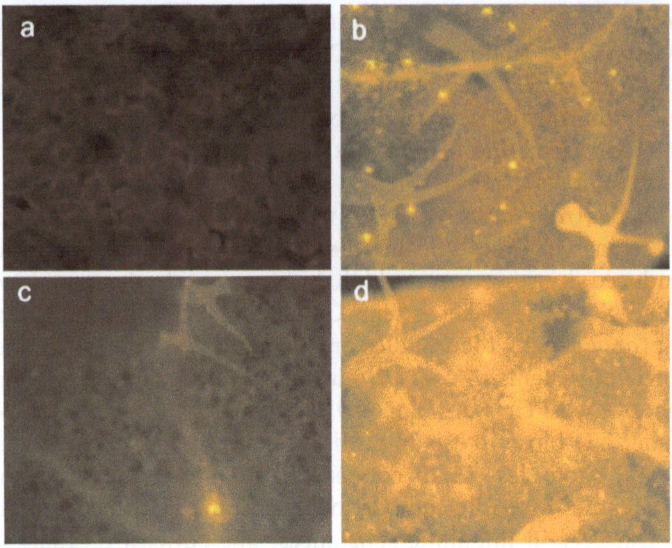

Fig. 3 Tissue distribution of NO-dependent fluorescence increase in leaves of *Arabidopsis* seedlings. Seedlings were grown on half-strength Murashige-Skoog and 2% sucrose on 1.2% agar containing: **a** no DAR-4M AM, **b–d** 5 μM DAR-4M AM and no zeatin in **b**, 0.5 μM zeatin in **c** and 5 μM zeatin in **d.** Plants grew in the light (16/8 h light/dark) for 14 days. For further details see Tun et al. (2006). A general increase in all tissues was seen but veins and trichomes showed especially high fluorescence increases due to zeatin

seedlings exhibited strongly enhanced NO-dependent endogenous fluorescence, even without addition of zeatin. Also, in cotyledons the fluorescence in untreated *nia1,2* seedlings was stronger than in wild-type seedlings and was already enhanced at 0.5 μM zeatin.

Endogenous NO release from *nia1,2* seedlings was equally rapidly enhanced by 20 μM zeatin after a 2–3 min lag phase, showing that NR is not the zeatin-regulated NO-generating enzyme. Short-term initial kinetics showed no apparent differences between the two genotypes, but in experiments of longer duration the amount of NO released by *nia1,2* seedlings was slightly higher during the first hours but clearly higher after 24 h in *nia1,2* seedlings, which corresponds to higher NO-induced fluorescence in *nia1,2* seedlings.

Remarkably, *nia1,2* plants exhibited a different morphological phenotype in the absence of exogenous hormone but an exaggerated phenotype in the presence of increasing zeatin concentrations. The *nia1,2* seedlings had a bushy root phenotype without hormone. Leaf areas of primary leaves were larger, even without hormone, but leaf growth was more strongly inhibited by zeatin than in the wild type. Moreover, in response to zeatin, the *nia1,2* cotyledons became larger and fleshier than in wild type. In addition, hypocotyls became thicker and longer than in wild type. In effect, *nia1,2* mutants behaved like zeatin-response mutants. Overall, this might be characterized as an "exaggerated" zeatin response but not in all aspects (e.g., main root growth) so that the term higher hormone sensitivity is not an exact description. Since cytokinin up-regulates NIA genes and nitrate affects some response regulator genes the relationship cannot be a simple "upstream" or "downstream" of *NIA* genes and cytokinin-induced NO biosynthesis (Samuelson et al. 1995; Miyata et al. 1998a,b; Taniguchi et al. 1998; Takei et al. 2002).

AET and the NO scavenger PTIO inhibited the endogenous and the zeatin-induced fluorescence increase in *nia1,2* seedlings, verifying that the effects were NO-specific. Inhibition by AET supports the notion that the zeatin-regulated enzyme was not NR since arginine analogs like AET did not inhibit NR (Rockel et al. 2002).

Two mutants of cytokinin signal components were available, a receptor knockout *cre1/ahk4* for one of the three histidine kinase receptor genes in *Arabidopsis* (Inoue et al. 2001; Ueguchi et al. 2001), and a triple knockout *ahp1,2,3* for three out of five *AHP* genes in *Arabidopsis* (Grefen and Harter, 2004). A remarkable difference was found in the mature central zone of *cre1/ahk4* plants, where the zeatin-induced NO-dependent fluorescence increase was absent although it was as strong in the basal hypocotyl zone and in other parts of the plants as in the wild type. This zone of the bundle in the mature part of the root is the site of strongest expression of the CRE1/AHK4 receptor (Mähönen et al. 2000; Nishimura et al. 2004). Moreover, wild-type and *cre1/ahk4* seedlings responded similarly to zeatin, except that a transient inhibition of root growth was observed in *cre1/ahk4* seedlings approximately 10 days after germination at higher zeatin (1 μM, 5 μM) concentration. Because of the only small overall response differences seen in *cre1/ahk4* we did not investigate this line with respect to NO release. The conclusion is that *cre1/ahk4* mutants made less NO in response at the site of highest (missing) expression of the *CRE1/AHK4* gene, corresponding to the response phenotype.

Up-regulation of NO biosynthesis by zeatin could be a function upstream or downstream of the AHPs. To test this we studied the tissue distribution and response to zeatin of NO-dependent fluorescence in a triple knockout line *ahp1,2,3* (gift of Dr. J.J. Kieber, UNC Chapel Hill, USA). The *Arabidopsis* genome contains five *AHP* genes so that only *AHP4* and *AHP5* were active in this mutant. When grown without zeatin, the tissue distribution of fluorescence in *ahp1,2,3* was very similar to wild-type seedlings. However, when grown in the presence of 5 μM zeatin, fluorescence in the primary leaves of the triple mutant was stronger than in wild type. In contrast, we observed no zeatin response in the roots, hypocotyls, and the cotyledons of the *ahp1,2,3*. The remaining active *AHP4* and *AHP5* genes are both expressed in the leaves. *AHP1*, *AHP2*, and *AHP3* are all expressed in the root and shoot (Tanaka et al. 2004), which could help to explain the aberrant response of NO tissue accumulation. When the NO release from seedlings was quantified, it was similar in both genotypes, probably because the balance of NO emission in all organs of *ahp1,2,3* was similar to the wild type.

The triple knockout mutant *ahp1,2,3* also had a zeatin-response phenotype. On minimal medium, without hormone the roots were shorter and bushier. Primary root growth and lateral root formation were less sensitive to zeatin in *ahp1,2,3* compared to wild type. The lower sensitivity to zeatin in NO accumulation and in the morphological response to zeatin indicate that these genes are necessary for general responses to zeatin and for the induction of NO in some organs so that at least one of the three mutant *AHP* genes is necessary for NO induction. This provided good evidence that *CRE1/AHK4* and one or several *AHP* genes are necessary for a complete induction of NO by zeatin and that this NO induction could be necessary for the physiological response.

Finally, if cytokinin-induced NO acts downstream of the type-B ARR proteins to regulate transcription of type-A *ARRs* then chemical NO donors

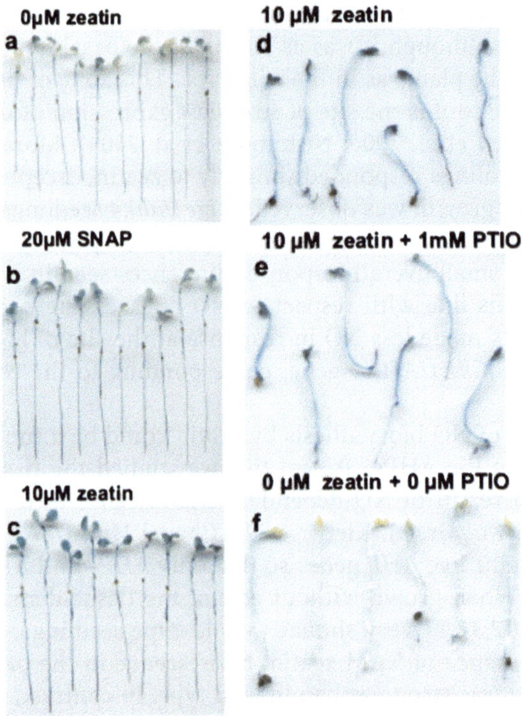

Fig. 4 Induction of the cytokinin-regulated promoter *ARR5* in *Arabidopsis* seedlings by mock treatment (**a**), by the NO donor 20 µM SNAP (**b**), and 10 µM zeatin (**c**). Induction by 10 µm zeatin (**d**), 10 µM zeatin and 1 mM PTIO (**e**), mock treatment (**f**). Seedlings were grown for 7 days in the light then treated for 2 h by 0 or 10 µM zeatin, 20 µM SNAP, or by the NO scavenger 1 mM PTIO or combinations

should stimulate transcription of the promoter of the type-A *ARR5* gene. *ARR5* is up-regulated by cytokinin within 10 min, reaching a peak at 40 min (D'Agostino et al. 2002). When the donors SNAP (*N*-(β-D-glucopyranosyl)-N^2-acetyl-*S*-nitroso-D,L-penicillaminamide) and NOC-12 (*N*-ethyl-2-(1-ethyl-2-hydroxy-2-nitrosohydrazino)ethanamine, data not shown) were applied to *ARR5-GUS* seedlings for 2 h, only marginal positive effects were observed (Fig. 4a–c). However, this response was partially suppressed by PTIO (Fig. 4d–f) so that NO specificity is indicated, not an unspecific response to NO stress. Since indirect effects could occur within 2 h treatments, this experiment does not yet give a clear answer to the question of to what extent NO is directly involved in triggering the transcription of this primary cytokinin-activated gene. Indirect ways of regulation could play a role in such a time span. A better answer will be provided once the relevant cytokinin-induced NO-generating enzyme(s) has/have been identified so that a more precise test of its downstream effects on NO signaling can be made.

2.4 Cytokinin Signal Transduction Models and NO

Current signal transduction models for cytokinin suggest a biochemical phosphate transfer chain starting with phosphorylation of histidine kinase receptors, phosphate transfer onto a histidine phosphotransfer protein, and then to A-type regulator response proteins. These act as transcriptional cofactors, to regulate the expression of the B-type response regulators (Hwang et al. 2001; Hutchison and Kieber 2002; Kakimoto 2003). However, most authors do not exclude the possibility of other mediators or second messengers in cytokinin action besides this phosphorylation cascade (Hutchison and Kieber 2002, Kakimoto 2003; Grefen and Harter 2004; Kiba et al. 2005). Perhaps, proteins other than B-type ARR proteins could also be phosphorylated. Because our results suggest that at least one out of three AHP proteins are needed in an *ahp1,2,3* triple knockout mutant for correct NO biosynthesis stimulation, conceivably, phosphorylation by an AHP protein leading to NO induction could be the critical step missing in this mutant.

The NO generated in a plant cell could transmit the signal through activation of a guanylate cyclase, which would be similar to animal models since cGMP (Durner et al. 1998; Wendehenne et al. 2001) and calcium-activated channels (Gould et al. 2003; Lamotte et al. 2004; Sokolovski et al. 2005), possibly activated by cADPR (Allen et al. 1995), could be involved. Alternative models of animal NO-mediated signal transduction assume that a set of plasma membrane calcium channels are regulated by reversible nitrosylation at defined cysteines by a NOS activity in direct vicinity of the channel (Stamler et al. 2001; Mannick and Schonhoff 2004; Martinez-Moreno et al. 2005), without participation of cGMP. These authors stress that strictly local NO sources regulate nearby proteins in their activity by reversible nitrosylation. This cannot be easily mimicked by a general change in NO level, such as by addition of NO donors. Also, nitrosylation of cysteines is not random but occurs only at cysteines surrounded by suitable neighboring amino acids (Stamler et al. 2001). The first nitrosylated protein to be described in plants was an enzyme involved in ethylene biosynthesis. An effect of S-nitrosylation on the enzyme activity was demonstrated (Lindermayr et al. 2005, 2006). Nitrosylation in plant channels was suggested by pharmacologyical criteria (Sokolovski and Blatt 2004). An enzyme capable of removing nitrosylation at cysteines, to make nitrosylation reversible, was also identified in plants (Liu et al. 2001; Sakamoto et al. 2002) so that reversible nitrosylation as a new mechanism for plant enzyme regulation is a clear prospect.

2.5 Polyamines and NO Functions in Plants

Motivated by the observation that polyamines and cytokinin have some overlapping functions in senescence inhibition (Cohen et al. 1979; Shi et al. 1982; Greenland and Lewis 1984; Laxalt et al. 1997; Lester 2000) and embryogenesis (Kakkar et al. 2000) we pursued an investigation of polyamines acting on NO biosynthesis. Polyamines indeed proved to be interesting in this respect. Most remarkably, when *Arabidopsis* seedlings were used to measure the response to polyamines, a very rapid increase of NO release by 1 mM spermine into the medium above endogenous levels was observed with no apparent lag (Fig. 5).

When spermine, spermidine, putrescine, and arginine effects were compared, only polyamines were able to stimulate NO release in a statistically significant manner. This was not the case for 1 mM arginine ($122 \pm 25\%$, $n = 8$) even though in individual tests stimulation by arginine reached 48%. Spermidine was consistently the most active polyamine (1 mM spermidine $185 \pm 46\%$, $n = 15$; 1 mM spermidine $147 \pm 28\%$, $n = 10$; 1 mM putrescine $130 \pm 15\%$, $n = 7$ in a 4 h assay). The non-biological amines Tris and imidazole were inactive. Polyamines are also known to be oxidized by polyamine oxidases whereby H_2O_2 originates (Binda et al. 2002). Hydrogen peroxide might also influence DAR-4M fluorescence but failed to do so in a chemical test with 1 mM H_2O_2 (Tun et al. 2006).

AET, an inhibitor of cytokinin-induced NO release (Tun et al. 2001) inhibited both endogenous and spermine-induced NO release. The NO-specific scavenger PTIO also decreased polyamine-induced NO fluorescence in assays with *Arabidopsis* seedlings when tested at 1 μM DAR-4M, proving that polyamine-induced fluorescence was due to NO release. Similar inhibition by PTIO was obtained when spermidine and putrescine were tested.

When we investigated the tissue distribution of accumulated NO the overall pattern distribution was very similar to that observed for zeatin-induced NO accumulation. We obtained the best differences in increase of

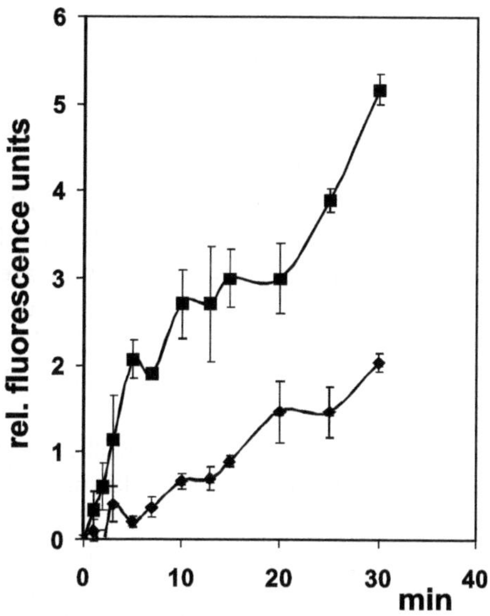

Fig. 5 Rapid kinetics of NO emission from *Arabidopsis* seedlings after addition of 1 mM spermidine at $T = 0$. *Diamonds* endogeneous NO emission. *Squares* with added 1 mM spermidine. *Bars* S.D., $n = 3$ per sample. Fluorescence was recorded with DAR-4M in the medium. For method details, see Tun et al. (2006)

NO-induced fluorescence in primary leaves treated by spermidine. Trichomes light up most dramatically within the cells and at their cuticles, revealing a characteristic sculpture. Guard cells remained rather dark but the cuticles around the opening cleft became bright by polyamine treatment and veins were highlighted. Again, as with zeatin, certain lipophilic structures such as cuticles may attract the dye. Otherwise, the dye seems to be concentrated in the vacuoles as one would expect from a cationic dye.

When different tissues and several polyamines were analyzed, spermidine and spermine induced higher NO-dependent fluorescence in *Arabidopsis* in the elongation zone of the root tip as compared to control, arginine, or putrescine treatment. This was the only clear difference as compared to the zeatin response tissue pattern of NO accumulation. In individual organs such as primary leaves the quantitative trends were similar to the seedling NO release, and spermine was the most active polyamine tested. In the cotyledons no increase in NO-induced fluorescence was found or only weakly so. PTIO quenched the polyamine-induced fluorescence increase in the leaves and roots. These data suggest a tissue specificity in the root of polyamine-induced NO biosynthesis in *Arabidopsis*, according to the polyamine supplied. Possibly this indicates different enzymatic activities interacting with different polyamines and different localizations of these as-yet unknown activities.

2.6 Polyamine Physiology and NO Biosynthesis

One important result was that polyamines very rapidly stimulate NO biosynthesis in *Arabidopsis* seedlings with clearly higher efficiency than arginine. The main and new finding is that polyamines are intimately linked to plant NO physiology. This opens new perspectives in the field of plant defence where polyamines have a functional role; but how they are linked to defence processes remains largely unclear. The second aspect is abiotic stresses, especially salt stress and potassium deficiency, and the third is the anti-senescence function of polyamines. In all three functional areas NO also plays a role. Last but not least, there is a known functional overlap of polyamines and cytokinin. This is highlighted by the intriguingly similar patterns of NO tissue distribution evoked by polyamines and cytokinin (compare micrographs shown here and in Tun et al. 2006).

That polyamines could be an indirect or direct source of NO opens up a new avenue of NO biology. Comparing, on the one hand, that polyamine addition to plants generates NO and, on the other hand, the other known NO-generating processes or signals in plants raises the question: could NO be a common theme of, or a link in, polyamine-mediated physiological responses?

2.6.1 Pathogen Defence, Polyamines, and NO Biosynthesis Regulation

An important process employing NO as a probable second messenger is pathogen defence (Delledonne et al. 1998, 2001; Durner et al. 1998; Wendehenne et al. 2001; Lamotte et al. 2004; Zeidler et al. 2004). NO can be generated within a few minutes upon elicitor application (Foissner et al. 2000; Huang et al. 2004; Zeidler et al. 2004). An increase in polyamine synthesis in response to various pathogens was observed earlier (Mo and Pua 2002) and resistance to virus infection was enhanced by polyamines (Yamakawa et al. 1998). Spermine, but not putrescine or spermidine, stimulated the activity of defence gene expression and two important MAP kinases involved in plant defence, WIPK and SIPK, after 6 h (Takahashi et al. 2003, 2004). Reactive oxygen species (ROS) and Ca^{2+} influx were upstream components of polyamine action on MAP kinases in these experiments. The far more rapid time course of polyamine-induced NO biosynthesis observed in our experiments indicates that NO should be upstream of the activation of defence MAP kinases, if it is involved in MAP kinase activation at all, and/or that two waves of NO biosynthesis can occur in pathogen defence. A rapid burst could have a function as a second messenger and a second late wave could function rather as a chemical weapon. Stimulation of expression of defence genes by NO was shown earlier (Huang et al. 2002; Polverari et al. 2003) so polyamine-induced

NO as an element of both polyamine-mediated defence reactions and defence gene induction is not a contradiction.

2.6.2
Abiotic Stress and Polyamines and NO

Several forms of abiotic stress lead to enhanced polyamine biosynthesis and can be ameliorated by polyamines (review: Bouchereau et al. 1999). Polyamine increases were found in several species such as rice, sorghum, maize, and tomato under salt stress and under osmotic stress (Flores and Galston 1984; Krishnamurthy and Bhagnat 1984; Prakash and Prathapsenan 1988; Erdei et al. 1996; Willadiano et al. 1996; Santa-Cruz et al. 1997). It is remarkable that the osmotic stress and, to some extent, salt stress-mediating hormone abscisic acid also seems to use NO as a second messenger (review: Neill 2003). It has long been known that polyamines accumulate under potassium deficiency (Richards and Coleman 1952; Watson and Malmberg 1996; Geny et al. 1997). Another stress common for plants is iron deficiency. Recently, it was found that NO has a function in iron deficiency and stress (Murgia et al. 2002, Graziano et al. 2002, Graziano and Lamattina 2005). Hypoxia is a stress that generates NO and involves nitrite (Rockel et al. 2002). Hypoxia leads to polyamine accumulation in rice, wheat, and zucchini (Reggiani et al. 1990, 1989a,b; Racz et al. 1996) and also leads to NO accumulation (Dordas et al. 2003), so that NO generated during hypoxia might, in part, originate from this accumulating polyamine. Again, NO may be a link between polyamine-mediated stress responses and other stress mediators using NO as intermediate. Conceivably, the source enzyme for stress-induced NO biosynthesis will be decisive for how different stresses can be regulated by one second messenger or diffusible mediator. As pointed out above, subcellular localization and the linkage to individual receptors will determine specificity of such pathways using one mediator: NO provided by different enzymes ought not to have the same function. In conclusion, several biotic and abiotic stresses in which polyamine are involved may use NO as a mediator.

2.6.3
Hormones, Polyamines and NO

Our findings on the polyamine-stimulated NO release may have a bearing on the anti-senescence effects of polyamines (Shi et al. 1992; Lester, 2000; Bregoli et al. 2002) since NO counteracts the senescence-enhancing effects of ethylene (Leshem et al. 1998; Leshem and Pinchasov 2000). Most recently, Durner's group showed that S-nitrosylation inhibits one of the three methionine adenosyltransferases generating the precursor to ethylene biosynthesis. Also, cytokinin has an anti-senescence effect (Hajouj et al. 2000; Li et al. 2000) and was shown to induce rapid release of NO (Tun et al. 2001, submitted). Recent findings explain the anti-senescence effect as well as the classical "nutrient-attracting" effect of cytokinin, such as the "green islands" on leaves after cytokinin treatment, as a stimulation of invertase transcription (Roitsch and Gonzales 2005). This allows hexoses to be used for sustaining the longer life of senescing leaves.

2.6.4 Embryogenesis and NO

Embryogenesis was the impetus for studies on polyamines and NO biosynthesis (Moura-Costa et al. 1993; Viana and Mantell 1999; Santa-Catarina et al. 2003, 2004; Silveira et al. 2004). Polyamines were described as mediators of embryogenesis (Galston et al. 1995; Shoeb et al. 2001; Bertoldi et al. 2004) and polyamine-induced NO could be involved in this function. We found that embryogenic cells of *Ocotea catharinensis* responded to polyamines by synthesizing more NO and that the round and embryogenic cells of *Araucaria angustifolia* accumulated more NO than the suspensor-like elongated cells (Silveira et al. 2006), suggesting that embryogenic cells might have a distinct NO physiology. Our findings should have bearings on future research on polyamines in embryogenesis.

3 Conclusions and Further Questions

NO biosynthesis is reported to be arginine-specific (Guo et al. 2003, 2005; Corpas et al. 2006). However, polyamine addition was never tested by other authors so that the polyamine-relevant enzyme has not been identified. So, the source enzyme for rapid polyamine- and cytokinin-induced NO is one big question, since it is neither NR nor *At*NOS1 for cytokinin and the similar consequences of both signals on NO biosynthesis are obvious. Is there a new group of enzymes using polyamines as a signal or even as a substrate? Or is it still one other protein linked to the *AtNOS* having different properties to *AtNOS1*? The second major question is: is protein nitrosylation the major mechanism for propagation of NO-dependent signals? This could be true for both cytokinin and polyamine and provide an explanation of their functional overlaps.

References

Allen GJ, Muir SR, Sanders D (1995) Release of Ca^{2+} from individual plant vacuoles by both $InsP_3$ and cyclic ADP ribose. Science 268:735–737

Arita NO, Cohen MF, Tokuda G, Yamasaki H (2007) Fluorometric detection of nitric oxide with diaminofluoresceins: applications and limitations for plant no research. In:

Lamattina L, Polacco JC (eds) Nitric oxide in plant growth. Plant Cell Monographs, vol 6. Springer, Berlin Heidelberg New York (in press)
Bertoldi D, Tassoni A, Martinelli L, Bagni N (2004) Polyamines and somatic embryogenesis in two *Vitis vinifera* cultivars. Physiol Plant 120:657–666
Bethke PC, Badger MR, Jones RL (2004) Apoplastic synthesis of nitric oxide by plant tissues. Plant Cell 16:332–341
Binda C, Mattevi A, Edmondson DE (2002) Structure-function relationships in flavoenzyme-dependent amine oxidations: a comparison of polyamine oxidase and monoamine oxidase. J Biol Chem 277:23973–23976
Bouchereau A, Aziz A, Larher F, Martin-Tanguy J (1999) Polyamines and environmental challenges: recent developments. Plant Sci 140:103–125
Bregoli AM, Scaramagli S, Costa G, Sabatini E, Ziosi V, Biondi S, Torrigiani P (2002) Peach (*Prunus persica*) fruit ripening: aminoethoxyvinylglycine (AVG) and exogenous polyamines affect ethylene emission and flesh firmness. Physiol Plant 114:472–481
Broillet M-C, Randin O, Chatton J-Y (2001) Photoactivation and calcium sensitivity of the fluorescent NO indicator 4,5-diamniofluorescein (DAF-2): implications for cellular NO imaging. FEBS Lett 491:227–232
Butt YK, Lum JH, Lo SC (2003) Proteomic identification of plant proteins probed by mammalian nitric oxide synthase antibodies. Planta 216:762–771
Carimi F, Zottini M, Costa A, Cattalani I, De Michele M, Terzi M, Lo Schiavo F (2005) NO signaling in cytokinin-induced programmed cell death. Plant Cell Environ 28:1171–1178
Cohen AS, Popovic RB, Zalk S (1979) Effects of polyamines on chlorophyll and protein content, photochemical activity, and chloroplast ultrastructure of barley discs during senescence. Plant Physiol 64:7171–720
Conrath U, Amoroso G, Köhle H, Sültemeier DF (2004) Non-invasive on-line detection of nitric oxide from plants and some other organisms by mass spectrophotometry. Plant J 38:1015–1022
Corpas FJ, Barroso JB, Carreras A, Quiros M, Leon AM, Romero-Puertas MC, Esteban FJ, Valderrama R, Palma JM, Sandalio LM, Gomez M, del Rio LA (2004) Cellular and subcellular localization of endogenous nitric oxide in young and senescent pea plants. Plant Physiol 136:2722–2733
Corpas FJ, Barroso JB, Carreras A, Valderrama R, Palma JM, León AM, Sandalio LM, del Río LA (2006) Constitutive arginine-dependent nitric oxide synthase activity in different organs of pea seedlings during plant development. Planta 224:246–254
Correa-Aragunde N, Graziano M, Lamattina L (2004) Nitric oxide plays a central role in determining lateral root development in tomato. Planta 218:900–905
D'Agostino IB, Deruere J, Kieber JJ (2000) Characterization of the response of the *Arabidopsis* response regulator gene family to cytokinin. Plant Physiol 124:1706–1717
Delledonne M (2005) NO news is good news for plants. Curr Opin Plant Biol 8:390–396
Delledone M, Xia Y, Dixon RA, Lamb C (1998) Nitric oxide functions as a signal in plant disease resistance. Nature 394:585–588
Desikan R, Griffiths R, Hancock J, Neill S (2002) A new role for an old enzyme: nitrate reductase-mediated nitric oxide generation is required for abscisic acid-induced stomatal closure in *Arabidospsis thaliana*. Proc Natl Acad Sci USA 99:16314–16318
Dordas C, Hasinoff BB, Igamberdiev AU, Manac'h N, al Rivo J, Hill RD (2003) Expression of a stress-induced hemoglobin affects NO levels produced by alfalfa root cultures under hypoxic stress. Plant J 35:763–770
Durner J, Gow AJ, Stamler JS, Glazebrook J (1999) Ancient origins of nitric oxide signaling in biological systems. Proc Natl Acad Sci USA 96:14206–14207

Durner J, Wendehenne D, Klessig DF (1998) Defense gene induction in tobacco by nitric oxide, cyclic GMP, and cyclic ADP-ribose. Proc Natl Acad Sci USA 95:10328–10333
Erdei L, Szegeltes Z, Barabas K, Pestenacz A (1996) Response in polyamine titer under osmotic and salt stress in sorghum and maize seedlings. J Plant Physiol 147:599–603
Flores HE, Galston AW (1984) Osmotic stress-induced polyamine content in cereal leaves. I. Physiological parameters of the response. Plant Physiol 75:102–109
Foissner I, Wendehenne D, Langebartels C, Durner J (2000) In vivo imaging of an elicitor-induced nitric oxide burst in tobacco. Plant J 23:817–824
Gabaldón C, Gómez Ros LV, Pedreno MA, Ros Barceló A (2005) Nitric oxide production by the differentiating xylem of Zinnia elegans. New Phytol 165:121–130
Galston AW, Kauw-Shawney R, Altabella T, Tiburcio AF (1995) Plant polyamines in reproductive activity and response to abiotic stress. Bot Acta 110:197–207
Geny L, Broquedis M, Martin-Tanguy J, Soyer JP, Bouard J (1997) Effects of potassium nutrition on polyamine content of various organs of fruiting cuttings of *Vitis vinifera* L. cv. Cabernet Sauvignon. Am J Enol Vitic 48:85–91
Gerber IB, Zeidler D, Durner J, Dubery IA (2004) Early perception responses of *Nicotiana tabacum* cells in response to lipopolysaccharides from *Burkholderia cepacia*. Planta 218:647–657
Graziano M, Beligni V, Lamattina L (2002) Nitric Oxide improves the iron deficiency phenotype in maize. Plant Physiol 130:1852–1859
Graziano M, Lamattina L (2005) Nitric oxide and iron in plants: an emerging and converging story. Trends Plant Sci 10:4–8
Greenland AJ, Lewis DH (1984) Amines in barley leaves infected by brown rust and their possible relevance to formation of for the green islands. New Phytol 96:283–291
Grefen C, Harter K (2004) Plant two-component systems: principles, functions, complexity and cross talk. Planta 219:733–742
Gould KS, Lamotte O, Klinguer A, Pugin A, Wendehenne D (2003) Nitric oxide production in tobacco leaf cells: a generalized stress response? Plant Cell Environ 26:1851–1862
Guo FQ, Crawford NM (2005) *Arabidopsis* nitric oxide synthase1 is targeted to mitochondria and protects against oxidative damage and dark-induced senescence. Plant Cell 17:3436–3450
Guo F-Q, Okamoto M, Crawford NM (2003) Identification of a plant nitric oxide synthase gene involved in hormonal signaling. Science 302:100–103
Hajouj T, Michelis R, Gepstein S (2000) Cloning and characterization of a receptor-like protein kinase gene associated with senescence. Plant Physiol 124:1305–1314
Harper JE (1981) Evolution of nitrogen oxide(s) during in vivo nitrate reductase assay of soybean leaves. Plant Physiol 68:1488–1493
Huang X, von Rad U, Durner J (2002) Nitric oxide induces transcriptional activation of the nitric oxide-tolerant alternative oxidase in *Arabidopsis* suspension cells. Planta 215:914–923
Huang X, Stettmaier K, Michel C, Hutzler P, Mueller MJ, Durner J (2004) Nitric oxide is induced by wounding and influences jasmonic acid signaling in *Arabidopsis thaliana*. Planta 218:938–946
Hutchison CE, Kieber JJ (2002) Cytokinin signaling in *Arabidopsis*. Plant Cell 14:S47-S59
Hwang I, Sheen J (2001) Two-component circuitry in *Arabidopsis* cytokinin signal transduction. Nature 413:383–389
Inoue T, Higuchi M, Hashimoto Y, Seki M, Kobayashi M, Kato T, Tabata S, Shinozaki K, Kakimoto T (2001) Identification of CRE1 as a cytokinin receptor from *Arabidopsis*. Nature 409:1060–1063

Jaffrey SR, Snyder SH (2001) The biotin switch method for the detection of S-nitrosylated proteins. Sci STKE 2001:PL1

Kakimoto T (2003) Perception and signal transduction of cytokinins. Annu Rev Plant Biol 54:605–627

Kakkar RK, Nagar PK, Ahuja PS, Rai VK (2000) Polyamines and plant morphogenesis. Physiol Plant 43:1–11

Kiba T, Naitou T, Koizumi N, Yamashino T, Sakakibara H, Mizuno T (2005) Combinatorial microarray analysis revealing *Arabidopsis* genes implicated in cytokinin responses through the his→asp phosphorelay circuitry. Plant Cell Physiol 46:339–355

Kojima H, Hirotani M, Nakatsubo N, Kikuchi K, Urano Y, Higuchi T, Hirata Y, Nagano T (2001) Bioimaging of nitric oxide with fluorescent indicators based on the rhodamine chromophore. Anal Chem 73:1967–1973

Kojima H, Hirotani M, Urano Y, Kikuchi K, Higuchi T, Nagano T (2000) Fluorescence indicators for nitric oxide based on rhodamine chromophore. Tetrahedron Lett 41:69–72

Krishnamurthy R, Bhagnat KA (1984) Polyamines as modulators of salt tolerance in rice cultivars. Plant Physiol 91:500–504

Lamattina L, Garcia-Mata C, Graziano M, Pagnussat G (2003) Nitric oxide: the versatility of an extensive signal molecule. Annu Rev Plant Biol 54:109–136

Lamotte O, Courtois C, Barnavon L, Pugin A, Wendehenne D (2005) Nitric oxide in plants: the biosynthesis and cell signaling properties of a fascinating molecule. Planta 221:1–4

Lamotte O, Gould K, Lecourieux D, Sequeira-Legrand A, Lebrun-Garcia A, Durner J, Pugin A, Wendehenne D (2004) Analysis of nitric oxide signaling functions in tobacco cells challenged by the elicitor cryptogein. Plant Physiol 135:516–529

Laxalt A, Beligni MV, Lamatina L (1997) Nitric oxide preserves the level of chlorophyll in potato leaves infected by *Phythophthora infestans*. Eur J Plant Pathol 73:643–651

Leshem YY (1996) Nitric oxide in biological systems. Plant Growth Regul 18:155–159

Leshem YY Pinchasov Y (2000) Noninvasive photoacoustic spectroscopic determination of relative endogenous nitric oxide and ethylene content stoichiometry during the ripening of strawberries *Fragaria annanasa* (Duch.) and avocados *Persea americana* (Mill.). J Exp Bot 51:1471–1473

Leshem YY, Wils RBH, Ku VVV (1998) Evidence for the function of the free radical gas – nitric oxide – as an endogenous maturation and senescence regulating factor in higher plants. Plant Physiol Biochem 36:825–833

Lester GE (2000) Polyamines and their cellular anti-senescence properties in honey dew muskmelon fruit. Plant Sci 160:105–112

Li Q, Bettany AJ, Donnison I, Griffiths CM, Thomas H, Scott IM (2000) Characterisation of a cysteine protease cDNA from *Lolium multiflorum* leaves and its expression during senescence and cytokinin treatment. Biochim Biophys Acta 1492:233–236

Lindermayr C, Saalbach G, Bahnweg G, Durner J (2006) Differential inhibition of *Arabidopsis* methionine adenosyltransferases by protein S-nitrosylation. J Biol Chem 281:4285–4291

Lindermayr C, Saalbach G, Durner J (2005) Proteomic identification of S-nitrosylated proteins in *Arabidopsis*. Plant Physiol 137:921–930

Liu L, Hausladen A, Zeng M, Que L, Heitman J, Stamler JS (2001) A metabolic enzyme for S-nitrosothiol conserved from bacteria to humans. Nature 410:490–494

Mähönen AP, Bonke M, Kauppinen L, Riikonen M, Benfey PN, Helariutta Y (2000) A novel two-component hybrid molecule regulates vascular morphogenesis of the *Arabidopsis* root. Genes Dev 14:2938–2943

Mannick JB, Schonhoff CM (2004) NO means no and yes: regulation of cell signaling by protein nitrosylation. Free Radic Res 38:1–7

Martinez-Moreno M, Alvarez-Barrientos A, Roncal F, Albar JP, Gavilanes F, Sokolovski S, Hills A, Gay R, Garcia-Mata C, Lamattina L, Blatt MR (2005) Protein phosphorylation is a prerequisite for intracellular Ca release and ion channel control by nitric oxide and abscisic acid in guard cells. Plant J 43:520–529

Miyata S, Urao T, Yamaguchi-Shinozaki K, Shinozaki K (1998a) Expression of *Arabidopsis* regulator homologs is induced by cytokinins and nitrate. FEBS Lett 429:259–262

Miyata S, Urao T, Yamaguchi-Shinozaki K, Shinozaki K (1998b) Characterization of genes for two-component phosphorelay mediators with a single HPt domain in *Arabidopsis*

Mo H, Pua EC (2002) Up-regulation of arginine decarboxylase gene expression and accumulation of polyamines in mustard (*Brassica juncea*) in response to stress. Physiol Plant 114:439–449

Morot-Gaudry-Talarmain Y, Rockel P, Moureaux T, Quilerè I, Leydecker MT, Kaiser WM, Morot-Gaudry JF (2002) Nitrite accumulation and nitric oxide emission in relation to signaling in nitrite reductase antisense plants. Planta 215:708–715

Moura-Costa PH, Viana AM, Mantell SH (1993) In vitro plantlet regeneration of *Ocotea catharinensis*, an endangered Brasilian hardwood forest tree. Plant Cell Tiss Org Cult 35:279–286

Mur LA, Santosa IE, Laarhoven LJ, Holton NJ, Harren FJ, Smith AR (2005) Laser photoacoustic detection allows in planta detection of nitric oxide in tobacco following challenge with avirulent and virulent *Pseudomonas syringae* pathovars. Plant Physiol 138:1247–1258

Murgia I, Delledonne M, Soave C (2002) Nitric oxide mediates iron-induced ferritin accumulation in *Arabidopsis*. Plant J 30:521–532

Neill SJ, Desikan R, Hancock J (2003) Nitric oxide signaling in plants. New Phytol 159:11–35

Nishimura C, Ohashi Y, Sato S, Kato T, Tabato S, Ueguchi C (2004) Histidine kinase homologs that act as cytokinin receptors possess overlapping functions in the regulation of shoot and root growth in *Arabidopsis*. Plant Cell 16:1365–1377

Polverari A, Molesini B, Pezzotti M, Buonoaurio R, Marte M, Delledone M (2003) Nitric oxide-mediated transcriptional changes in *Arabidopsis* thaliana. Mol Plant Microbe Interact 16:1094–1105

Prakash L, Prathapsenan G (1988) Effect of NaCl on salinity and putrescine on shoot growth, tissue ion concentration, and yield of rice (*Oryza sativa*). J Agron Crop Sci 160:325–334

Racz I, Kovacs M, Lasztity D, Veisz O, Szalai G, Paldi E (1996) Effects of short-term and long-term low temperature stress on polyamine biosynthesis in wheat genotypes with varying degrees of frost tolerance. J Plant Physiol 148:368–373

Reggiani R, Bertani A (1989a) Effect of decreasing oxygen concentration of polyamine metabolism in rice and wheat shoots. J Plant Physiol 135:375–377

Reggiani R, Guissani P, Bertani A (1990) Relationship of the accumulation of putrescine and the tolerance to oxygen deficit stress in Gramineae seedlings. Plant Cell Physiol 31:489–494

Reggiani R, Hochkoeppler A, Bertani A (1989b) Polyamines and anaerobic elongation in rice coleoptile. Plant Cell Physiol 30:893–898

Richards FJ, Coleman EG (1952) Occurrence of putrescine in potassium deficient barley. Nature 170:460–461

Rockel P, Strube F, Rockel A, Wildt J, Kaiser WM (2002) Regulation of nitric oxide (NO) production by plant nitrate reductase in vivo and in vitro. J Exp Bot 53:103–110

Roitsch T, Gonzalez MC (2004) Function and regulation of plant invertases: sweet sensations. Trends Plant Sci 9:606–613

Roychowdhury S, Luthe A, Keilhoff G, Wolf G, Horn TFW (2002) Oxidative stress in glial cultures: detection by DAF-2 fluorescence used as a tool to measure peroxynitrite rather than nitric oxide. Glia 38:103–114

Sakamoto A, Ueda M, Morikawa H (2002) *Arabidopsis* glutathione dependent formaldehyde dehydrogenase is an S-nitrosoglutathione reductase. FEBS Lett 515:20–24

Sakihama Y, Nakamura S, Yamasaki H (2002) Nitric oxide production mediated by nitrate reductase in the green alga *Chlamydomonas reinhardtii*: an alternative NO production pathway in photosynthesis organisms. Plant Cell Physiol 43:290–297

Samuelson ME, Campbell WH, Larsson CM (1995) The influence of cytokinins in nitrate regulation of nitrate reductase activity and expression in barley. Physiol Plant 93:533–539

Santa-Catarina C, Hanai LR, Dornelas MC, Viana AM, Floh EIS (2004) Serk gene homolog expression, polyamines and amino acids associated with somatic embryogenic competence of *Ocotea catharinensis* Mez. (Lauraceae). Plant Cell Tiss Org Cult 90R1:1–9

Santa-Catarina C, Randi AM, Viana AM (2003) Growth and accumulation of storage reserves by somatic embryos of *Ocotea catharinensis* Mez. (Lauraceae). Plant Cell Tiss Org Cult 74:67–71

Santa-Cruz A, Perez-Alfocea MA, Bolarin C (1997) Changes in free polyamine levels induced by salt stress in leaves of cultivated and wild tomato species. Physiol Plant 101:341–346

Silveira V, Floh EIS, Handro W, Guerra M (2004) Effect of plant growth regulators on the cellular growth and levels of intracellular protein, starch and polyamines in embryogenic suspension cultures of *Pinus* taeda. Plant Cell Tiss Org Cult 76:53–60

Silveira V, Santa-Catarina C, Tun NN, Handro W, Scherer GFE, Handro W, Guerra M, Floh EIS (2006) Polyamine effects on the endogenous polyamine contents, nitric oxide release, growth and differentiation of embryonic suspension cultures of *Araucaria angustifolia* (Bert.) O.KTZE. Plant Sci 171:91–98

Shi LM, Kaur-Shawney R, Fuhrer J, Samanat S, Galston A (1982) Effects of exogeneous 1,3-diaminepropane and spermidine on senescence on oat leaves. Plant Physiol 70:1592–1596

Shoeb F, Yadav JS, Bajaj S, Rajam MV (2001) Polyamines as biomarkers for plant regeneration capacity: improvement of regeneration by modulation of polyamine metabolism in different genotypes of indica rice. Plant Sci 160:1229–1235

Sokolovski S, Blatt MR (2004) Nitric oxide block of outward-rectifying K^+ channels indicates direct control by protein nitrosylation in guard cells. Plant Physiol 136:4275–4284

Sokolovski S, Hills A, Gay R, Garcia-Mata C, Lamattina L, Blatt MR (2005) Protein phosphorylation is a prerequisite for intracellular Ca release and ion channel control by nitric oxide and abscisic acid in guard cells. Plant J 43:520–529

Stamler JS, Lamas S, Fang FC (2001) Nitrosylation. the prototypic redox-based signaling mechanism. Cell 106:675–683

Stöhr C, Strube F, Marx G, Ullrich WR, Rockel P (2001) A plasma membrane-bound enzyme of tobacco roots catalyses the formation of nitric oxide from nitrite. Planta 212:835–841

Takahashi Y, Berberich T, Miyazaki A, Seo S, Ohashi Y, Kusano T (2003) Spermine signaling in tobacco: activation of mitogen-activated protein kinases by spermine is mediated through mitochondrial dysfunction. Plant J 36:820–829

Takahashi Y, Uehara Y, Berberich T, Ito A, Saitoh H, Miyazaki A, Terauchi R, Kusano T (2004) A subset of hypersensitive response marker genes, including HSR203J, is the downstream target of a spermine signal transduction pathway in tobacco. Plant J 40:586–595

Takei K, Takahashi T, Sugiyama T, Yamaya T, Sakakibara H (2002) Multiple routes communicating nitrogen availability from roots to shoots: a signal transduction pathway mediated by cytokinin. J Exp Bot 53:971–977

Tanaka S, Suzuki T, Yamashino T, Mizuno T (2004) Comparative studies on the AHP histidine-containing phosphotransmitters implicated in his-to-asp phosphorelay in *Arabidopsis thaliana*. Biosci Biotechnol Biochem 68:462–465

Taniguch M, Kiba T, Sakakibara H, Ueguchi C, Mizuno T, Sugiyama T (1998) Expression of *Arabidopsis* response regulator homologs is induced by cytokinins and nitrate. FEBS Lett 429:259–262

Tischner R, Planchet E, Kaiser WM (2004) Mitochondrial electron transport as a source for nitric oxide in the unicellular green alga *Chlorella sorokiniana*. FEBS Lett 576:151–155

Tun NN, Holk A, Scherer GFE (2001) Rapid increase of NO release in plant cell cultures induced by cytokinin. FEBS Lett 509:174–176

Tun NN, Santa-Catarina C, Begum T, Silveira V, Handro W, Floh EIS, Scherer GFE (2006) Polyamines induce rapid biosynthesis of nitric oxide (NO) in *Arabidopsis thaliana* seedlings. Plant Cell Physiol 47:346–354

Ueguchi C, Koizumi H, Suzuki T, Mizuno T (2001) Novel family of sensor histidine kinase genes in *Arabidopsis thaliana*. Plant Cell Physiol 42:231–235

Viana AM, Mantell SH (1999) Somatic embryogenesis of *Ocotea catharinensis*: an endangered tree of the Mata Atlantica (S. Brasil). In: Jain SM, Gupta PK, Newton RJ (eds) Somatic embryogenesis in woody plants, vol 5. Kluwer, Dordrecht, pp 3–30

Watson MB, Malmberg RL (1996) Regulation of *Arabidopsis thaliana* (L.) Heynh arginine decarboxylase by potassium deficiency stress. Plant Physiol 111:1077–1083

Wendehenne D, Durner J, Klessig DF (2004) Nitric oxide: a new player in plant signaling and defence responses. Curr Opin Plant Biol 7:449–455

Wendehenne D, Pugin A, Klessig DF, Durner J (2001) Nitric oxide: comparative synthesis and signaling in animal and plant cells. Trends Plant Sci 4:177–183

Wilkinson JQ, Crawford NM (1993) Identification and characterisation of a chlorate-resistant mutant of *Arabidopsis thaliana* with mutations in both nitrate reductase structural genes *NIA1* and *NIA2*. Mol Gen Genet 239:289–297

Willidiano L, Camara T, Boget N, Claparols I, Santos M, Torne JM (1996) Polyamine and free amino acid variations in NaCl-treated embryogenic maize callus from sensitive and resistant cultivars. J Plant Physiol 149:179–185

Yamakawa H Kamada H, Satoh M, Ohashi Y (1998) Spermine is a salicylate-independent endogenous inducer for both tobacco acidic pathogenesis-related proteins and resistance against tobacco mosaic virus infection. Plant Physiol 118:1213–1222

Yamasaki H (2005) The NO world for plants: achieving balance in an open system. Plant Cell Environ 28:78–84

Yamasaki H, Sakihama Y Takahashi S (1999) An alternative pathway for nitric oxide production in plants: new features of an old enzyme. Trends Plant Sci 4:128–129

Zeidler D, Zähringer U, Gerber I, Dubery I, Hartung T, Bors W, Hutzler P, Durner J (2004) Innate immunity in *Arabidopsis* thaliana: lipopolysaccharides activate nitric oxide synthase (NOS) and induce defense genes. Proc Natl Acad Sci USA 101:15811–15816

Plant Cell Monogr (6)
L. Lamattina · J. C. Polacco: Nitric Oxide in Plant Growth
DOI 10.1007/7089_2006_089/Published online: 14 October 2006

Nitric Oxide and Plant Ion Channel Control

Sergei G. Sokolovski · Michael R. Blatt (✉)

Laboratory of Plant Physiology and Biophysics, Institute of Biomedical and Life Sciences, University of Glasgow, Bower Building, Glasgow G12 8QQ, UK
m.blatt@bio.gla.ac.uk

Abstract Nitric oxide (NO) has profound effects on the regulation of ion channels in plants. Although direct evidence to date comes exclusively from electrophysiological studies of guard cells, there is good reason to expect similar patterns of action in other plant cell types as well. As in animals, NO appears to act through two distinct mechanisms. One mechanism is mediated via stimulation of guanylate cyclase, which leads to a rise in cyclic ADP-ribose and, in turn, an increase in the efficacy of Ca^{2+} release triggered by Ca^{2+} entry across the plasma membrane. This signal cascade underpins intracellular Ca^{2+} release and the elevation of cytosolic-free $[Ca^{2+}]$ by the water-stress hormone abscisic acid and leads to profound changes in K^+ and Cl^- channel activities, to facilitate the ion fluxes for stomatal closure. The second mechanism appears to arise from direct, covalent modification of ion channels by NO, notably of the outward-rectifying K^+ channel at the guard cell plasma membrane. The physiological significance of this process of S-nitrosylation has yet to be explored in depth, but almost certainly is allied to plant cell responses to pathogen attack and apoptosis. Both processes, and ion transport in guard cells generally, are now sufficiently well-defined for a full description with accurate kinetics and flux equations in which all of the key parameters are constrained by experimental data. Thus, guard cells are now a prime focus for integrative (so-called systems biology) approaches. Applications of integrative analysis have already demonstrated the potential for accurately predicting physiological behaviours and signal interactions with membrane ion transport.

1 Introduction

Nitric oxide (NO) has long been associated with ion transport and the regulation of ion channels in mammalian tissues, although its relevance to plant ion channel control has come to the fore only recently. Interest in NO was heightened over one decade ago when it was found to hyperpolarise the K^+ channels of guinea-pig tracheae (Allen and Sanders 1996). In the intervening years, its role as a channel regulator has been firmly established in a wide variety of animal tissues and numerous physiological and diseased states (Brune et al. 1998; Stamler et al. 2001; Demple 2002; Ahern et al. 2002) as well as in tissue development, notably in synaptic plasticity and neurogeneration and apoptosis (Li et al. 1997; Zufall et al. 1997; Melino et al. 1997). NO in animals derives from distinct isoforms of nitric oxide synthase, including inducible activities, and functions to stimulate soluble guanylate cyclase and production of cGMP. Four principal targets of cGMP are protein kinase G (White et al. 1999), cyclic-nucleotide-gated channels (Zufall et al. 1997), cyclic-nucleotide phosphodiesterase and synthesis of cyclic-ADP-ribose (cADPR) leading to activation of ryanodine-sensitive Ca^{2+} channels (Lee et al. 1993; Clementi et al. 1996; Lukyanenko et al. 2001). Of these targets, Protein kinase G and cADPR provide especially broad control of ion channels, modifying the gating of Ca^{2+}-dependent K^+ channels, Na^+ channels, N- and T-type Ca^{2+} channels (Ahern et al. 2002; Moreno et al. 2001; Stamler and Meissner 2001; Willmott et al. 1996; Yuan et al. 1996).

In the past decade, NO has also come to be recognised in signal transduction events that occur independent of cGMP and, instead, depend on direct, post-translational modification of target proteins by covalent modification of cysteine thiol side-chains (Mannick and Schonhoff 2004; Ahern et al. 2002; Stamler et al. 2001). This so-called S-nitrosylation of proteins generally occurs at significantly higher concentrations of NO than does activation of guanylate cyclases and tends to proceed with slower kinetics than cGMP-mediated reactions, but plays equally important physiological roles in ion channel regulation (Mannick and Schonhoff 2004; Ahern et al. 2002).

By contrast, recognition of NO in plant physiology has come very late – especially in comparison to another well-known gaseous hormone (Ashley et al. 2006; Kieber and Ecker 1993; Yang and Hoffman 1984) – and despite a knowledge of its effects on mitochondrial and chloroplast electron transport (Shapiro 2005) and germination (Bhatia and Sybenga 1965). The spectrum of physiological functions for NO in plants has now grown rapidly to include plant defense responses (Delledonne et al. 1998; Foissner et al. 2000), drought resistance and guard cell signalling (Neill et al. 2002; Garcia-Mata and Lamattina 2001), light and gravitropic responses (Zhang et al. 2003; Hu et al. 2005) among others. While many of these are known to be associated with ion transport, for example with changes in cytosolic-free $[Ca^{2+}]$ ($[Ca^{2+}]_i$), there is still remarkably little known about the ion flux mechanisms involved outside of the singular example of the guard cell. Here we summarise the roles for NO in the control of ion channels in guard cells and recent evidence for both cGMP-dependent and S-nitrosylation-dependent signalling pathways in this plant cell model. The origins of NO in guard cells, and in plants generally, are subjects of other chapters in this volume and will not be dealt with here.

2
Guard Cells and Ion Channels

Guard cells surround pores (stomata) within the epidermis of all aerial parts of most plants to control gas exchange for photosynthesis and to prevent water vapour loss from plant tissues to the environment. Because the demands for CO_2 in photosynthesis and for water retention are frequently at odds, guard cells must integrate these and other signals to give a fine balance between the open and closed states of the stoma. Since the 1980s, work on ion transport and the control of guard cell ion channels has provided a wealth of information that is still unparallelled in plant biology, driven primarily by electrophysiological studies and more recently by molecular genetics. We know now sufficient detail of all of the major transport pathways at the plasma membrane to describe these fully with accurate kinetics and flux equations in which all of the key parameters are constrained by experimental data. Thus, guard cells are now a prime focus for integrative (so-called "systems biology") approaches, applications of which have already demonstrated the potential for accurately predicting physiological behaviours and signal interactions with membrane ion transport (Gradmann et al. 1993; Blatt 2000b).

Opening and closing of the stomata are driven by changes in guard cell turgor and solute content, and they offer a particularly useful handle for analysis of transport and its control in guard cells. Between open and closed states guard cells of *Vicia*, for example, take up or release 2–4 pmol of KCl – on a cell volume basis 200–300 mOsM in solute content – all of which must pass across the plasma membrane since mature guard cells lack plasmodesmata (Wille and Lucas 1984). The plasma membrane of guard cells, like that of plant cells generally, is energised by H^+-ATPases that drive H^+ out of the cell, generating an electrochemical gradient for H^+ ($\Delta\mu_H$) directed inward across the membrane. This $\Delta\mu_H$, in turn, facilitates solute uptake, especially during stomatal opening. Analyses of the *Vicia* guard cell H^+-ATPase has shown that one H^+ is transported for each molecule of ATP hydrolysed (Blatt 1987). This activity maintains a gradient of 2–3 pH units and a membrane voltage of − 150 mV to − 200 mV (inside negative) under most conditions. In fact, electrophysiological analyses have confirmed that the guard cell H^+-ATPase is more than sufficient to account for solute uptake during stomatal opening (Lohse and Hedrich 1992; Blatt 1987). There is evidence that *Vicia* guard cells express unusually high levels of two distinct H^+-ATPases at the plasma membrane (Hentzen et al. 1996; Becker et al. 1993; Villalba et al. 1991), consistent with the higher demand expected for membrane energisation in the guard cells.

The predominant inorganic solutes contributing to the guard cell osmotica are K^+ and Cl^-. Transport of these ions is mediated by several ion channels, both at the plasma membrane and at the tonoplast. Comparisons of these channels between species, in those instances for which comparable data are available, indicate some (albeit subtle) differences in kinetic and/or regulatory characteristics. However, the general patterns are largely the same (Very and Sentenac 2003; Hetherington 2001; Schroeder et al. 2001; Blatt 2000b; Dreyer et al. 2004a). At the plasma membrane, three major channel currents prevail that are separable on the bases of their biophysical and pharmacological properties. Genomic and functional studies have confirmed that these currents associate with different channel protein populations. At the functional level, notable features include differences in voltage dependence (rectification) as well as separate regulatory sensitivities to $[Ca^{2+}]_i$ and cytosolic pH (pH_i).

Current through inward-rectifying K^+ channels ($I_{K,in}$) facilitates K^+ uptake during stomatal opening and, in guard cells of *Arabidopsis*, is identified principally with current through the KAT1 K^+ channel (Nakamura et al. 1995; Very and Sentenac 2003; Hetherington 2001; Blatt 2000b; Dreyer et al. 2004a) [but see also Pilot, et al. (2001)]. KAT1 and related K^+ currents in plants show a requirement for millimolar $[K^+]$ outside (Thiel et al. 1992; Hertel et al. 2005; Blatt 1992), but gating otherwise is essentially independent of $[K^+]$ (Blatt 1992; Schroeder 1988). The guard cell $I_{K,in}$ is strongly suppressed by cytosolic-free $[Ca^{2+}]$ ($[Ca^{2+}]_i$). In *Vicia* guard cells, $I_{K,in}$ activity declines steeply with increasing $[Ca^{2+}]_i$ above resting values near 100 nM, showing an apparent K_i around 300 nM and a high (fourfold) degree of cooperativity. By contrast, this current is only weakly dependent on cytosolic pH (pH_i) (Grabov and Blatt 1997, 1999).

Current carried by outward-rectifying K^+ channels ($I_{K,out}$) provides the major pathway for K^+ efflux during stomatal closure (Clint and Blatt 1989; Schroeder et al. 2001; Blatt 2000b) and, in *Arabidopsis*, is identified exclusively with the GORK K^+ channel (Hosy et al. 2003). Characteristics of this K^+ channel and its counterparts in the guard cells of other species include a voltage sensitivity that coordinates with $[K^+]$ outside (Blatt 1988; Blatt and Gradmann 1997; Roelfsema and Prins 1997). Other outward-rectifying K^+ channels in plants show a similar dependence on extracellular K^+, including the phloem-associated SKOR channel of *Arabidopsis*, but this subfamily of eukaryotic K^+ channels appears otherwise unique (Hosy et al. 2003; Gaymard et al. 1998; Johansson et al. 2006). Activation of $I_{K,out}$ is unaffected by $[Ca^{2+}]_i$, by contrast with $I_{K,in}$, but the current is strongly promoted with increasing pH_i (Grabov and Blatt 1997; Blatt and Armstrong 1993; Blatt 1992; Miedema and Assmann 1996).

The third class of channels at the guard cell plasma membrane carry Cl^- as well as other anions. Anion efflux is important for stomatal closure to bring the membrane voltage positive of E_K for net loss of K^+ through $I_{K,out}$. Two Cl^- (anion) channels have been identified with different macroscopic current kinetics (Linder and Raschke 1992; Schroeder and Keller 1992; Hedrich et al. 1990) and may constitute different functional "modes" of the same channel protein (Dietrich and Hedrich 1994; Thomine et al. 1995). Of these, one current S-type (Schroeder and Keller 1992) or SLAC current (Linder and Raschke 1992), hereafter identified as I_{Cl} activates and deactivates slowly (halftimes, 5–30 s) on voltage steps, and exhibits a significant conductance at voltages between − 150 and − 200 mV. These characteristics are consistent with the prolonged (20–30 min) ion flux that occurs during stomatal closure in ABA (Grabov et al. 1997; Pei et al. 1997; Roelfsema et al. 2004).

As in virtually all mature plant cells, the vacuole in guard cells makes up 80–90% of the total cell volume. Thus, the greater proportion of solutes that pass across the plasma membrane during stomatal movements must also traverse the tonoplast, and in a coordinated manner (MacRobbie 1995, 1999). Furthermore, the vacuole is an important source and sink for Ca^{2+} and H^+, and therefore is expected to contribute to signalling events that lead to changes in the free concentration of these ions in the cytosol (Frohnmeyer et al. 1998; Leckie et al. 1998).By contrast with events at the plasma membrane, our understanding of transport across the tonoplast is relatively poor. Nonetheless, it is clear that the guard cell tonoplast harbours a complement of ion channels and pumps in common with most plant cells. These include at least two different cation channels that are capable of carrying K^+ and Ca^{2+} (Schulzlessdorf and Hedrich 1995; Ward and Schroeder 1994; Tikhonova et al. 1997) and an anion channel with high selectivity for Cl^- over K^+, and dependent on protein phosphorylation for activity (Pei et al. 1996). Of these, the so-called slow-vacuolar (SV) channel was recently identified with the Ca^{2+}-permeant TPC1 channel subunit (Peiter et al. 2005) and the vacuolar-K^+ (VK) channel has been suggested to correspond with the TPK1 (=KCO1) cation channel (Bihler et al. 2005). Both of these cation channels are clearly important for solute loss and stomatal closure, although their respective contributions need to be explored in detail.

3 Abscisic Acid, $[Ca^{2+}]_i$ and NO

Consider for a moment the three channel populations at the guard cell plasma membrane. Clearly all three channels, and the H^+-ATPase, locate within the same membrane and therefore interact through the common "intermediate" of the membrane voltage. So, it is no surprise that the voltage sensitivities of both K^+ channels and the Cl^- channel currents at the plasma membrane contribute significantly to their regulation under free-running (that is, in non-voltage-clamp) conditions (Gradmann et al. 1993). Indeed, the interplay between ion channels inhabiting a common membrane reflects the unique nature of the membrane voltage as the kinetic analogue of both substrate and product for all charge-carrying transport, as well as its role as a regulatory factor in controlling channel activity (Blatt 2004). Flux through each of these channels carries electrical charge associated with the ion and, therefore, affects the distribution of charge across the membrane (the membrane voltage). By the same token, charges that move across a membrane necessarily do so through an electric field (at the macroscopic level, the membrane voltage) and therefore will be affected both in rate and direction by changes in membrane voltage. All three channels are coupled independently of membrane voltage as well, and this additional layer of control accounts for the ability of stomata to integrate a wide range of hormonal and environmental factors (Willmer and Fricker 1996). By far the best documented example of this integration comes from studies of guard cell response to abscisic acid (ABA) and it is in these events that NO services a fundamental role.

During water-stress conditions, ABA accumulates in leaf tissues, where it triggers the closure of stomata to reduce transpirational water loss (Davies and Jones 1991). In the simplest of schemes, ABA triggers stomatal closing through a cascade of events that effectively bias the plasma membrane for K^+ and Cl^- loss (see Fig. 1):

1. ABA activates an inward-directed current, mediated at least in part by I_{Cl}, which depolarises the membrane to generate a driving force for K^+ efflux
2. ABA inactivates $I_{K,in}$, which normally mediates K^+ uptake
3. ABA activates current through $I_{K,out}$ which, together with I_{Cl}, facilitates salt loss from the cells

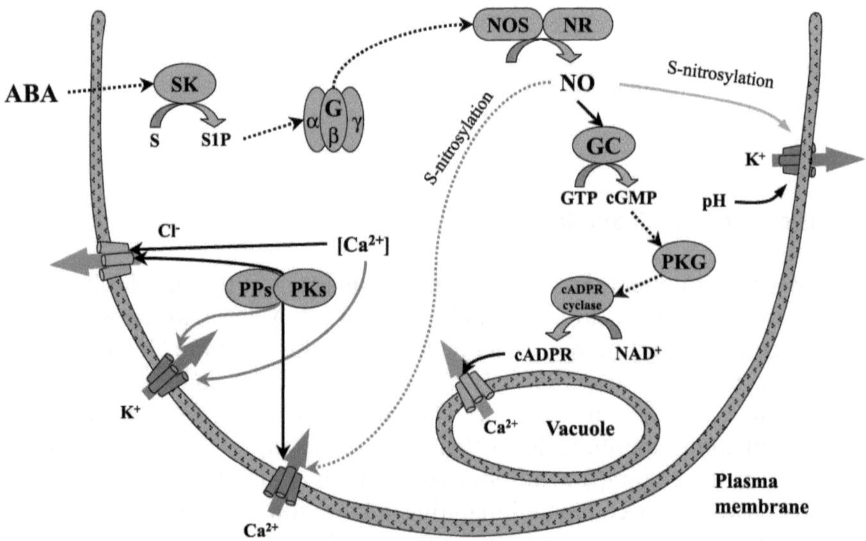

Fig. 1 Nitric oxide-dependent cascades in guard cell signalling. *SK* sphingosine kinase, *S* sphingosine, *S1P* sphingosine 1-phosphate, *G* heterotrimeric GTP-binding protein, *PPs* protein phosphatases, *PKs* protein kinases, *cADPR* cyclic ADP-ribose, *NOS* nitric oxide synthase, *NR* nitrite reductase, *NO* nitric oxide, *GC* guanylate cyclase, *PGK* cGMP-dependent protein kinase. *Black* and *grey arrows* show established positive and negative (respectively) regulatory relationships. *Solid arrows* known in plants; *dashed arrows* known in other eukaryotic systems but as yet undemonstrated in plants. Details can be found in the text and several of the references cited

Critical to these events is the interplay between the different channel activities and voltage as well as inputs from a number of cellular signals, including G proteins and sphingolipids (Coursol et al. 2005; Wang et al. 2001; Ng et al. 2001), inositol phosphates (Lemtiri-Chlieh et al. 2000; Hunt et al. 2003; Lee et al. 1996), protein (de-)phosphorylation (Assmann 2003; Mustilli et al. 2002; Li et al. 2000; Armstrong et al. 1995), reactive oxygen species (Desikan et al. 2005; Kwak et al. 2003) and pH_i (Grabov and Blatt 1997; Blatt and Armstrong 1993; Irving et al. 1992). These signals notwithstanding, a recurring theme centres around changes in $[Ca^{2+}]_i$. With the addition of NO as a component in these events (Sokolovski et al. 2005; Garcia-Mata et al. 2003) we can now reconstruct major control elements and their downstream responses in a "functional cassette".

4 The Ca^{2+} Theme

Resting $[Ca^{2+}]_i$ in guard cells, as in other eukaryotic cells, is situated near 100 nM but may be elevated to micromolar free concentrations (Allen et al. 2001; Grabov and Blatt 1997, 1999; Gilroy et al. 1990; McAinsh et al. 1990). Above resting $[Ca^{2+}]_i$, small increases in free divalent reduce $I_{K,in}$ with an apparent K_i of 300 nM and a cooperativity coefficient of 4, such that at 500 nM $[Ca^{2+}]_i$ the K^+ channels are essentially inactive (Grabov and Blatt 1999). Detailed analysis of the kinetics has shown that increasing $[Ca^{2+}]_i$ displaces the voltage sensitivity of K^+ channel gating out of the the normal physiological voltage range (approx. – 50 mV to – 200 mV) so that $I_{K,in}$ is not active under free running (non-voltage-clamp) conditions (Grabov and Blatt 1997, 1999). Increasing $[Ca^{2+}]_i$ also promotes I_{Cl} (Hedrich et al. 1990; Schroeder and Keller 1992), although quantitative kinetic detail is still incomplete. Significantly, ABA itself evokes qualitatively similar changes in both $I_{K,in}$ and I_{Cl} (Thiel et al. 1992; Lemtiri-Chlieh and MacRobbie 1994; Grabov et al. 1997; Pei et al. 1997) that have been correlated with $[Ca^{2+}]_i$ increases, in some cases to values above 1 μM (Irving et al. 1992; McAinsh et al. 1992; Fricker et al. 1991; Allan et al. 1994). However, even a relatively small rise in $[Ca^{2+}]_i$ is clearly sufficient to suppress $I_{K,in}$ (Grabov and Blatt 1999) and probably to promote I_{Cl}, thereby depolarising the membrane to bias it for solute efflux.

It is clear now that other factors also contribute to the "poise" of the cell, to the efficacy and extent of $[Ca^{2+}]_i$ changes that may be evoked by ABA. This point was proposed on the basis of studies from this laboratory implicating an adaptive feedback of $[Ca^{2+}]_i$ on its elevation in ABA (Grabov and Blatt 1998). For example, $[Ca^{2+}]_i$ and $[Ca^{2+}]_i$-dependent responses are affected by the redox state of the cell (Kwak et al. 2003; Kohler et al. 2003) as well as protein (de)phosphorylation both upstream and downstream of any changes in $[Ca^{2+}]_i$ (Kwak et al. 2002; Sokolovski et al. 2005; Köhler and Blatt

2002). These processes, too, are almost certainly subject to the background of $[Ca^{2+}]_i$ (Mustilli et al. 2002; Li et al. 1998; Mori and Muto 1997; Cheng et al. 2002) and may explain why guard cell response to ABA may be achieved without measurable changes in $[Ca^{2+}]_i$ (Romano et al. 2000; Armstrong et al. 1995; Grabov and Blatt 1999). In short, the efficacy of a stimulus to trigger $[Ca^{2+}]_i$ increases also depends on the prevailing $[Ca^{2+}]_i$ at rest, implying feedback control of $[Ca^{2+}]_i$ on Ca^{2+} entry and release within the guard cell.

5 NO and the "$[Ca^{2+}]_i$ Cassette"

In general, increases in $[Ca^{2+}]_i$ in guard cells arise both from Ca^{2+} entry across the plasma membrane and release from intracellular stores (Hetherington 2001; Blatt 2000b). However, the details of functional specialisation – and location within the cell – of each of these elements leading to $[Ca^{2+}]_i$ changes is almost certainly important for signal encoding (Clapham 2003; Sanders et al. 2002; Berridge 1998).

Thus, differences in the relative contributions of various pathways may reflect an inherent redundancy (and, hence, plasticity) to the Ca^{2+} signal. For ABA, Ca^{2+} entry across the guard cell plasma membrane appears to be mediated by a single population of Ca^{2+} channels that are kinetically distinct from K^+ channels, show a high selectivity for Ca^{2+} and Ba^{2+} and, by contrast with mammalian Ca^{2+} channels, are activated at negative membrane voltages (Hamilton et al. 2000, 2001). Significantly, Hamilton, et al. (2000) found that this Ca^{2+} current (I_{Ca}) was potentiated by ABA, consistent with the effects of ABA on the voltage-dependence for $[Ca^{2+}]_i$ increases. It was strongly suppressed by micromolar $[Ca^{2+}]_i$, indicating a self-regulating feedback mechanism appropriate for triggering Ca^{2+}-induced intracellular Ca^{2+} release in ABA. Intriguingly, I_{Ca} was promoted by ABA even in excised membrane patches, an observation that may relate to subsequent observations that both protein phosphorylation (Köhler and Blatt 2002) and NADPH oxidases (Kwak et al. 2003) affect Ca^{2+} channel activities in guard cells.

By contrast, intracellular Ca^{2+} release in guard cells appears to proceed via at least three different pathways, including stores that are subject to activation by inositol-1,4,5-trisphosphate (Blatt et al. 1990; Gilroy et al. 1990; Parmar and Brearley 1995; Lee et al. 1996), by inositol hexakisphosphate (Lemtiri-Chlieh et al. 2000), and by cyclic ADP-ribose (cADPR), a metabolite of nicotinamide adenine dinucleotide (Leckie et al. 1998; Grabov and Blatt 1998). Of these, the cADPR-activated (ryanodine-sensitive) Ca^{2+}-release channels are evidently the most important for ABA signalling. Leckie, et al. (1998) provided evidence for their presence in guard cell vacuoles. They observed $[Ca^{2+}]_i$ to rise following injections of guard cells with cADPR, and they noted that stomatal closure in ABA was slowed when guard cells were preloaded with the cADPR antagonist 8-NH_2-cADPR. Grabov and Blatt (1998; 1999) and Hamilton, et al. (2000) found that the ABA-evoked $[Ca^{2+}]_i$ signals were sensitive to ryanodine and that ABA strongly influenced the kinetics for the $[Ca^{2+}]_i$ rise and its recovery, as well as the voltage threshold triggering I_{Ca} and the $[Ca^{2+}]_i$ rise. These results indicated that ABA must act both on Ca^{2+} entry across the plasma membrane and on its release from cADPR-dependent Ca^{2+} stores.

The discoveries that NO could promote drought tolerance and that NO scavengers suppressed stomatal closure (Garcia-Mata and Lamattina 2001, 2002; Neill et al. 2002) yielded another piece of the puzzle of $[Ca^{2+}]_i$-mediated signalling in ABA. Garcia-Mata, et al. (2003) added essential molecular detail, showing that a low nanomolar level (< 10 nM) of NO was essential for normal inactivation of $I_{K,in}$ and activation of I_{Cl} by ABA, and that these events depended on NO-mediated elevation of $[Ca^{2+}]_i$. The $[Ca^{2+}]_i$ rise evoked by NO was sensitive to the antagonist of guanylate cyclase ODQ (1-*H*-(1,2,4)-oxadiazole-[4,3-*a*]quinolxalin-1-one), and to ryanodine, consistent with a signal cascade analogous to the canonical pathways mediated via cGMP, cADPR and ryanodine-sensitive Ca^{2+} channels in animals. Furthermore, Garcia-Mata, et al. could clearly separate ABA actions on I_{Ca} at the plasma membrane and on intracellular Ca^{2+} release through NO: whereas the enhancement of $[Ca^{2+}]_i$ by NO, like that of ABA, was evident when the membrane was clamped to negative voltages, NO had no (or even a moderate antagonistic) effect on the voltage threshold triggering the I_{Ca} and the $[Ca^{2+}]_i$ rise, unlike ABA (Grabov and Blatt 1998; Hamilton et al. 2000; Köhler and Blatt 2002). At these levels, NO action downstream was restricted to the Ca^{2+}-dependent currents and was without effect on the Ca^{2+}-independent K^+ channels and $I_{K,out}$. In short, these studies map out two different signalling pathways evoked by ABA that converge on the $[Ca^{2+}]_i$ signal intermediate, one acting at the plasma membrane on I_{Ca}, the second mediated independently via NO and affecting intracellular Ca^{2+} release.

Subsequent studies demonstrated that NO-mediated Ca^{2+} release is modulated by protein phosphorylation. Sokolovski, et al. (2005) reported that the $[Ca^{2+}]_i$ elevation in NO could be suppressed by protein kinase 1/2A antagonists staurosporine and K252A, leading to a loss of response to ABA of the Ca^{2+}-dependent currents $I_{K,in}$ and I_{Cl}. By contrast, these protein kinase antagonists did not influence ABA action on current through $I_{K,out}$, which is Ca^{2+}-insensitive but pH_i-dependent (see Sect. 2). The juxtaposition of these observations and evidence for cGMP as a downstream component in the NO-induced $[Ca^{2+}]_i$ rise does suggest an additional role for the cGMP-dependent protein kinase G, but no substantive evidence is yet forthcoming. Nonetheless, Sokolovski, et al. (2005) were able to rule out the action of protein tyrosine kinases, given that the NO-dependent rise in $[Ca^{2+}]_i$ and activities of $I_{K,in}$ and I_{Cl} were insensitive to the protein tyrosine kinase antagonist genestein.

These findings add substance to the "$[Ca^{2+}]_i$ cassette" of signalling and response elements in the guard cell (Blatt 2000a,b). In ABA, stomatal closure follows on solute loss from the guard cells and depends on an alteration in the time-averaged balance of ionic currents to favour the efflux of K^+ and Cl^-. As one means to achieving this end, ABA affects two separate pathways that combine to promote a rise in $[Ca^{2+}]_i$. One of these pathways is closely associated with Ca^{2+} channels at the plasma membrane and leads to a positive shift in the voltage-sensitivity of the channels, thereby favouring Ca^{2+} channel activation (Hamilton et al. 2000). The second pathway leads through NO, guanylate cyclase and cADPR cyclase to ryanodine-sensitive Ca^{2+} channels at one or more endomembranes. In turn, $[Ca^{2+}]_i$ promotes the exchange of inward current between $I_{K,in}$ and I_{Cl}, in effect suppressing K^+ influx while favouring Cl^- efflux and membrane depolarisation.

Finally, we must not forget the added dimension to this "$[Ca^{2+}]_i$ cassette" associated with the membrane voltage (Blatt 2000a). As noted previously, all three currents – I_{Ca}, $I_{K,in}$ and I_{Cl} – integrate across a common membrane. Thus, not only $[Ca^{2+}]_i$ elevation, but also activation of I_{Cl} and the resulting membrane depolarisation, serve to shut down Ca^{2+} entry at the plasma membrane. This self-limiting effect may be seen to ensure the guard cell is able to re-sequester the free Ca^{2+} released to the cytosol. It also implies an oscillation in $[Ca^{2+}]_i$ and, as a consequence, in the activities of both $I_{K,in}$ and I_{Cl}. In fact, guard cells show two states of membrane voltage, one close to the K^+ equilibrium voltage (E_K), and the second largely K^+-insensitive and typified by voltages well-negative of E_K (Thiel et al. 1992; Gradmann et al. 1993). Oscillations between these states occur in response to stimuli, including ABA, with a time course similar to that of the oscillations in $[Ca^{2+}]_i$ (Thiel et al. 1992; Blatt and Thiel 1994; Blatt and Armstrong 1993). These voltage oscillations have since been shown, in turn, to drive $[Ca^{2+}]_i$ oscillations (Grabov and Blatt 1998). This interplay between membrane voltage and $[Ca^{2+}]_i$ oscillations has been suggested to function in "tuning" the $[Ca^{2+}]_i$-dependent currents, $I_{K,in}$ and I_{Cl}, to the prevailing transport status of the guard cell (Blatt 2000a; Grabov and Blatt 1998) and it provides an obvious starting point for a "systems biology" approach to predictive modelling of entrainment behaviours such as the "closure memory" of guard cells (Allen et al. 2001).

6 S-Nitrosylation as an Alternative NO Signalling Pathway

NO can bond covalently with cysteine sulfhydryl residues to form *S*-nitrosothiols. This reaction, like that of protein phosphorylation, is able to affect protein conformation, substrate and protein–protein interactions that are elements of many regulatory cascades (Hess et al. 2001; Stamler et al. 1997, 2001; Ahern et al. 2002). S-Nitrosylation can also interact in Ca^{2+} signal cascades (Lai et al. 2001). In skeletal muscle, it plays a couterpart role to the cGMP- and cADPR-mediated pathway in modifying the efficacy of ryanodine-sensitive Ca^{2+} channels that facilitate intracellular Ca^{2+} release (Sun et al. 2003; Eu et al. 2000). In neurons it is known to activate cyclic-nucleotide (cGMP)-dependent cation channels (Broillet 2000) and to suppress persistent Na^+ channel current (Renganathan et al. 2002). The cellular mechanics of *S*-nitrosothiol homeostasis is still poorly understood, but it is reversible and is likely to engage glutathione transfer reactions, peroxidases (Mannick and Schonhoff 2004; Abu-Soud and Hazen 2000; Liu et al. 2004) and, in plants, also haemoglobins (Perazzolli et al. 2006).

This second pathway for NO-mediated signalling has attracted much less attention in plant research to date. Nonetheless, S-nitrosylation is likely to serve analogous functions, not only on the short-term in physiological stress responses (Liu et al. 2004) but also in control of gene expression (Bogdan 2001; Marshall et al. 2000) and apoptosis (Melino et al. 1997). There is good evidence for effects of S-nitrosylation on ion channels in vivo, at least in guard cells. Work from this laboratory (Sokolovski and Blatt 2004) has shown that NO suppresses guard cell K^+ channels that mediate $I_{K,out}$. The NO levels required for this action are only marginally higher than those required to potentiate $[Ca^{2+}]_i$ increases and inactivate $I_{K,in}$, but the mechanism is entirely different since $I_{K,out}$ is demonstrably insensitive to $[Ca^{2+}]_i$. Analysis of $I_{K,out}$ kinetics and voltage dependence showed that NO acts in a cooperative manner to remove channels from the functional pool in the membrane, as NO bound with an apparent K_i of 20 nM. Studies with redox reagents indicated that the process was facilitated by the presence of vicinal cysteines accessible for reaction inside the guard cell. Significantly, NO action on $I_{K,out}$ was reversibly blocked by reducing reagents, especially the British anti-Lewisite reagent BAL (2,3-dimercapto-1-propanol), and was mimicked by the oxidising reagent phenylarsine oxide that targets vicinal SH residues. These findings indicate that the $I_{K,out}$ is profoundly sensitive to nitrosative stress and, on the assumption that the K^+ channels are comprised of four homologous subunits (Dreyer et al. 2004b), are consistent with inactivation effected by NO binding to at least two of these subunits. They also suggest that the K^+ channels can be "locked down", an observation underscored by work with H_2O_2 (Kohler et al. 2003) and evidence of stomatal sensitivity to oxidative stress (Willmer and Fricker 1996; Black and Black 1979).

An important question remains: what is the physiological function of the NO sensitivity of $I_{K,out}$? At present, we can offer two conjectures. First, on the assumption that NO action is not specifically targeted to $I_{K,out}$, its inactivation may represent a broad-range response to oxidative signals associated with plant responses, for example during pathogen attack (Khurana et al. 2005; Feechan et al. 2005; Klusener et al. 2002). Indeed, the K^+ channels are known to be sensitive to concentrations of H_2O_2 only marginally higher than the steady-state levels of NO required for inactivation (Kohler et al. 2003).

Second, on the assumption that NO action is targeted to $I_{K,out}$, its inactivation may reflect an imbalance between nitrosylation and denitrosylation (Liu et al. 2001), superimposed by the experimental addition of NO. This explanation would imply that the K^+ channels are finely tuned to the balance of S-nitrosylation in vivo.

7
Conclusions

NO plays an important role as one element of a"$[Ca^{2+}]_i$ cassette" that mediates ABA signalling in guard cells. Although some components are still a matter of conjecture – and targets for experimentation – the different transport mechanisms occurring at the plasma membrane (pumps, channels and carriers, or simply diffusion through the bilayer) are all known and at a sufficient depth to provide detailed and accurate flux equations with parameters fully constrained by experimental data. Many of the binding and buffering interactions between solutes are understood in quantitative detail. Furthermore, the physico-chemical relationships that constrain the behaviour of homeostatic variables (e.g. conservation of mass; conservation of charge; electroneutrality of intracellular and extracellular solutions; osmotic transport of water linked to all solute concentrations and their gradients; membrane potential linked to all ion gradients and permeabilities) are well-understood and are simple quantitative relationships easily contained mathematically. Thus, the guard cell is now prime target for a "systems biology" approach to quantitative predictive modelling, especially in relation to signalling and control of ion channels by $[Ca^{2+}]_i$. There are also significant gaps in our understanding of NO signalling, especially in relation to the molecular mechanisms behind S-nitrosylation and its molecular targets. Nonetheless, the tools for such work are available, and it can be expected therefore that the next 5 years will see a quantum leap in our knowledge of these events.

References

Abu-Soud HM, Hazen SL (2000) Nitric oxide is a physiological substrate for mammalian peroxidases. J Biol Chem 275:37524–37532

Ahern GP, Klyachko VA, Jackson MB (2002) cGMP and S-nitrosylation: two routes for modulation of neuronal excitability by NO. Trends Neurosci 25:510–517

Allan AC, Fricker MD, Ward JL, Beale MH, Trewavas AJ (1994) Two transduction pathways mediate rapid effects of abscisic acid in *Commelina* guard cells. Plant Cell 6:1319–1328

Allen GJ, Chu SP, Harrington CL, Schumacher K, Hoffman T, Tang YY, Grill E, Schroeder JI (2001) A defined range of guard cell calcium oscillation parameters encodes stomatal movements. Nature 411:1053–1057

Allen GJ, Sanders D (1996) Control of ionic currents in guard cell vacuoles by cytosolic and luminal calcium. Plant J 10:1055–1069

Armstrong F, Leung J, Grabov A, Brearley J, Giraudat J, Blatt MR (1995) Sensitivity to abscisic acid of guard cell K^+ channels is suppressed by *abi1-1*, a mutant *Arabidopsis* gene encoding a putative protein phosphatase. Proc Natl Acad Sci USA 92:9520–9524

Ashley MK, Grant M, Grabov A (2006) Plant responses to potassium deficiencies: a role for potassium transport proteins. J Exp Bot 57:425–436

Assmann SM (2003) OPEN STOMATA1 opens the door to ABA signaling in Arabidopsis guard cells. Trends Plant Sci 8:151–153

Becker D, Zeilinger C, Lohse G, Depta H, Hedrich R (1993) Identification and biochemical characterization of the plasma membrane H^+-ATPase in guard cells of *Vicia faba* L. Planta 190:44–50

Berridge MJ (1998) Neuronal calcium signaling. Neuron 21:13–26

Bhatia CR, Sybenga J (1965) Effect of nitric oxide on X-ray sensitivity of *Crotalaria intermedia* seeds. Mutat Res 2:332–338

Bihler H, Eing C, Hebeisen S, Roller A, Czempinski K, Bertl A (2005) TPK1 is a vacuolar ion channel different from the slow-vacuolar cation channel. Plant Physiol 139:417–424

Black CR, Black VJ (1979) The effects of low concentrations of sulphur dioxide on stomatal conductance and epidermal cell survival in field bean (*Vicia faba* L). J Exp Bot 30:291–298

Blatt MR (1987) Electrical characteristics of stomatal guard cells: the contribution of ATP-dependent, electrogenic transport revealed by current-voltage and difference-current-voltage analysis. J Membr Biol 98:257–274

Blatt MR (1988) Potassium-dependent bipolar gating of potassium channels in guard cells. J Membr Biol 102:235–246

Blatt MR (1992) K^+ channels of stomatal guard cells: characteristics of the inward rectifier and its control by pH. J Gen Physiol 99:615–644

Blatt MR (2000a) Ca^{2+} signaling and control of guard-cell volume in stomatal movements. Curr Opin Plant Biol 3:196–204

Blatt MR (2000b) Cellular signaling and volume control in stomatal movements in plants. Ann Rev Cell Dev Biol 16:221–241

Blatt MR (2004) Concepts and techniques in plant membrane physiology. In: Blatt MR (ed) Membrane transport in plants. Blackwell, Oxford, pp 1–39

Blatt MR, Armstrong F (1993) K^+ channels of stomatal guard cells: abscisic acid-evoked control of the outward rectifier mediated by cytoplasmic pH. Planta 191:330–341

Blatt MR, Gradmann D (1997) K^+-sensitive gating of the K^+ outward rectifier in *Vicia* guard cells. J Membr Biol 158:241–256

Blatt MR, Thiel G (1994) K^+ channels of stomatal guard cells: bimodal control of the K^+ inward-rectifier evoked by auxin. Plant J 5:55–68

Blatt MR, Thiel G, Trentham DR (1990) Reversible inactivation of K^+ channels of *Vicia* stomatal guard cells following the photolysis of caged inositol 1,4,5-trisphosphate. Nature 346:766–769

Bogdan C (2001) Nitric oxide and the regulation of gene expression. Trends Cell Biol 11:66–75

Broillet MC (2000) A single intracellular cysteine residue is responsible for the activation of the olfactory cyclic nucleotide-gated channel by NO. J Biol Chem 275:15135–15141

Brune B, von Knethen A, Sandau KB (1998) Nitric oxide and its role in apoptosis. Eur J Pharmacol 351:261–272

Cheng SH, Willmann MR, Chen HC, Sheen J (2002) Calcium signaling through protein kinases. The *Arabidopsis* calcium-dependent protein kinase gene family. Plant Physiol 129:469–485

Clapham DE (2003) TRP channels as cellular sensors. Nature 426:517–524

Clementi E, Riccio M, Sciorati C, Nistico G, Meldolesi J (1996) The type 2 ryanodine receptor of neurosecretory PC12 cells is activated by cyclic ADP-ribose – Role of the nitric oxide cGMP pathway. J Biol Chem 271:17739–17745

Clint GM, Blatt MR (1989) Mechanisms of fusicoccin action: evidence for concerted modulations of secondary K^+ transport in a higher-plant cell. Planta 178:495–508

Coursol S, Le Stunff H, Lynch DV, Gilroy S, Assmann SM, Spiegel S (2005) *Arabidonsis* sphingosine kinase and the effects of phytosphingosine-1-phosphate on stomatal aperture. Plant Physiol 137:724–737

Davies WJ, Jones HG (eds)(1991) Abscisic acid: physiology and biochemistry. Bios Scientific, Oxford

Delledonne M, Xia YJ, Dixon RA, Lamb C (1998) Nitric oxide functions as a signal in plant disease resistance. Nature 394:585–588

Demple B (2002) Signal transduction by nitric oxide in cellular stress responses. Mol Cell Biochem 234:11–18

Desikan R, Hancock JT, Bright J, Harrison J, Weir L, Hooley R, Neill SJ (2005) A role for ETR1 in hydrogen peroxide signaling in stomatal guard cells. Plant Physiol 137:831–834

Dietrich P, Hedrich R (1994) Interconversion of fast and slow gating modes of GCAC1, a guard cell anion channel. Planta 195:301–304

Dreyer I, MullerRober B, Kohler B (2004a) Voltage-gated ion channels. In: Blatt MR (ed) Membrane transport in plants. Blackwell, Oxford, pp 150–192

Dreyer I, Poree F, Schneider A, Mittelstadt J, Bertl A, Sentenac H, Thibaud JB, Mueller-Roeber B (2004b) Assembly of plant Shaker-like K-out channels requires two distinct sites of the channel alpha-subunit. Biophys J 87:858–872

Eu JP, Sun JH, Xu L, Stamler JS, Meissner G (2000) The skeletal muscle calcium release channel: Coupled O-2 sensor and NO signaling functions. Cell 102:499–509

Feechan A, Kwon E, Yun BW, Wang YQ, Pallas JA, Loake GJ (2005) A central role for S-nitrosothiols in plant disease resistance. Proc Natl Acad Sci USA 102:8054–8059

Foissner I, Wendehenne D, Langebartels C, Durner J (2000) In vivo imaging of an elicitor-induced nitric oxide burst in tobacco. Plant J 23:817–824

Fricker MD, Gilroy S, Read ND, Trewavas AJ (1991) Visualisation and measurement of the calcium message in guard cells. In: Schuch W, Jenkins G (eds) Molecular biology of plant development. Cambridge University Press, Cambridge, pp 177–190

Frohnmeyer H, Grabov A, Blatt MR (1998) A role for the vacuole in auxin-mediated control of cytosolic pH by *Vicia* mesophyll and guard cells. Plant J 13:109–116

Garcia-Mata C, Gay R, Sokolovski S, Hills A, Lamattina L, Blatt MR (2003) Nitric oxide regulates K^+ and Cl^- channels in guard cells through a subset of abscisic acid-evoked signaling pathways. Proc Natl Acad Sci USA 100:11116–11121

Garcia-Mata C, Lamattina L (2001) Nitric oxide induces stomatal closure and enhances the adaptive plant responses against drought stress. Plant Physiol 126:1196–1204

Garcia-Mata C, Lamattina L (2002) Nitric oxide and abscisic acid cross talk in guard cells. Plant Physiol 128:790–792

Gaymard F, Pilot G, Lacombe B, Bouchez D, Bruneau D, Boucherez J, MichauxFerriere N, Thibaud JB, Sentenac H (1998) Identification and disruption of a plant shaker-like outward channel involved in K^+ release into the xylem sap. Cell 94:647–655

Gilroy S, Read ND, Trewavas AJ (1990) Elevation of cytoplasmic calcium by caged calcium or caged inositol trisphosphate initiates stomatal closure. Nature 346:769–771

Grabov A, Blatt MR (1997) Parallel control of the inward-rectifier K^+ channel by cytosolic-free Ca^{2+} and pH in *Vicia* guard cells. Planta 201:84–95

Grabov A, Blatt MR (1998) Membrane voltage initiates Ca^{2+} waves and potentiates Ca^{2+} increases with abscisic acid in stomatal guard cells. Proc Natl Acad Sci USA 95:4778–4783

Grabov A, Blatt MR (1999) A steep dependence of inward-rectifying potassium channels on cytosolic free calcium concentration increase evoked by hyperpolarization in guard cells. Plant Physiol 119:277–287

Grabov A, Leung J, Giraudat J, Blatt MR (1997) Alteration of anion channel kinetics in wild-type and *abi1-1* transgenic *Nicotiana benthamiana* guard cells by abscisic acid. Plant J 12:203–213

Gradmann D, Blatt MR, Thiel G (1993) Electrocoupling of ion transporters in plants. J Membr Biol 136:327–332

Hamilton DWA, Hills A, Blatt MR (2001) Extracellular Ba^{2+} and voltage interact to gate Ca^{2+} channels at the plasma membrane of stomatal guard cells. FEBS Lett 491:99–103

Hamilton DWA, Hills A, Kohler B, Blatt MR (2000) Ca^{2+} channels at the plasma membrane of stomatal guard cells are activated by hyperpolarization and abscisic acid. Proc Natl Acad Sci USA 97:4967–4972

Hedrich R, Busch H, Raschke K (1990) Ca^{2+} and nucleotide dependent regulation of voltage dependent anion channels in the plasma membrane of guard cells. EMBO J 9:3889–3892

Hentzen AE, Smart LB, Wimmers LE, Fang HH, Schroeder JI, Bennett AB (1996) Two plasma-membrane H^+-ATPase genes expressed in guard-cells of *Vicia faba* are also expressed throughout the plant. Plant Cell Physiol 37:650–659

Hertel B, Horvath F, Wodala B, Hurst A, Moroni A, Thiel G (2005) KAT1 inactivates at sub-millimolar concentrations of external potassium. J Exp Bot 56:3103–3110

Hess DT, Matsumoto A, Nudelman R, Stamler JS (2001) S-Nitrosylation: spectrum and specificity. Nat Cell Biol 3:E46-E49

Hetherington AM (2001) Guard cell signaling. Cell 107:711–714

Hosy E, Vavasseur A, Mouline K, Dreyer I, Gaymard F, Poree F, Boucherez J, Lebaudy A, Bouchez D, Very A-A, Simonneau T, Thibaud J-P, Sentenac H (2003) The *Arabidopsis* outward K^+ channel GORK is involved in regulation of stomatal movements and plant transpiration. Proc Nat Acad Sci USA 100:5549–5554

Hu XY, Neill SJ, Tang ZC, Cai WM (2005) Nitric oxide mediates gravitropic bending in soybean roots. Plant Physiol 137:663–670

Hunt L, Mills LN, Pical C, Leckie CP, Aitken FL, Kopka J, Mueller-Roeber B, McAinsh MR, Hetherington AM, Gray JE (2003) Phospholipase C is required for the control of stomatal aperture by ABA. Plant J 34:47–55

Irving HR, Gehring CA, Parish RW (1992) Changes in cytosolic pH and calcium of guard cells precede stomatal movements. Proc Natl Acad Sci USA 89:1790–1794

Johansson I, Wulfetange K, Poree F, Michard E, Gajdanowicz P, Lacombe B, Sentenac H, Thibaud JB, Mueller-Roeber B, Blatt MR, Dreyer I (2006) External K^+ modulates the activity of the *Arabidopsis* potassium channel SKOR via an unusual mechanism. Plant J 46:269–281

Khurana SMP, Pandey SK, Sarkar D, Chanemougasoundharam A (2005) Apoptosis in plant disease response: a close encounter of the pathogen kind. Curr Sci 88:740–752

Kieber JJ, Ecker JR (1993) Ethylene gas – its not just for ripening any more. Trends Genet 9:356–362

Klusener B, Young JJ, Murata Y, Allen GJ, Mori IC, Hugouvieux V, Schroeder JI (2002) Convergence of calcium signaling pathways of pathogenic elicitors and abscisic acid in Arabidopsis guard cells. Plant Physiol 130:2152–2163

Köhler B, Blatt MR (2002) Protein phosphorylation activates the guard cell Ca^{2+} channel and is a prerequisite for gating by abscisic acid. Plant J 32:185–194

Kohler B, Hills A, Blatt MR (2003) Control of guard cell ion channels by hydrogen peroxide and abscisic acid indicates their action through alternate signaling pathways. Plant Physiol 131:385–388

Kwak JM, Moon JH, Murata Y, Kuchitsu K, Leonhardt N, Delong A, Schroeder JI (2002) Disruption of a guard cell-expressed protein phosphatase 2A regulatory subunit, RCN1, confers abscisic acid insensitivity in *Arabidopsis*. Plant Cell 14:2849–2861

Kwak JM, Mori IC, Pei ZM, Leonhardt N, Torres MA, Dangl JL, Bloom RE, Bodde S, Jones JDG, Schroeder JI (2003) NADPH oxidase AtrbohD and AtrbohF genes function in ROS-dependent ABA signaling in *Arabidopsis*. EMBO J 22:2623–2633

Lai TS, Hausladen A, Slaughter TF, Eu JP, Stamler JS, Greenberg CS (2001) Calcium regulates S-nitrosylation, denitrosylation, and activity of tissue transglutaminase. Biochemistry 40:4904–4910

Leckie CP, McAinsh MR, Allen GJ, Sanders D, Hetherington AM (1998) Abscisic acid-induced stomatal closure mediated by cyclic ADP-ribose. Proc Natl Acad Sci USA 95:15837–15842

Lee HJ, Tucker EB, Crain RC, Lee Y (1993) Stomatal opening is induced in epidermal peels of *Commelina communis* L. by GTP analogs or pertussis toxin. Plant Physiol 102:95–100

Lee YS, Choi YB, Suh S, Lee J, Assmann SM, Joe CO, Kelleher JF, Crain RC (1996) Abscisic acid-induced phosphoinositide turnover in guard-cell protoplasts of *Vicia faba*. Plant Physiol 110:987–996

Lemtiri-Chlieh F, MacRobbie EAC (1994) Role of calcium in the modulation of *Vicia* guard cell potassium channels by abscisic acid: a patch-clamp study. J Membr Biol 137:99–107

Lemtiri-Chlieh F, MacRobbie EAC, Brearley CA (2000) Inositol hexakisphosphate is a physiological signal regulating the K^+-inward rectifying conductance in guard cells. Proc Natl Acad Sci USA 97:8687–8692

Li JX, Lee YRJ, Assmann SM (1998) Guard cells possess a calcium-dependent protein kinase that phosphorylates the KAT1 potassium channel. Plant Physiol 116:785–795

Li JX, Wang XQ, Watson MB, Assmann SM (2000) Regulation of abscisic acid-induced stomatal closure and anion channels by guard cell AAPK kinase. Science 287:300–303

Li WH, Schultz C, Llopis J, Tsien RY (1997) Membrane-permeant esters of inositol polyphosphates, chemical syntheses and biological applications. Tetrahedron 53:12017–12040

Linder B, Raschke K (1992) A slow anion channel in guard cells, activating at large hyperpolarization, may be principal for stomatal closing. FEBS Lett 313:27–30

Liu L, Yan Y, Zeng M, Zhang J, Hanes MA, Ahearn G, McMahon TJ, Dickfeld T, Marshall HE, Que LG, Stamler JS (2004) Essential roles of S-nitrosothiols in vascular horneostasis and endotoxic shock. Cell 116:617–628

Liu LM, Hausladen A, Zeng M, Que L, Heitman J, Stamler JS (2001) A metabolic enzyme for S-nitrosothiol conserved from bacteria to humans. Nature 410:490–494

Lohse G, Hedrich R (1992) Characterization of the plasma-membrane H^+-ATPase from *Vicia faba* guard cells. Planta 188:206–214

Lukyanenko V, Gyorke I, Wiesner TF, Gyorke S (2001) Potentiation of Ca^{2+} release by cADP-ribose in the heart is mediated by enhanced SR Ca^{2+} uptake into the sarcoplasmic reticulum. Circ Res 89:614–622

MacRobbie EAC (1995) ABA-induced ion efflux in stomatal guard-cells - multiple actions of ABA inside and outside the cell. Plant J 7:565–576

MacRobbie EAC (1999) Vesicle trafficking: a role in trans-tonoplast ion movements? J Exp Bot 50:925–934

Mannick JB, Schonhoff CM (2004) NO means no and yes: regulation of cell signaling by protein nitrosylation. Free Radic Res 38:1–7

Marshall HE, Merchant K, Stamler JS (2000) Nitrosation and oxidation in the regulation of gene expression. Faseb J 14:1889–1900

McAinsh MR, Brownlee C, Hetherington AM (1990) Abscisic acid-induced elevation of guard cell cytosolic Ca^{2+} precedes stomatal closure. Nature 343:186–188

McAinsh MR, Brownlee C, Hetherington AM (1992) Visualizing changes in cytosolic-free Ca^{2+} during the response of stomatal guard cells to abscisic acid. Plant Cell 4:1113–1122

Melino G, Bernassola F, Knight RA, Corasaniti MT, Nistico G, FinazziAgro A (1997) S-Nitrosylation regulates apoptosis. Nature 388:432–433

Miedema H, Assmann SM (1996) A membrane-delimited effect of internal pH on the K^+ outward rectifier of *Vicia faba* guard cells. J Membr Biol 154:227–237

Moreno K, de Miera EVS, Nadal MS, Amarillo Y, Rudy B (2001) Modulation of Kv3 potassium channels expressed in CHO cells by a nitric oxide-activated phosphatase. J Physiol 530:345–358

Mori IC, Muto S (1997) Abscisic acid activates a 48-kilodalton protein kinase in guard cell protoplasts. Plant Physiol 113:833–839

Mustilli AC, Merlot S, Vavasseur A, Fenzi F, Giraudat J (2002) *Arabidopsis* OST1 protein kinase mediates the regulation of stomatal aperture by abscisic acid and acts upstream of reactive oxygen species production. Plant Cell 14:3089–3099

Nakamura RL, Mckendree WL, Hirsch RE, Sedbrook JC, Gaber RF, Sussman MR (1995) Expression of an *Arabidopsis* potassium channel gene in guard cells. Plant Physiol 109:371–374

Neill SJ, Desikan R, Clarke A, Hancock JT (2002) Nitric oxide is a novel component of abscisic acid signaling in stomatal guard cells. Plant Physiol 128:13–16

Ng CKY, Carr K, McAinsh MR, Powell B, Hetherington AM (2001) Drought-induced guard cell signal transduction involves sphingosine-1-phosphate. Nature 410:596–599

Parmar PN, Brearley CA (1995) Metabolism of 3-phosphorylated and 4-phosphorylated phosphatidylinositols in stomatal guard cells of *Commelina communis* l. Plant J 8:425–433

Pei ZM, Kuchitsu K, Ward JM, Schwarz M, Schroeder JI (1997) Differential abscisic acid regulation of guard cell slow anion channels in *Arabidopsis* wild-type and *abi1* and *abi2* mutants. Plant Cell 9:409–423

Pei ZM, Ward JM, Harper JF, Schroeder JI (1996) A novel chloride channel in *Vicia faba* guard cell vacuoles activated by the serine/threonine kinase, CDPK. EMBO J 15:6564–6574

Peiter E, Maathuis FJM, Mills LN, Knight H, Pelloux M, Hetherington AM, Sanders D (2005) The vacuolar Ca^{2+}-activated channel TPC1 regulates germination and stomatal movement. Nature 434:404–408

Perazzolli M, Romero-Puertas MC, Delledonne M (2006) Modulation of nitric oxide bioactivity by plant haemoglobins. J Exp Bot 57:479–488

Pilot G, Lacombe B, Gaymard F, Cherel I, Boucherez J, Thibaud JB, Sentenac H (2001) Guard cell inward K^+ channel activity in *Arabidopsis* involves expression of the twin channel subunits KAT1 and KAT2. J Biol Chem 276:3215–3221

Renganathan M, Cummins TR, Waxman SG (2002) Nitric oxide blocks fast, slow, and persistent Na^+ channels in C-type DRG neurons by S-nitrosylation. J Neurophysiol 87:761–775

Roelfsema MRG, Levchenko V, Hedrich R (2004) ABA depolarizes guard cells in intact plants, through a transient activation of R- and S-type anion channels. Plant J 37:578–588

Roelfsema MRG, Prins HBA (1997) Ion channels in guard cells of *Arabidopsis thaliana* (L.) Heynh. Planta 202:18–27

Romano LA, Jacob T, Gilroy S, Assmann SM (2000) Increases in cytosolic Ca^{2+} are not required for abscisic acid-inhibition of inward K^+ currents in guard cells of *Vicia faba* L. Planta 211:209–217

Sanders D, Pelloux J, Brownlee C, Harper JF (2002) Calcium at the crossroads of signaling. Plant Cell 14:S401-S417

Schroeder JI (1988) K^+ transport properties of K^+ channels in the plasma membrane of *Vicia faba* guard cells. J Gen Physiol 92:667–683

Schroeder JI, Allen GJ, Hugouvieux V, Kwak JM, Waner D (2001) Guard cell signal transduction. Ann Rev Plant Physiol Mol Biol 52:627–658

Schroeder JI, Keller BU (1992) Two types of anion channel currents in guard cells with distinct voltage regulation. Proc Natl Acad Sci USA 89:5025–5029

Schulzlessdorf B, Hedrich R (1995) Protons and calcium modulate SV-type channels in the vacuolar lysosomal compartment – channel interaction with calmodulin inhibitors. Planta 197:655–671

Shapiro AD (2005) Nitric oxide signaling in plants. Plant Hormones 72:339–398

Sokolovski S, Blatt MR (2004) Nitric oxide block of outward-rectifying K^+ channels indicates direct control by protein nitrosylation in guard cells. Plant Physiol 136:4275–4284

Sokolovski S, Hills A, Gay R, Garcia-Mata C, Lamattina L, Blatt MR (2005) Protein phosphorylation is a prerequisite for intracellular Ca^{2+} release and ion channel control by nitric oxide and abscisic acid in guard cells. Plant J 43:520–529

Stamler JS, Lamas S, Fang FC (2001) Nitrosylation: The prototypic redox-based signaling mechanism. Cell 106:675–683

Stamler JS, Meissner G (2001) Physiology of nitric oxide in skeletal muscle. Physiol Rev 81:209–237

Stamler JS, Toone EJ, Lipton SA, Sucher NJ (1997) (S)NO signals: translocation, regulation, and a consensus motif. Neuron 18:691–696

Sun JH, Xu L, Eu JP, Stamler JS, Meissner G (2003) Nitric oxide, NOC-12, and S-nitrosoglutathione modulate the skeletal muscle calcium release channel/ryanodine receptor by different mechanisms – an allosteric function for O-2 in S-nitrosylation of the channel. J Biol Chem 278:8184–8189

Thiel G, MacRobbie EAC, Blatt MR (1992) Membrane transport in stomatal guard cells: the importance of voltage control. J Membr Biol 126:1–18

Thomine S, Zimmermann S, Guern J, Barbierbrygoo H (1995) ATP-dependent regulation of an anion channel at the plasma membrane of protoplasts from epidermal cells of *Arabidopsis* hypocotyls. Plant Cell 7:2091–2100

Tikhonova LI, Pottosin II, Dietz KJ, Schonknecht G (1997) Fast-activating cation channel in barley mesophyll vacuoles: inhibition by calcium. Plant J 11:1059–1070

Very AA, Sentenac H (2003) Molecular mechanisms and regulation of K^+ transport in higher plants. Annu Rev Plant Biol 54:575–603

Villalba JM, Lutzelschwab M, Serrano R (1991) Immunocytolocalization of plasma membrane H^+-ATPase in maize coleoptiles and enclosed leaves. Planta 185:458–461

Wang XQ, Ullah H, Jones AM, Assmann SM (2001) G protein regulation of ion channels and abscisic acid signaling in *Arabidopsis* guard cells. Science 292:2070–2072

Ward JM, Schroeder JI (1994) Calcium-activated K^+ channels and calcium-induced calcium release by slow vacuolar ion channels in guard-cell vacuoles implicated in the control of stomatal closure. Plant Cell 6:669–683

White PJ, Biskup B, Elzenga JTM, Homann U, Thiel G, Wissing F, Maathuis FJM (1999) Advanced patch-clamp techniques and single-channel analysis. J Exp Bot 50:1037–1054

Wille A, Lucas W (1984) Ultrastructural and histochemical studies on guard cells. Planta 160:129–142

Willmer C, Fricker MD (1996) Stomata. Chapman and Hall, London, pp 1–375

Willmott N, Sethi JK, Walseth TF, Lee HC, White AM, Galione A (1996) Nitric oxide-induced mobilization of intracellular calcium via the cyclic ADP-ribose signaling pathway. J Biol Chem 271:3699–3705

Yang SF, Hoffman NE (1984) Ethylene biosynthesis and its regulation in higher plants. Ann Rev Plant Physiol 35:155–182

Yuan XJ, Tod ML, Rubin LJ, Blaustein MP (1996) NO hyperpolarizes pulmonary artery smooth muscle cells and decreases the intracellular Ca^{2+} concentration by activating voltage-gated K^+ channels. Proc Natl Acad Sci USA 93:10489–10494

Zhang MX, An LZ, Feng HY, Chen T, Chen K, Liu YH, Tang HG, Chang JF, Wang XL (2003) The cascade mechanisms of nitric oxide as a second messenger of ultraviolet B in inhibiting mesocotyl elongations. Photochem Photobiol 77:219–225

Zufall F, Shepherd GM, Barnstable CJ (1997) Cyclic nucleotide gated channels as regulators of CNS development and plasticity. Curr Opin Neurobiol 7:404–412

Plant Cell Monogr (6)
L. Lamattina · J. C. Polacco: Nitric Oxide in Plant Growth
DOI 10.1007/7089_2006_090/Published online: 25 October 2006

Nitric Oxide in Nitrogen-Fixing Symbiosis

Emmanuel Baudouin · Nicolas Pauly · Alain Puppo (✉)

Interactions Plantes-Microorganismes et Santé Végétale,
UMR INRA 1064/Université de Nice-Sophia Antipolis/CNRS 6192, 400, route des Chappes, BP 167, 06903 Sophia Antipolis, France
puppo@unice.fr

Abstract The establishment and functioning of nitrogen-fixing symbioses between legumes and rhizobia rely on a succession of infectious, developmental, and metabolic processes, which end in the formation of root nodules and the acquisition of nitrogen-fixing capacity by the bacteria. A tight regulation is required for the establishment of a successful interaction, and the identification of the regulatory network operating is therefore a major challenge for a better understanding of symbiotic associations. It is now well established that nitric oxide (NO) is a major signaling molecule controlling not only plant interactions with pathogens, but also plant development. This review presents our current knowledge on the metabolism and possible roles of NO during symbiotic interactions. In particular, the involvement of NO in nitrogen-fixation regulation, O_2-deprivation sensing, and nodule development are discussed.

1 Introduction

Dinitrogen (N_2) constitutes about 80% of the earth's atmosphere, but the stability of the triple bond renders it inert and metabolically inaccessible to most organisms. Nevertheless, whereas eukaryotes cannot utilize N_2, some prokaryotes are able to catalyze the enzymatic reduction of this compound to ammonia. This reaction is catalyzed by a unique enzyme called nitrogenase and occurs at ambient temperatures and atmospheric pressure (Buchanan et al. 2000). In a certain number of species, N_2-fixation capacity is acquired upon a symbiotic interaction with plants. This is the case for a phylogenetically diverse group of Gram-negative bacteria (termed the rhizobia, which include bacteria of the genera *Azorhizobium*, *Bradyrhizobium*, *Mesorhizobium*, *Rhizobium*, and *Sinorhizobium*) that interact with legumes under nitrogen-limiting conditions (Perret et al. 2000). The symbiotic interaction leads to the formation of a new root organ, the so-called nodule, where bacteria differentiate into their symbiotic forms, the bacteroids, able to reduce atmospheric N_2 into ammonia (NH_4^+). Nodulation is a highly host-specific process in which, with few exceptions, specific rhizobial strains infect a limited range of plant hosts (Perret et al. 2000). Such specificity is achieved through a cross-recognition of particular signals produced by the two partners. Indeed, plants secrete (iso)flavonoids that are recognized by the compatible bacteria, resulting in the induction of nodulation genes responsible for the synthesis of nodulation factors (Nod factors).

Upon recognition by the plant roots, Nod factors provoke the profound morphological changes required for proper symbiosis (Gage 2004). The establishment of a nitrogen-fixing symbiosis involves an infection process in which the rhizobia penetrate root hairs via infection threads and trigger the formation of a nodule by reprogramming root cortical cell development. A successful interaction requires the strict coordination of these two processes. Depending on the plant species, two major types of nodules are formed. The indeterminate nodules maintain an apical meristem and are therefore elongated (e.g., pea, alfalfa, etc), as opposed to the determinate nodules, which do not have a persistent meristem and are rounder (e.g., soybean, bean, etc.).

In a mature nodule, the acquisition of N_2-fixing capacity, which characterizes the differentiation of bacteria into bacteroids, is correlated with the novo synthesis of nitrogenase. This enzyme is irreversibly inactivated by oxygen and requires microaerobic conditions to function. Firstly, the entry of oxygen is controlled by a variable-permeability barrier in the nodule parenchyma. Secondly, leghemoglobin, an oxygen-binding plant protein, plays an active role in regulating and delivering oxygen to the infected cells. It increases the flux of oxygen moving through the plant cytoplasm to the bacteroids while controlling the concentration of free oxygen (Buchanan et al. 2000). This allows the bacteroids to respire actively, therefore supporting the ATP synthesis necessary for the nitrogen-fixation process.

Thus, it appears that N_2-fixing symbiotic interactions involve both infection and developmental processes. On the other hand, such interactions are often considered as controlled incompatible pathogenic interactions (Rolfe and Gresshoff 1988), and symbiosis and pathogenesis appear to be variations of a common theme (Baron and Zambryski 1995). The involvement of nitric oxide (NO) in plant–pathogen interactions and developmental processes is well described elsewhere in this book (see chapters by Shapiro, De Stefano et al., and Salgado et al., in this volume). Moreover, there is compelling evidence that NO plays an important role in the symbiotic interaction (Pauly et al. 2006). The aim of this review is to present relevant data on NO metabolism and possible functions during symbiotic interactions and to open new fields for future investigations. It will principally focus on the symbiosis between legumes and rhizobia, which are to date the best documented among the relevant biological systems.

2
Metabolism of NO During Symbiotic Interactions

The most recent reports have illustrated that, as for any cell signal, the biological outcome of NO production is largely influenced by where, when, at which concentration, how long, and possibly through which mechanisms NO is evolved. The following paragraphs summarize our current knowledge about these questions in the framework of plant–symbiont interactions. A particular emphasis is put on how the symbiotic partners may modulate the levels of NO during the interaction.

2.1 Detection of NO Production

Several lines of evidence have demonstrated the occurrence of NO production during symbiotic interactions. Until recently, such evidence only relied on the detection of NO complexed to leghemoglobin, the major hemoprotein of legume nodules, using electron paramagnetic resonance (EPR). The presence of such nitrosylleghemoglobin (LbNO) complexes was reported in several legumes including cowpea (Maskall et al. 1977), soybean (Maskall et al. 1977; Mathieu et al. 1998), and pea (Kanayama and Yamamoto 1991) thus representing determinate and indeterminate nodule types. The abundance of LbNO was greatly enhanced in nitrate-treated plants (Kanayama and Yamamoto 1991), but significant levels were also detected in field-grown plants (Maskall et al. 1977) or in soybeans plants in nitrate-free conditions (Mathieu et al. 1998). Moreover, Mathieu et al. (1998) presented evidence that LbNO was accumulated mainly in young nodules but decreased with nodule age, which pointed out a relationship between LbNO abundance and nodule development and/or functioning. Based on the recent development of permeable NO-sensitive probes, the production of nitric oxide was re-examined in *Medicago truncatula* nodules formed upon infection with *Sinorhizobium meliloti* (Baudouin et al. 2006). Using the 4,5 diamino fluorescein diacetate (DAF2-DA) probe, which was reported as a specific and sensitive tool for NO detection in plants (Foissner et al. 2000), we could characterize the pattern of NO production in nodules. Nitric oxide detection was restricted to the fixing zone of the nodule and could be impaired with the NO scavenger 2-(4-carboxyphenyl)-4,4,5,5-tetramethyl imidazoline-1-oxyl-3-oxide (cPTIO). Moreover, NO detection was only achieved in a large part of the bacteroid-containing cells, which suggested that particular stages of differentiation and/or metabolism have to be completed for NO to be produced (Baudouin et al. 2006).

Although the presence of NO has mainly been examined in functioning nodules, it is tempting to assume that it may also participate in the early signaling network that controls recognition and infection by the microsymbiont and/or the development of the new organ. Recent data from Creus et al. (2005) evidenced that NO is synthesized during the interaction of tomato seedlings and the diazotroph bacterium *Azospirillum brasilense*. Combining EPR and DAF2-DA-based NO detection, a production of NO was observed in free-living *A. brasilense* as well as in *A. brasilense*-infected tomato seedlings. In this last case, NO production was located at the vascular tissues and subepidermal cells (Creus et al. 2005). Production of NO during the early steps of legume–rhizobia symbiosis has also been reported by Shimoda et al. (2005). The authors detected DAF2-DA-related fluorescence in *Lotus japonicus* roots 4 h after inoculation with *Mesorhizobium loti* (Shimoda et al. 2005). In contrast, we could not correlate NO detection with early responses of *M. truncatula* to *S. meliloti*, observed between 12 and 72 h (Baudouin et al. 2006). The observations of Shimoda et al. (2005) may therefore reflect a highly transient production of NO in infected roots, the kinetics and role of which remains to be studied. In particular, as symbiotic and pathogen interactions share common features, it would be of interest to compare how NO production is regulated in these situations and to determine whether it plays different roles in these two types of interactions.

2.2 Sources of NO

Plants and microorganisms possess multiple mechanisms to generate NO, relying on enzymatic and non-enzymatic processes (for review, Crawford 2006). Our current knowledge strongly suggests that diverse NO sources may independently or concomitantly contribute to NO production in response to a particular signal, as illustrated in other chapters. Because symbiosis results from the interaction between two organisms, there are multiple potential sources for NO, depending on the biological partner on the one hand, and on the enzymatic system on the other hand.

Investigations in fixing nodules established that NO accumulation is not a by-product of symbiotic nitrogen fixation (Baudouin et al. 2006). Similarly, NO detection in *M. truncatula* nodules formed upon infection with bacterial mutants impaired in the denitrification pathway excluded denitrification as a NO source in nodules (Baudouin et al. 2006). On the other hand, current reports point out a possible role of NO synthase (NOS)-like enzymes to form NO in symbiotic tissues. A NOS-like activity, which was inhibited by the mammalian NOS inhibitor N^G-monomethyl-L-arginine (l-NMMA), was detected in *Lupinus albus* nodules (Cueto et al. 1996). The same inhibitor efficiently impaired NO production in *M. truncatula* nodules (Baudouin et al. 2006).

Whether the NO synthase activity is of plant or bacterial origin is still unclear. Creus et al. (2005) recently reported that *A. brasilense* synthesized NO in pure culture, through a process stimulated by L-arginine but insensitive to mammalian NOS inhibitors. We could not detect a similar production in *S. meliloti*, which may corroborate the lack of an identifiable gene in the completely sequenced *S. meliloti* genome, showing homology with eukaryotic and prokaryotic NOS (unpublished data). On the other hand, homologs of AtNOS1, which participates in NO synthesis in *Arabidopsis thaliana* (Guo et al. 2003), are present in legumes. Nevertheless, the *M. truncatula* homolog did not exhibit NO synthase activity when expressed in *Escherichia coli*, which makes it unlikely that it catalyzes NO synthesis in nodules. Further investigations are therefore necessary to identify the possible NOS operating in nodules.

The possible participation of additional enzymes unrelated to NOS should not be discounted, as NOS inhibitors were only partially effective. Nitrate reductases may be good candidates. A nitrate reductase isoform expressed independently of nitrate supply, was recently characterized in *L. japonicus* nodules (Kato et al. 2003). Moreover, Becana et al. (1985) presented evidence that a sufficient nitrite concentration is present in bacteroids extracted from nitrate-starved alfalfa nodules to sustain NO synthesis from nitrate reductases (Becana et al. 1985).

2.3 NO Detoxification

A crucial element for NO bioactivity is its turnover. Recent work has highlighted modifications of the NO scavenging capacities of plants and/or bacteria that have a dramatic impact on the issue of plant–pathogen interactions (Perazzolli et al. 2006). In this respect, both microbes and plants possess enzymatic systems to detoxify NO and may therefore participate in NO homeostasis during plant–microbe interactions. It is therefore worth considering the putative role of NO "removal" during symbiotic interaction, even though its role has not yet been reported in this biological context. Most of the enzymes involved in NO scavenging are heme-containing proteins and metabolize NO to NO_3^- in the presence of dioxygen and NAD(P)H (for review see Frey and Kallio 2003; Kundu et al. 2003; Perazzolli et al. 2006). The best studied enzymes are the bacterial flavohemoglobins which participate in nitrosative stress response. In some species, flavohemoglobin activity is an important determinant of pathogenicity as it enables the bacteria to cope with NO produced as an antimicrobial agent by the infected cells (Favey et al. 1995; Stevanin et al. 2002; Boccara et al. 2005). A putative flavohemoglobin designated SmfHb has been identified in the *S. meliloti* genome (Lira-Ruan et al. 2003). It shares highest similarities with *Ralstonia eutropha* FHP, which presents a low NO dioxygenase activity compared to other flavohemoglobins (Cramm et al. 1994; Frey et al. 2002).

We have constructed an *S. meliloti* SmfHb mutant strain (D. Touati, E. Baudouin and A. Puppo, unpublished). The mutant has growth parameters comparable to the wild type and does not exhibit an increased sensitivity towards NO donors. On the other hand, the overexpression of *Erwinia chrysanthemi* HmpX flavohemoglobin in *S. meliloti* did not enhance its resistance towards exogenous NO (B. Demple, E. Baudouin and A. Puppo, unpublished). These data suggest that other enzymes participate in NO detoxification in *S. meliloti*. They may include the NO reductase activity that operates in the

S. meliloti denitrification pathway. Several reports have presented evidence for the activation of denitrification-related genes in nodules and suggested that it could play a role in NO turnover (Mesa et al. 2002, 2004). Interestingly, the *SmfHb* gene is located on the pSymA megaplasmid, close to the genes involved in denitrification processes (Lira-Ruan et al. 2003). Although the mechanisms remain unknown, flavohemoglobins have been associated with denitrification activity in some species, including *R. eutropha* (Cramm et al. 1994; Takaya et al. 1997). It is therefore possible that a coordinated regulation of NO reductase and *SmfHb* genes leads to NO modulation during symbiosis. We did not observe significant differences in the symbiotic capacities of the *SmfHb* mutant strain (E. Baudouin and A. Puppo, unpublished). Similarly, a *NorC* mutant impaired in NO reductase was unaffected in its symbiotic capacities, and the NO level was unmodified in the nodules obtained (Baudouin et al. 2006). Further research, including the construction of bacterial strains mutated in more than one NO scavenging system, is required to elucidate whether and how symbiotic bacteria may interfere with NO metabolism when infecting plant cells.

Recent reports have shed light on how NO can be eliminated in plant cells and have pointed out the major role of non-symbiotic hemoglobins as NO scavengers (Dordas et al. 2003; Perazzolli et al. 2004). Whereas the function of symbiotic hemoglobins, and in particular leghemoglobins, is well established as an O_2 carrier in the symbiotic tissues, it is only recently that a function for non-symbiotic hemoglobins has been demonstrated. In plants establishing nitrogen fixing symbiosis, both symbiotic and non-symbiotic hemoglobins are present. If providing O_2 to bacteroids is the role of leghemoglobins, then the NO scavenging activity may be shared by both classes of hemoglobins. Herold and Puppo (2005) have provided evidence for oxyleghemoglobin capacity to scavenge NO and peroxinitrite, which suggests that it participates in the protection of bacteroids against the inhibition of nitrogenase by NO (Herold and Puppo 2005). On the other hand, several genes encoding non-symbiotic hemoglobins have been isolated in legumes or related symbiotic plants.

In *M. truncatula*, the *MtTrHb1* and *MtTrHb2* genes, which encode truncated hemoglobins, were found to be highly induced in nodules compared to roots (Vieweg et al. 2005). *MtTrHb1* was activated in the infected cells of the fixing zone whereas *MtTrHb2* was predominantly expressed in the nodule vasculature (Vieweg et al. 2005). Analysis of two *L. japonicus* non-symbiotic hemoglobins *LjHb1* and *LjHb2* suggested that their function during symbiosis may not be restricted to nodule tissues but may also take place during the early steps of the plant–bacteria interaction (Shimoda et al. 2005). Indeed, *LjHb1* is highly expressed in nodules compared to roots, but its expression is also rapidly and transiently increased 4 h after *M. loti* inoculation (Shimoda et al. 2005). Moreover, NO production was visualized over the same time period and *LjHb1* transcription could be induced with exogenous NO.

Taken together, these data suggest that LjHb1 is involved in NO scavenging during establishment of the symbiosis.

Recent data speak for the non-symbiotic hemoglobins being conserved elements of plant–symbiont interactions. Indeed, *AfHb1*, an *Alnus firma* homolog of *LjHb1*, is strongly expressed in nodules formed upon infection with the actinomycete *Frankia* (Sasakura et al. 2006). In this context, AfHb1 may restrain NO production in functioning nodules, preserving a fully functional nitrogen fixing capacity. Beyond nitrogen fixing symbiosis, non-symbiotic hemoglobins have also been associated to arbuscular mycorrhizal symbiosis. Upon infection of *M. truncatula* with the fungus *Glomus intraradices*, *MtTrHb2* gene expression was induced in response to infection, and the expression pattern co-localized with the presence of the arbuscular structures (Vieweg et al. 2005). Whether MtTrHb2 actually functions as a NO scavenger is still unknown, as is the occurrence and role of NO during mycorrhization.

3
Outcomes of NO Production During Symbiosis

To date, nitric oxide has only been considered as a deleterious species, negatively interfering with nodule nitrogen fixation. Nevertheless, the recent data described above, which confirm the presence of NO in fully functional nitrogen-fixing cells and maybe even during the early stages of symbiosis establishment, lead us to consider that NO may play important regulatory functions during symbiotic interactions. The following paragraphs give an overview of some current hypothesis for NO participation in plant–symbiotic interactions.

3.1
Nitrogen Metabolism

Nitric oxide has been known for more than two decades as a potent inhibitor of nitrogenase activity. Using nitrogenase purified from soybean bacteroids, Trinchant and Rigaud (1982) showed that NO inhibited C_2H_2 reduction by a non-competitive mechanism (Trinchant and Rigaud 1982). Moreover, when compared to nitrite, the K_i value (56 μM) for NO inhibition was very low. The exogenous application of NO donors (sodium nitroprussiate, SNP; *S*-nitroso-*N*-acetyl-D,L-penicillamine, SNAP) on nodulated roots also led to a dramatic decrease of nitrogen-fixation capacities (E. Baudouin and A. Puppo, unpublished data). It is therefore probable that, if NO concentrations sufficient to interfere with nitrogen fixation are reached in fixing cells, active mechanisms have been developed to protect nitrogenase from NO inhibition in functional bacteroids. In this view, activation of the denitrification pathway, including the nitric oxide reductase activity, in a context where no nitrate is available, may be related to such a protective requirement (Mesa et al. 2004; Meakin et al. 2006). Nevertheless, the in vivo significance of nitrogenase inhibition by NO remains elusive. It has been proposed that NO may participate in the inhibition of nitrogen fixation observed in response to nitrate (Streeter 1988). Indeed, plants can use nitrate to generate NO in a nitrate reductase-dependent process (Desikan et al. 2002; Rockel et al. 2002). Mesa et al. (2004) have shown that, although soybean plants inoculated with a *Bradyrhizobium japonicum* strain impaired in NO reductase activity presented a reduced nodule number in the presence of 4 mM nitrate, their nitrogen fixation efficiency remained unchanged. Nevertheless, additional experiments are required to analyze the production of NO in nodules upon nitrate treatment, and to study the possible consequences of such a production on nitrogen fixation.

So far, the impact of NO on nitrogen metabolism in fixing cells has only been focused on interactions with nitrogenase. Recent data suggest that it may also have an impact on the metabolism of NH_4^+ following N_2 reduction. Glutamine synthase, which catalyzes the incorporation of NH_4^+ into carbon backbones, was identified as an S-nitrosylated protein in vivo in *A. thaliana* (Lindermayr et al. 2005). We also observed that several key genes of NH_4^+ metabolism, including the nodule-specific glutamine synthase isoform, are induced by NO donor treatment (M. De Stefano, E. Baudouin, A. Puppo and M. Delledonne, unpublished data). These preliminary data open new perspectives for studying the relationship between NO and nitrogen metabolism through genomic and postgenomic approaches.

3.2
O_2 Deprivation Sensing

As mentioned in the introduction, although microaerobic conditions are needed in nodules for proper nitrogen fixation, the intense metabolic activity of the fixing cells also requires that large amounts of O_2 must be provided to bacteroids. Therefore, Lb, as an O_2 carrier, plays a crucial role in the nitrogen-fixing capacities of the nodules. Obviously NO, which readily reacts with Lb, may interfere with O_2 fixation and delivery and, therefore, limit nodule metabolism. The formation of nitrosylLb ($LbFe^{II}NO$), prevents both Lb autoxidation and its conversion to oxyLb in the presence of O_2 (Maskall et al. 1977; Kanayama et al. 1990). On the other hand, oxyLb is able to efficiently scavenge NO (Herold and Puppo 2005). Several reports indicate that the level of $LbFe^{II}NO$ is increased by nitrate supply (Kanayama et al. 1990; Kanayama and Yamamoto 1990, 1991). It is also influenced by the age of the nodules (Mathieu et al. 1998). Since the young nodules have the highest $LbFe^{II}NO$ content and since a decline is observed with aging, it may suggest that $LbFe^{II}NO$ is not only an inactivation form of Lb but also sustains additional functions not yet elucidated. Moreover, a direct impact of the presence of $LbFe^{II}NO$ in nodules on nitrogen fixation has not yet been demonstrated.

O_2 deprivation, on the other hand, is sensed as a signal by the invading bacteria and triggers fundamental differentiation steps ending in the formation of the nitrogen-fixing bacteroids. Of particular interest is the FixLJ complex, which acts as a sensor and transducer of the low pO_2 signal and controls the activation of genes of microaerobic respiration, through the transcription factor FixK, and of nitrogen fixation, through the transcription factor NifA (Fischer 1994). Several studies have evidenced that FixL, the O_2 sensor, binds NO with a ten- to 100-fold greater affinity than O_2 (Winkler et al. 1996; Gong et al. 2000; Rodgers et al. 2000). As FixLJ downstream signaling is triggered by the activation of FixL autophosphorylation and subsequent phosphotransfer to FixJ upon O_2 deprivation, several studies have examined the effect of NO binding on FixL kinase activity (Akimoto et al. 2003; Dunham et al. 2003). Although conformational changes clearly occur in the kinase domain upon NO binding, the consequences of such changes on kinase activity remain unclear. Nevertheless, NO binding did not mimic O_2 binding and only partially inhibited kinase activity (Akimoto et al. 2003). Therefore, one can speculate that NO binding may prevent O_2 binding and lock FixLJ in an activated state. Interaction of NO with the FixLJ-FixK cascade could be even more subtle and complex. Mesa et al. (2003) identified a new transcription factor NnrR, which acts downstream of FixK to activate *nir* and *nor* genes in *B. japonicum*. They showed that *nir* and *nor* genes can be activated upon NO treatment in an NnrR-dependent manner and suggested that NnrR could integrate the NO signal for a maximal induction of denitrification genes (Mesa et al. 2003). If so, the identification of additional NnrR-regulated genes may reveal unsuspected regulatory functions of NO in *Rhizobia*.

Recent reports have illustrated the importance of NO in plants exposed to O_2 deprivation (for review, Igamberdiev and Hill 2004). Hypoxic conditions triggered NO accumulation, and the manipulation of NO levels through the modification of non-symbiotic hemoglobin expression in transgenic alfalfa modulated the plant metabolic response to hypoxia (Dordas et al. 2003). In nodules, fixing cells face microaerobic conditions. In this context, NO may induce specific plant responses reminiscent of those observed during hypoxia. Indeed, we isolated a series of hypoxia-related genes, such as pyruvate decarboxylase and alcohol dehydrogenase, that were induced upon NO donor treatment in *M. truncatula* roots (M. De Stefano, E. Baudouin, A. Puppo and M. Delledonne, unpublished data). Interestingly, Colebatch et al. (2004) reported that these genes are also induced in functional *L. japonicus* nodules (Colebatch et al. 2004). Although a direct link between NO production and hypoxia-related genes in nodules has not been established, these data suggest that NO is a central actor of the O_2-regulated processes occurring in nodules.

3.3 Development

Creus et al. (2005) have provided clear evidence for the impact on plant development of NO produced during symbiotic interactions. Indeed, lateral root formation (LRF), which is mediated by association with *A. brasilense*, relies on the production of NO. Previous reports demonstrated the involvement of NO in the auxin-induced LRF and growth in several species (Pagnussat et al. 2002, 2003; Correa-Aragunde et al. 2004). It is therefore probable that a similar auxin-dependent pathway involving NO is responsible for LRF in response to *A. brasilense*. This result raises the question of a general role for an auxin/NO pathway in organogenesis during symbiotic interactions. Mathesius et al. (1998) suggested that LRF and nodule development share a similar auxin requirement in order to proceed. In particular, it has been evidenced that modifications of the polar auxin transport triggered by bacterial nodulation factors, but also by plant flavonoids, are determinant for proper nodule development (Mathesius et al. 1998; Wasson et al. 2006). Whether NO is required for auxin-mediated nodule development remains to be explored. It is noteworthy that NO is involved in organ primordia formation rather than in the subsequent organ growth as observed for LRF (Correa-Aragunde et al. 2006), which may also be the case for nodules.

A crucial step for nodule organogenesis is the re-initiation of root cortical cell division that gives rise to the nodule primordia. Recent works have illustrated that NO is a potent regulator of the cell cycle in plants (Otvos et al. 2005; Correa-Aragunde et al. 2006). In particular, Otvos et al. (2005) suggested that NO is involved in the activation, but not the progression, of the cell cycle. This hypothesis is strengthened by the analysis of cell cycle-related genes. Indeed, NO induces genes such as *CycD3* that stimulates the G1-to-S phase transition, and represses genes such as *KRP2* that prevent the G1-to-S phase transition (Correa-Aragunde et al. 2006). These data shed light on a new potential aspect of NO regulation during symbiosis, which has to be further explored in legume–rhizobia symbiosis.

4 Concluding Remarks

Beside its deleterious effects on nodule activity, which have been known for years, recent data suggest that NO is an actor in the regulation of plant–bacteria symbiosis. The development of techniques to visualize NO in vivo will bring additional clues on the spatio-temporal fluctuations of NO concentration during symbiotic interactions. Of particular interest will be the comparison between different symbiotic systems, such as rhizobial and mycorrhizal symbioses, which share common signaling elements. Indeed, no information has been provided so far concerning the relationship between mycorrhizal symbioses and NO. Together with the use of symbiotic mutants, it will highlight the potential role of NO as a general signal of symbiotic interaction establishment/functioning.

The identification of enzymes controlling the production and removal of NO in plants gives new opportunities to investigate the physiological processes regulated by the variations of NO concentration during symbiosis. On the basis of these data, transgenic plants with modified capacities to accumulate NO will be inestimable tools for investigating our current hypothesis of NO being a regulatory element of nodule development and/or of nodular metabolic activity. Moreover, in association with genomic and proteomic approaches, they will permit the isolation and characterization of target genes and proteins regulated through a NO signaling pathway during symbiotic interactions. Such information will help evaluate the universality of NO-regulated mechanisms during plant–symbiont interactions, and also during plant–pathogen interactions.

References

Akimoto S, Tanaka A, Nakamura K, Shiro Y, Nakamura H (2003) O_2-specific regulation of the ferrous heme-based sensor kinase FixL from Sinorhizobium meliloti and its aberrant inactivation in the ferric form. Biochem Biophys Res Commun 304:136–142

Baron C, Zambryski PC (1995) The plant response in pathogenesis, symbiosis, and wounding: variations on a common theme? Annu Rev Genet 29:107–129

Baudouin E, Pieuchot L, Engler G, Pauly N, Puppo A (2006) Nitric oxide is formed in Medicago truncatula-Sinorhizobium meliloti functional nodules. Mol Plant Microbe Interact 19:970–975

Becana M, Aparicio-Tejo PM, Sanchez-Diaz M (1985) Nitrate and nitrite reduction in the plant fraction of alfalfa root nodules. Physiol Plant 65:185–188

Boccara M, Mills CE, Zeier J, Anzi C, Lamb C, Poole RK, Delledonne M (2005) Flavohaemoglobin HmpX from Erwinia chrysanthemi confers nitrosative stress tolerance and affects the plant hypersensitive reaction by intercepting nitric oxide produced by the host. Plant J 43:226–237

Buchanan BB, Gruissem W, Jones RL (eds) (2000) Biochemistry and molecular biology of plants. American Society of Plant Physiologists, Rockville, Maryland

Colebatch G, Desbrosses G, Ott T, Krusell L, Montanari O, Kloska S, Kopka J, Udvardi MK (2004) Global changes in transcription orchestrate metabolic differentiation during symbiotic nitrogen fixation in Lotus japonicus. Plant J 39:487–512

Correa-Aragunde N, Graziano M, Chevalier C, Lamattina L (2006) Nitric oxide modulates the expression of cell cycle regulatory genes during lateral root formation in tomato. J Exp Bot 57:581–588

Correa-Aragunde N, Graziano M, Lamattina L (2004) Nitric oxide plays a central role in determining lateral root development in tomato. Planta 218:900–905

Cramm R, Siddiqui RA, Friedrich B (1994) Primary sequence and evidence for a physiological function of the flavohemoprotein of Alcaligenes eutrophus. J Biol Chem 269:7349–7354

Crawford NM (2006) Mechanisms for nitric oxide synthesis in plants. J Exp Bot 57:471–478

Creus CM, Graziano M, Casanovas EM, Pereyra MA, Simontacchi M, Puntarulo S, Barassi CA, Lamattina L (2005) Nitric oxide is involved in the Azospirillum brasilense-induced lateral root formation in tomato. Planta 221:297–303

Cueto M, Hernandez-Perera O, Martin R, Bentura ML, Rodrigo J, Lamas S, Golvano MP (1996) Presence of nitric oxide synthase activity in roots and nodules of Lupinus albus. FEBS Lett 398:159–164

De Stefano M, Vandelle E, Polverari A, Ferrarini A, Delledonne M (2007) Nitric oxide-mediated signaling functions during the plant hypersensitive response. In: Lamattina L, Polacco JC (eds) Nitric oxide in plant growth. Plant Cell Monographs, vol 6. Springer, Berlin Heidelberg New York (in press)

Desikan R, Griffiths R, Hancock J, Neill S (2002) A new role for an old enzyme: nitrate reductase-mediated nitric oxide generation is required for abscisic acid-induced stomatal closure in Arabidopsis thaliana. Proc Natl Acad Sci USA 99:16314–16318

Dordas C, Rivoal J, Hill RD (2003) Plant haemoglobins, nitric oxide and hypoxic stress. Ann Bot (Lond) 91 Spec No:173–178

Dunham CM, Dioum EM, Tuckerman JR, Gonzalez G, Scott WG, Gilles-Gonzalez MA (2003) A distal arginine in oxygen-sensing heme-PAS domains is essential to ligand binding, signal transduction, and structure. Biochemistry 42:7701–7708

Favey S, Labesse G, Vouille V, Boccara M (1995) Flavohaemoglobin HmpX: a new pathogenicity determinant in Erwinia chrysanthemi strain 3937. Microbiology 141(4):863–871

Fischer HM (1994) Genetic regulation of nitrogen fixation in rhizobia. Microbiol Rev 58:352–386

Foissner I, Wendehenne D, Langebartels C, Durner J (2000) In vivo imaging of an elicitor-induced nitric oxide burst in tobacco. Plant J 23:817–824

Frey AD, Farres J, Bollinger CJ, Kallio PT (2002) Bacterial hemoglobins and flavohemoglobins for alleviation of nitrosative stress in Escherichia coli. Appl Environ Microbiol 68:4835–4840

Frey AD, Kallio PT (2003) Bacterial hemoglobins and flavohemoglobins: versatile proteins and their impact on microbiology and biotechnology. FEMS Microbiol Rev 27:525–545

Gage DJ (2004) Infection and invasion of roots by symbiotic, nitrogen-fixing rhizobia during nodulation of temperate legumes. Microbiol Mol Biol Rev 68:280–300

Gong W, Hao B, Chan MK (2000) New mechanistic insights from structural studies of the oxygen-sensing domain of Bradyrhizobium japonicum FixL. Biochemistry 39:3955–3962

Guo FQ, Okamoto M, Crawford NM (2003) Identification of a plant nitric oxide synthase gene involved in hormonal signaling. Science 302:100–103

Herold S, Puppo A (2005) Oxyleghemoglobin scavenges nitrogen monoxide and peroxynitrite: a possible role in functioning nodules? J Biol Inorg Chem 10:935–945

Igamberdiev AU, Hill RD (2004) Nitrate NO and haemoglobin in plant adaptation to hypoxia: an alternative to classic fermentation pathways. J Exp Bot 55:2473–2482

Kanayama Y, Watanabe I, Yamamoto Y (1990) Inhibition of nitrogen fixation in soybean plants supplied with nitrate I. Nitrite accumulation and formation of nitrosylleghemoglobin in nodules. Plant Cell Physiol 31:341–346

Kanayama Y, Yamamoto Y (1990) Inhibition of nitrogen fixation in soybean plantssupplied with nitrate II. Accumulation and properties of nitrosylleghemoglobin in nodules. Plant Cell Physiol 31:207–214

Kanayama Y, Yamamoto Y (1991) Formation of nitrosylleghemoglobin in nodules of nitrate-treated cowpea and pea plants. Plant Cell Physiol 32:19–24

Kato K, Okamura Y, Kanahama K, Kanayama Y (2003) Nitrate-independent expression of plant nitrate reductase in Lotus japonicus root nodules. J Exp Bot 54:1685–1690

Kundu S, Trent JT 3rd, Hargrove MS (2003) Plants, humans and hemoglobins. Trends Plant Sci 8:387–393

Lindermayr C, Saalbach G, Durner J (2005) Proteomic identification of S-nitrosylated proteins in Arabidopsis. Plant Physiol 137:921–930

Lira-Ruan V, Sarath G, Klucas RV, Arredondo-Peter R (2003) In silico analysis of a flavohemoglobin from Sinorhizobium meliloti strain 1021. Microbiol Res 158:215–227

Maskall CS, Gibson JF, Dart PJ (1977) Electron-paramagnetic-resonance studies of leghaemoglobins from soya-bean and cowpea root nodules. Identification of nitrosylleghaemoglobin in crude leghaemoglobin preparations. Biochem J 167:435–445

Mathesius U, Schlaman HR, Spaink HP, Of Sautter C, Rolfe BG, Djordjevic MA (1998) Auxin transport inhibition precedes root nodule formation in white clover roots and is regulated by flavonoids and derivatives of chitin oligosaccharides. Plant J 14:23–34

Mathieu C, Moreau S, Frendo P, Puppo A, Davies MJ (1998) Direct detection of radicals in intact soybean nodules: presence of nitric oxide leghemoglobin complexes. Free Radic Biol Med 24:1242–1249

Meakin GE, Jepson BJ, Richardson DJ, Bedmar EJ, Delgado MJ (2006) The role of Bradyrhizobium japonicum nitric oxide reductase in nitric oxide detoxification in soya bean root nodules. Biochem Soc Trans 34:195–196

Mesa S, Alche Jd J, Bedmar E, Delgado MJ (2004) Expression of nir, nor and nos denitrification genes from Bradyrhizobium japonicum in soybean root nodules. Physiol Plant 120:205–211

Mesa S, Bedmar EJ, Chanfon A, Hennecke H, Fischer HM (2003) Bradyrhizobium japonicum NnrR, a denitrification regulator, expands the FixLJ-FixK2 regulatory cascade. J Bacteriol 185:3978–3982

Mesa S, Velasco L, Manzanera ME, Delgado MJ, Bedmar EJ (2002) Characterization of the norCBQD genes, encoding nitric oxide reductase, in the nitrogen fixing bacterium Bradyrhizobium japonicum. Microbiology 148:3553–3560

Otvos K, Pasternak TP, Miskolczi P, Domoki M, Dorjgotov D, Szucs A, Bottka S, Dudits D, Feher A (2005) Nitric oxide is required for, and promotes auxin-mediated activation of, cell division and embryogenic cell formation but does not influence cell cycle progression in alfalfa cell cultures. Plant J 43:849–860

Pagnussat GC, Lanteri ML, Lamattina L (2003) Nitric oxide and cyclic GMP are messengers in the indole acetic acid-induced adventitious rooting process. Plant Physiol 132:1241–1248

Pagnussat GC, Simontacchi M, Puntarulo S, Lamattina L (2002) Nitric oxide is required for root organogenesis. Plant Physiol 129:954–956

Pauly N, Pucciariello C, Mandon K, Innocenti G, Jamet A, Baudouin E, Herouart D, Frendo P, Puppo A (2006) Reactive oxygen and nitrogen species and glutathione: key players in the legume-Rhizobium symbiosis. J Exp Bot 57:1769–1776

Perazzolli M, Dominici P, Romero-Puertas MC, Zago E, Zeier J, Sonoda M, Lamb C, Delledonne M (2004) Arabidopsis nonsymbiotic hemoglobin AHb1 modulates nitric oxide bioactivity. Plant Cell 16:2785–2794

Perazzolli M, Romero-Puertas MC, Delledonne M (2006) Modulation of nitric oxide bioactivity by plant haemoglobins. J Exp Bot 57:479–488

Perret X, Staehelin C, Broughton WJ (2000) Molecular basis of symbiotic promiscuity. Microbiol Mol Biol Rev 64:180–201

Rockel P, Strube F, Rockel A, Wildt J, Kaiser WM (2002) Regulation of nitric oxide (NO) production by plant nitrate reductase in vivo and in vitro. J Exp Bot 53:103–110
Rodgers KR, Lukat-Rodgers GS, Tang L (2000) Nitrosyl adducts of FixL as probes of heme environment. J Biol Inorg Chem 5:642–654
Rolfe BG, Gresshoff PG (1988) Genetic analysis of legume nodule initiation. Annu Rev Physiol Plant Mol Biol 39:297–319
Salgado I, Modolo LV, Augusto O, Bragav MR, Oliveira HC (2007) Mitochondrial nitric oxide synthesis during plant–pathogen interactions: role of nitrate reductase in providing substrates. In: Lamattina L, Polacco JC (eds) Nitric oxide in plant growth. Plant Cell Monographs, vol 6. Springer, Berlin Heidelberg New York (in press)
Sasakura F, Uchiumi T, Shimoda Y, Suzuki A, Takenouchi K, Higashi S, Abe M (2006) A class 1 hemoglobin gene from Alnus firma functions in symbiotic and nonsymbiotic tissues to detoxify nitric oxide. Mol Plant Microbe Interact 19:441–450
Shapiro AD (2007) Nitric oxide in cell-to-cell communication coordinating the plant hypersensitive response. In: Lamattina L, Polacco JC (eds) Nitric oxide in plant growth. Plant Cell Monographs, vol 6. Springer, Berlin Heidelberg New York (in press)
Shimoda Y, Nagata M, Suzuki A, Abe M, Sato S, Kato T, Tabata S, Higashi S, Uchiumi T (2005) Symbiotic rhizobium and nitric oxide induce gene expression of non-symbiotic hemoglobin in Lotus japonicus. Plant Cell Physiol 46:99–107
Stevanin TM, Poole RK, Demoncheaux EA, Read RC (2002) Flavohemoglobin Hmp protects Salmonella enterica serovar typhimurium from nitric oxide-related killing by human macrophages. Infect Immun 70:4399–4405
Streeter J (1988) Inhibition of legume nodule formation and N_2 fixation by nitrate. Crit Rev Plant Sci 7:1–23
Takaya N, Suzuki S, Matsuo M, Shoun H (1997) Purification and characterization of a flavohemoglobin from the denitrifying fungus Fusarium oxysporum. FEBS Lett 414:545–548
Trinchant JC, Rigaud J (1982) Nitrite and nitric oxide as inhibitors of nitrogenase from soybean bacteroids. Appl Environ Microbiol 44:1385–1388
Vieweg MF, Hohnjec N, Kuster H (2005) Two genes encoding different truncated hemoglobins are regulated during root nodule and arbuscular mycorrhiza symbioses of Medicago truncatula. Planta 220:757–766
Wasson AP, Pellerone FI, Mathesius U (2006) Silencing the flavonoid pathway in Medicago truncatula inhibits root nodule formation and prevents auxin transport regulation by rhizobia. Plant Cell 18:1617–1629
Winkler WC, Gonzalez G, Wittenberg JB, Hille R, Dakappagari N, Jacob A, Gonzalez LA, Gilles-Gonzalez MA (1996) Nonsteric factors dominate binding of nitric oxide, azide, imidazole, cyanide, and fluoride to the rhizobial heme-based oxygen sensor FixL. Chem Biol 3:841–850

Plant Cell Monogr (6)
L. Lamattina · J. C. Polacco: Nitric Oxide in Plant Growth
DOI 10.1007/7089_2006_091/Published online: 14 October 2006

Nitrosative Stress in Plants: A New Approach to Understand the Role of NO in Abiotic Stress

Francisco J. Corpas[1] (✉) · Juan B. Barroso[2] · Alfonso Carreras[2] · Raquel Valderrama[2] · José M. Palma[1] · Luis A. del Río[1]

[1]Departamento de Bioquímica, Biología Celular y Molecular de Plantas, Estación Experimental del Zaidín (EEZ), CSIC, Apartado 419, 18008 Granada, Spain
javier.corpas@eez.csic.es

[2]Grupo de Señalización Molecular y Sistemas Antioxidantes en Plantas, Unidad Asociada al CSIC (EEZ), Área de Bioquímica y Biología Molecular, Universidad de Jaén, Paraje Las Lagunillas, Edificio B-3, 23071 Jaén, Spain

Abstract Early reports that plants have the capacity to generate $NO^{\cdot}$ did not attract the attention of plant physiologists for many years, until 1987 when $NO^{\cdot}$ was identified by Prof. Moncada's group as the endothelium-derived relaxing factor in mammals. Plant physiologists and biochemists have started to pay attention to this gas and today $NO^{\cdot}$ is known as a versatile molecule with multiple functions in several complex processes such as seed germination, development, senescence, and defence against biotic/abiotic stress. This chapter presents an overview of the present knowledge on the involvement of $NO^{\cdot}$ and other reactive nitrogen species in plant abiotic stress, with special emphasis on nitrosative stress as a new component of plant stress physiology.

1 Introduction

The gaseous free radical nitric oxide ($NO^{\cdot}$) can have two antagonistic or opposite physiological roles. While a high cellular production of $NO^{\cdot}$ can bring about extensive cellular damage, at low levels it is involved as a signal molecule in many important physiological processes (del Río et al. 2004). In plants, intensive research is being done on the physiological function of $NO^{\cdot}$ under normal and stress conditions (Neill et al. 2003; Lamattina et al. 2003; Shapiro 2005), but comparatively much less information is available on other reactive nitrogen species like *S*-nitrosoglutathione, peroxynitrite, and nitrotyrosine. At present, it is well established that plant cells can generate $NO^{\cdot}$, but there is some controversy on the specific source of $NO^{\cdot}$ in a defined physiological process, as well as on how $NO^{\cdot}$ levels are modulated under normal and adverse conditions. In addition, the cellular and subcellular localization of $NO^{\cdot}$ is also under intensive investigation.

Parallel to the expression "reactive oxygen species" (ROS), the term "reactive nitrogen species" (RNS) was coined to designate nitric oxide ($NO^{\cdot}$)

and a family of related molecules such as peroxynitrite ($ONOO^-$), dinitrogen trioxide (N_2O_3), *S*-nitrosoglutathione (GSNO), nitrogen dioxide ($NO_2^{\cdot}$), nitrosonium cation (NO^+) etc. (Hausladen and Stamler 1999; Halliwell and Gutteridge 2000). In an analogous manner to the term "oxidative stress", in animal systems the expression "nitrosative stress" started to be employed to define the deregulated synthesis or overproduction of $NO^{\cdot}$ and $NO^{\cdot}$-derived products and its toxic physiological consequences (Klatt and Lamas 2000). Excessive amounts of reactive nitrogen species can damage DNA, lipids, proteins, and carbohydrates leading to impaired cellular functions. They have been shown to play a prominent role in the pathogenesis of various lung disorders (Ricciardolo et al. 2006), kidney and cardiovascular diseases (Redondo-Horcajo and Lamas 2005; Stevens 2005), and neurodegenerative disorders (Chung et al. 2005). For this reason, plasma nitrotyrosine, protein tyrosine nitration, and/or nitrite/nitrate levels have been used as markers of nitrosative stress.

However, in plant systems the information available on nitrosative stress is rather limited. Nitrosative stress could be involved in damages produced by different abiotic and biotic stress conditions with the direct participation of different RNS. As in the case of lipid peroxidation and protein oxidation, widely recognized markers of oxidative stress, it would be necessary to establish some biological markers characteristic of nitrosative stress. This chapter will focus on the description of the elements necessary to establish relationships between adverse environmental conditions, $NO^{\cdot}$ metabolism, and nitrosative stress.

2 Nitric Oxide, Reactive Nitrogen Species, and NO–Protein Interactions

Nitric oxide is a versatile molecule. Among the physical and chemical properties of $NO^{\cdot}$, some of them relevant to its biological action can be emphasized. Thus, $NO^{\cdot}$ is a lipophylic radical, which diffuses across cell membranes and through the cytoplasm. It diffuses at a rate of $50\,\mu m\,s^{-1}$. The solubility of $NO^{\cdot}$ is 1.9 mM in aqueous solutions at 1 atm pressure. In vivo, $NO^{\cdot}$ half-life is relatively short, less than 10 s (Pfeiffer et al. 1999). On the other hand, the term "reactive nitrogen species" (RNS) was coined to designate $NO^{\cdot}$ and a family of related molecules (Table 1) (Halliwell and Gutteridge 2000), in an attempt to keep an expression equivalent to "reactive oxygen species" (ROS) since RNS are generally considered to arise from the interaction of $NO^{\cdot}$ with other molecules. In plants, most of the studies have been focused on $NO^{\cdot}$ and little is known about the other RNS. However, special attention has been paid to peroxynitrite ($ONOO^-$). This is a potent oxidant formed by the non-enzymatic reaction between superoxide anion and $NO^{\cdot}$ in a one-to-one-stoichiometry. Peroxynitrite has been shown to react with virtually all classes

Table 1 NO-derived molecules designated as reactive nitrogen species (RNS) including radical and non-radical molecules

Radicals	Non-radicals
Nitric oxide ($NO^{\cdot}$)	Nitrous acid (HNO_2)
Nitrogen dioxide ($NO_2^{\cdot}$)	Dinitrogen trioxide (N_2O_3)
	Dinitrogen tetraoxide (N_2O_4)
	Nitronium (nitryl) ion (NO_2^+)
	Peroxynitrite ($ONOO^-$)
	Peroxynitrous acid (ONOOH)
	Alkyl peroxynitrites (ROONO)
	Nitroxyl anion (NO^-)
	Nitrosonium cation (NO^+)

of biomolecules in vitro (Pryor and Squadrito 1995). After discovering that peroxynitrite is generated in vivo and the demonstration that peroxynitrite in animal cells can modify a number of biological molecules, including proteins, lipids, and nucleic acids, considerable interest has been given to the role of this molecule in oxidative cellular damage (Alvarez and Radi 2003; Dalle-Donne et al. 2005). One of the changes produced by peroxynitrite in proteins is the nitration of tyrosine residues in the *ortho* position, used as a biomarker of nitrosative stress in human diseases (Ischiropoulos 1998; Dalle-Donne et al. 2005).

In these phenomena the nomenclature is at times somewhat confusing. For instance, the terms "nitrosation" and "nitrosylation" have been used indiscriminately. Nitrosation is the addition of a diatomic nitroso group (– NO), whilst nitrosylation is the addition of a nitrosyl radical group, stressing the concept of the addition of a chemical group that, if free, would be a radical. Finally, the term "nitration" describes the incorporation of a nitro triatomic group ($-NO_2$) and, in protein chemistry, is generally used to describe the incorporation of that group at position 3 of the phenolic ring of tyrosine residues (Martinez-Ruiz and Lamas 2004).

RNS are able to react with heme groups, thiols, or metal clusters and induce posttranslational modifications in the proteome, such as S-glutathionylation, cysteine oxidation, *S*-nitrosylation, metal nitrosylation, and tyrosine nitration, which can alter a great number of molecular targets in animal tissues. However, in plants, there are few data about these modifications.

Special interest has been paid to the interaction between $NO^{\cdot}$ and the thiol tripeptide, γ-glutamyl cysteinyl glycine (glutathione, GSH), which is one of the major low molecular weight soluble antioxidants in plant cells and is involved in the antioxidative ascorbate–glutathione cycle (Foyer 2001). But, GSH also has important functions apart from its antioxidative role, including the detoxification of xenobiotics and heavy metals (Steffens 1990; Cobbett and Goldsbrough 2002; Romero-Puertas et al. 2004a) and its involvement in signal transduction processes (Puppo et al. 2005). The *S*-nitrosylation reaction of NO˙ with GSH to form *S*-nitrosoglutathione (GSNO) has a significant physiological relevance in plants since GSNO is thought to function as a mobile reservoir of NO˙ bioactivity (Durner et al. 1999; Díaz et al. 2003), as reported in animals (Stamler et al. 2001). Besides, GSNO could also be formed by reaction of reduced GSH with peroxynitrite (Halliwell and Gutteridge 2000).

3
Recent Advances in the Metabolism of NO in Plant Cells

In animal cells, knowledge of the physiological and metabolic functions of NO˙ and other RNS has advanced significantly in recent years. As a result, researchers in plant biology started to pay attention to the interactions reported in animal cells between NO˙ and molecules like thiols, aminoacids, and proteins (Alvarez and Radi 2003). Some important physiological functions of NO˙ are performed through the so-called *S*-nitrosothiols (SNOs), which are formed by reaction of NO˙ with protein or non-protein sulfhydryl-containing compounds (Hogg 2000; Foster et al. 2003). SNOs carry out important indirect biological reactions, including NO˙ release, transnitrosation, *S*-thiolation, as well as having direct actions (Hogg 2000; Stamler et al. 2001). In animal cells, there is considerable evidence indicating that *S*-nitrosylation of the cysteine thiols of proteins is an important redox-based posttranslational modification (Stamler et al. 2001; Foster et al. 2003).

3.1
Enzymatic and Non-enzymatic Sources of NO

Nitric oxide can be produced in plants by non-enzymatic and enzymatic systems (for a review, see del Río et al. 2004). The enzyme nitrate reductase (NR) is a well-established generator of NO˙ (Dean and Harper 1988; Yamasaki et al. 1999; Rockel et al. 2002) and some authors have proposed that it is involved in the NO˙ generation which determines stomata closure (Desikan et al. 2002; García-Mata and Lamattina 2003). However, there is little information about the involvement of NR-derived NO˙ in plant stress (Kaiser and Huber 2001). NR does not produce NO˙ from L-arginine and, therefore, it cannot be considered a characteristic NOS. In the past decade, many plant biologists intensively searched for an NO-generating enzyme similar to the NOS identified in mammalian systems (Corpas et al. 2004c), and there are an increasing number of reports showing the presence of NOS-like activities in plants (for a review, see del Río et al. 2004). Recently, the presence of a constitutive arginine-dependent NOS activity in roots, stems, and leaves of pea plants was demonstrated (Corpas et al. 2006). At the subcellular level, a NOS activity sensitive to the characteristic inhibitors of animal NOS and with the same cofactor requirement (NADPH, BH_4, FAD, FMN, calcium, and calmodulin) has been identified in peroxisomes from pea leaves (Barroso et al. 1999; Corpas et al. 2004a,b). In parallel, a plant protein AtNOS1, which produces $NO^{\cdot}$ from L-arginine, has been identified in *Arabidopsis*, although this NOS activity does not depend on BH_4, FAD, and FMN as cofactors (Guo et al. 2003; Guo and Crawford 2005). To the best of our knowledge, peroxisomes and mitochondria are the only plant cell organelles where an arginine-dependent $NO^{\cdot}$ generation has been demonstrated (Corpas et al. 2001, 2004b; Guo and Crawford 2005). However, knowledge of the roles of these enzymatic $NO^{\cdot}$ sources in specific plant physiological processes is still limited.

3.2 Function of *S*-Nitroglutathione and *S*-Nitrosoglutathione Reductase

S-Nitroglutathione (GSNO) is considered to be a reservoir of $NO^{\cdot}$ and a long-distance signaling molecule, transporting glutathione-bound $NO^{\cdot}$ throughout the plant. In recent years, the glutathione-dependent enzyme formaldehyde dehydrogenase (FALDH; EC 1.2.1.1) has been demonstrated to have GSNO reductase (GSNOR) activity in bacteria, yeast, and mammals (Liu et al. 2001) and to be involved in the mechanism of protein *S*-nitrosation in mammalian cells (Haqqani et al. 2003). The enzyme GSNOR catalyzes the NADH-dependent reduction of GSNO to GSSG and NH_3 (Liu et al. 2001; Lamotte et al. 2005). Therefore, this enzyme is able to regulate the level of GSNO and indirectly the $NO^{\cdot}$ content and its availability. In plants, glutathione-dependent FALDH was also found to be very active in the reduction of GSNO in *Arabidopsis* (Sakamoto et al. 2002) and tobacco (Díaz et al. 2003). In tobacco plants, the gene encoding this enzyme was modulated in response to wounding, jasmonic acid, and salicylic acid (Díaz et al. 2003). The glutathione-dependent FALDH, also termed a class III alcohol dehydrogenase, has been purified and characterized in pea seeds (Uotila and Koivusalo 1979; Shafqat et al. 1996) and *Arabidopsis* plants (Martínez et al. 1996). Recently it has been reported that mutations of the *Arabidopsis* gene *AtGSNOR1*, encoding GSNOR, affect the extent of cellular *S*-nitrosothiol formation and turnover, and this seems to regulate multiple modes of plant disease resistance (Feechan et al. 2005). Therefore, GSNOR could be a new component in the regulation of $NO^{\cdot}$ metabolism.

In leaves of *Arabidopsis* plants treated with gaseous $NO^{\cdot}$, up to 52 proteins were identified that represented candidates for S-nitrosylation (Lindermayr et al. 2005). These included stress-related, redox-related, signaling/regulating, cytoskeleton and metabolic proteins (Lindermayr et al. 2005). In pea leaves, the GSNOR activity was reduced in plants grown with 200 mM NaCl, but treatment of pea leaves with the herbicide 2,4-dichlorophenoxyacetic acid

(2,4-D) produced an increase of the enzyme activity and transcripts (Barroso et al. unpublished results). All these data indicate that NO˙ and NO˙-derived molecules have a relevant physiological role in plants under normal and stress conditions. Moreover, it has been reported that NO˙ can also mediate gene regulation in plants (Grün et al. 2006).

4 RNS Metabolism and Abiotic Stress

Increased NO˙ production in response to certain abiotic stresses has been reported in different plant species. A summary of the main effects on NO˙ production, and on NOS activity of abiotic stresses produced by xenobiotics (paraquat, diquat, methylviologen), heavy metals, salinity, mechanical stress, ozone, environmental conditions (high light intensity, UV-B radiation, low and high temperature) etc. are presented in Table 2 However, many of the data available have been obtained by indirect studies using exogenous NO˙ donors and some of the results are contradictory, depending on the type of NO˙ donors (Murgia et al. 2004). Furthermore, in many cases it has not been clearly determined which is the specific endogenous source of NO˙ (e.g. non-enzymatic or enzymatic) and therefore it is difficult to establish a clear relationship between stress conditions and activation/inhibition of a specific NO˙ source.

Therefore, one of the goals of future research will be to establish how NO˙ and NO-related molecules vary under adverse environmental/abiotic stress conditions and to establish reliable markers of nitrosative stress.

4.1 Heavy Metals

Heavy metals occur naturally in small quantities in soil, but human activity has raised these to exceptionally high levels at many polluted land and water sites. Some of these metals can be micronutrients (Fe, Zn, Cu, Mo, and Mn) for plants, whereas others such as As, Hg, Cd, and Pb are very toxic even at low concentrations. There is very little information on the relationship between heavy metal toxicity and NO˙ metabolism.

That NO˙ donors were able to reverse the chlorotic phenotype of Fe-inefficient maize mutants has suggested a role for NO˙ in Fe nutrition (Graziano et al. 2002). Moreover, Fe and NO˙ can form Fe-nitrosyl complexes which could be involved in plant iron homeostasis (Graziano and Lamattina 2005).

On the other hand, *Cassia tora* plants pretreated for 12 h with 0.4 mM sodium nitroprusside (SNP), an NO˙ donor, and subsequently exposed for 24 h to 10 μM Al exhibited a significantly greater root elongation and a de-

Table 2 Nitric oxide and/or nitric oxide synthase activity in plants under several abiotic stress conditions

Plant species/ tissue	Abiotic stress agent	Effect on NO• production/ NOS activity/of NO• donors	Refs.
Glycine max			
Leaves	Herbicide treated	Increased NO• production	Klepper 1979
Solanum tuberosum			
Leaves	Methylviologen, diquat	NO• donors protect	Beligni and Lamattina 1999, 2002
Oryza sativa			
Leaves	Paraquat	NO• donors protect	Hung et al. 2002
Phaseolus aureus			
Leaves	H_2O_2	Increased NO• production	Lum et al. 2002
Medicago sativa			
Roots	Hypoxia	Increased NO• production	Dordas et al. 2003
Pisum sativum			
Leaves	Wilting	Increased NO• production	Leshem and Haramaty 1996
	Senescence	Decreased NO• and NOS activity	Corpas et al. 2004
	Continuous light or dark	No effect on NOS activity	Corpas et al., unpublished
	High light intensity	Increased NOS activity	Corpas et al., unpublished
	Low temperature	Increased NOS activity	Corpas et al., unpublished
	High temperature	No effect on NOS	Corpas et al., unpublished
	Mechanical wounding	Increased NOS activity	
Leaves and roots	Cadmium	Reduced NO• production	Corpas et al., unpublished
Helianthus annuus			Barroso et al. 2006
			Rodriguez-Serrano et al. 2006
Leaves	Cd	NO• donors protect	Laspina et al. 2005
Lupinus luteus			
Seeds and roots	Cd, lead, salinity	NO• donors protect	Kopyra and Gwózdz 2003
Cassia tora			
Roots	Aluminium	NO• donors protect	Wang and Yang 2005

Table 2 (continued)

Plant species/ tissue	Abiotic stress agent	Effect on $NO^{\cdot}$ production/ NOS activity/of $NO^{\cdot}$ donors	Refs.
Kalanchoe			
Leaves	Gravitational force	Increased $NO^{\cdot}$ production	Pedroso and Durzan 2000
Taxus cuspidate			
Suspension cells	Shear	Increased $NO^{\cdot}$ production	Gong and Yuan 2005
Phragmites australis			
Leaves	Ozone	Increased $NO^{\cdot}$ production	Velikova et al. 2005
Lycopersicon esculentum			
Leaves	Wounding	Defence genes inhibited by $NO^{\cdot}$	Orozco-Cárdenas and Ryan 2002
Arabidopsis thaliana			
Epidemal cells	Wounding	Increased $NO^{\cdot}$ production	Huang et al. 2004
Leaf cells	Mechanical stress	Increased $NO^{\cdot}$ production	Garces et al. 2001
	Ozone	Induced NOS activity	Rao and Davis 2001
	UV-B radiation	Induced NOS activity	Mackerness et al. 2001
Vicia faba			
Epidermal cells	UV-B radiation	Increased $NO^{\cdot}$ production	He et al. 2005
Phaseolus vulgaris			
Leaves	UV-B radiation	$NO^{\cdot}$ donors protect	Shi et al. 2005
Zea mays			
Leaves	UV-B radiation	Induced NOS activity	An et al. 2005
	Salinity	$NO^{\cdot}$ donors protect	Zhang et al. 2006
Olea europaea			
Leaves	Salinity	Induced NOS activity and $NO^{\cdot}$ production	Valderrama et al., unpublished

Table 2 (continued)

Plant species/ tissue	Abiotic stress agent	Effect on $NO^{\cdot}$ production/ NOS activity/of $NO^{\cdot}$ donors	Refs.
Phragmites communis			
Callus	Salinity	$NO^{\cdot}$ donors protect	Zhao et al. 2004
Nicotiana tabacum			
Leaf cells	Salt, osmotic and heat stress	Increased $NO^{\cdot}$ production	Gould et al. 2003
	Light stress and mechanical injury	No effect	
Triticum aestivum			
Leaves	Osmotic stress	$NO^{\cdot}$ donors protect	Xing et al. 2004
Detached leaves	Drought	$NO^{\cdot}$ donors protect	García-Mata and Lamattina 2001

crease in Al accumulation in root apexes as compared with plants without SNP treatment (Wang and Yang 2005).

In roots of *Pisum sativum* and *Brassica juncea* in the presence of 100 μM Cu, Zn, and Cd there was a time-dependent endogenous NO˙ production (Bartha et al. 2005). These data contrast with the results observed in leaves and roots of *Pisum sativum* grown with a toxic Cd concentration (50 μM), which produced growth inhibition and oxidative damages (Sandalio et al. 2001; Romero-Puertas et al. 2002, 2004b), and a drastic reduction of the endogenous NO˙ content (Rodríguez-Serrano 2006; Barroso et al. 2006). Furthermore, the decline of leaf NO˙ content was accompanied by a reduction of GSH, GSNO, and GSNOR activity and transcript (Barroso et al. 2006). In contrast, pretreatment of seedlings with 100 μM SNP protected sunflower leaves against Cd-induced oxidative stress (Laspina et al. 2005). A similar situation has been described in *Lupinus* roots grown with 50 μM Cd and 1.5 mM Pb, and it was proposed that the protective effect of NO˙ could involve the stimulation of superoxide dismutase activity to counteract the overproduction of superoxide radicals (Kopyra and Gwóźdz 2003).

All these data indicate the importance of exogenous NO˙ in the uptake of micronutrients and in the protection against deleterious effects of toxic heavy metals such as Cd or Al. However, the physiological implication of the plant endogenous NO˙ in the response to heavy metal stress is still not known.

4.2 Salinity

Salinity is one of the major abiotic stresses affecting plant productivity due to its negative effects on plant growth, ion balance, and water relations (Hasegawa and Bressan 2000). Application of NO˙ donors in callus of *Phragmites communis* exposed to 200 mM NaCl revealed that NO˙ affected the K/Na ratio by increasing a plasma membrane H^+-ATPase activity (Zhao et al. 2004). In the same way, application of NO˙ donors enhanced maize tolerance to salt stress by elevating the activities of proton-pump and Na^+/H^+ antiport of the tonoplast (Zhang et al. 2006). Germination of *Lupinus luteus* is inhibited by 200 mM NaCl, but preincubation of seeds with the NO˙ donor SNP, restored germination (Kopyra and Gwóźdz 2003).

In olive (*Olea europaea*) plants, it has been shown that salinity (200 mM NaCl) produced a 40% reduction in leaf fresh weight and induced oxidative stress (Valderrama et al. 2006). Salinity also caused an increase in the number of proteins that underwent tyrosine nitration, as evaluated by immunoblotting and confocal laser scanning microscopy analysis using a specific antibody against 3-nitrotyrosine (Fig. 1b–d). This effect was accompanied by an increase of other RNS such as endogenous NO˙ and GSNO (Valderrama et al., unpublished). Taken together, these data show that in olive plants salinity inhibits leaf growth and induces oxidative stress (Valderrama et al. 2006), and,

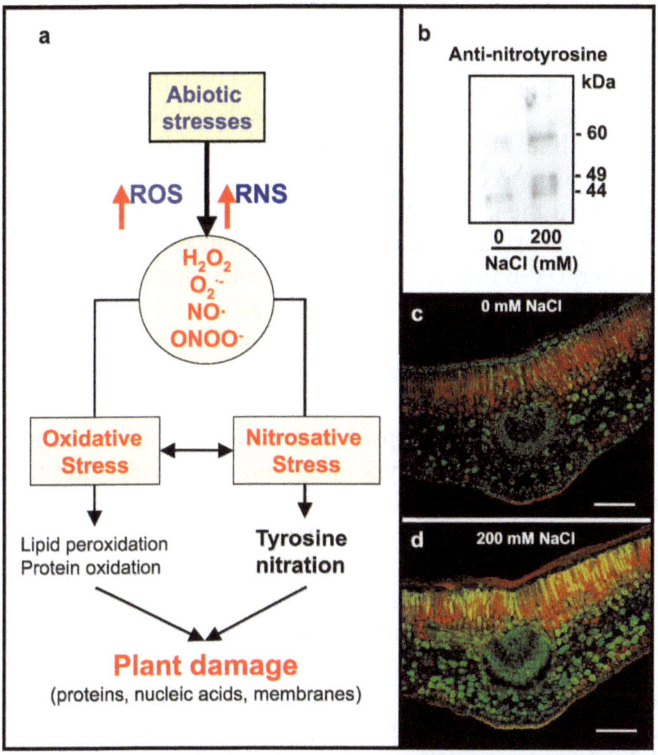

Fig. 1 **a** Model of the involvement of RNS and ROS in damage produced in plants by abiotic stress conditions. Overproduction of these species can cause the tyrosine nitration of proteins, lipid peroxidation and protein oxidation, which are markers of nitrosative and oxidative stress, respectively. **b** Immunoblot of leaf extracts from control and NaCl-treated olive plants probed with an antibody against 3-nitrotyrosine (nTyr) to detect proteins that underwent tyrosine nitration. **c,d** Representative images illustrating the confocal laser scanning microscopy (CLSM) detection of nTyr in olive leaf sections of plants grown under optimal conditions and in the presence of 200 mM NaCl, respectively. The *bright green fluorescence* corresponds to the detection of nTyr proteins. The *orange-yellow color* corresponds to the autofluorescence. Each picture was prepared from 30–40 cross-sections, which were analyzed by CLSM. The *scale bars* represent 200 μm

interestingly, this toxic situation is also accompanied by alterations in the RNS metabolism.

4.3 Atmospheric Pollutants and Other Abiotic Stresses

There are several studies dealing with the relationship between $NO^{\cdot}$ metabolism and UV-B radiation and ozone (Table 2). UV-B radiation (280–320 nm) has increased as consequence of the destruction of the ozone layer and this radiation clearly affects plant growth and usually induces oxidative stress. In *Arabidopsis thaliana* a NOS inhibitor and an NO˙ scavenger partially blocked the induction of the chalcone synthase gene by UV-B (Mackerness et al. 2001). In maize seedlings, UV-B radiation strongly induced NOS activity and caused a decrease of both leaf biomass and exo- and endo β-glucanase activity (An et al. 2005). In bean seedlings subjected to UV-B radiation, exogenous NO˙ partially alleviated the UV-B effect characterized by a decrease of the chlorophyll contents and oxidative damage to the thylakoid membrane (Shi et al. 2005). Moreover, UV-B induced stomatal closure which was mediated by NO˙ and H_2O_2, and the generation of NO˙ was produced by a NOS-like activity (He et al. 2005). However, other authors reported that the NO˙ generated in the guard cells was produced by the nitrate reductase activity (Bright et al. 2006).

Ozone treatment of *Arabidopsis* plants induced NOS activity and this preceded accumulation of salicylic acid and cell death. Additionally, NO˙ treatment has been shown to increase the levels of ozone-induced ethylene production and leaf injury (Rao and Davis 2001). In another report, using leaves of *Phragmites* exposed to ozone, it has been shown that endogenous isoprene was able to decrease the concentration of NO˙ (Velikova et al. 2005). Isoprene is one of the most abundant volatile organic compounds emitted by plants.

NO˙ has also been implicated in other stresses. Thus, centrifugation of either cell cultures or leaves from several species induced NO˙ production that led to cell death responses (Garcês et al. 2001; Pedroso and Durzan 2000; Pedroso et al. 2000). In *Taxus cuspidata* cells, NO˙ was generated after 8 h of shear stress. In this case, NO˙ formation was inhibited by N^{ω}-nitro-L-arginine (LNNA), a NOS inhibitor, indicating the involvement of this activity in the response to such a stressing agent. Moreover, the glutathione S-transferase (GST) activity, a principal enzyme responsible for detoxification of xenobiotics, decreased during shear stress, and this could be somehow mediated by tyrosine nitration of this protein. In fact, this inactivation was partially recovered when NOS inhibitors or NO˙ scavengers were added to cell cultures during shear stress (Gong and Yuan 2005).

Exogenous application of NO˙ induces stomatal closure in wheat and increased tolerance to drought stress, which was accompanied by an accumulation of a group of late embryogenesis abundant (LEA) proteins (García-Mata and Lamatina 2001).

In wheat seedlings, the osmotic stress produced by treatment with 0.4 M manitol reduced leaf water loss while increasing the leaf ABA content. These effects were partially reversed by NO˙ scavengers and NOS activity inhibitors (Xing et al. 2004). In tomato detached leaves, the application of NO˙ donors inhibited the synthesis of proteinase inhibitor I and the H_2O_2 generated in response to mechanical wounding (Orozco-Cárdenas and Ryan 2002).

Hypoxia induced the production of NO˙ in alfalfa roots and maize cell-suspension cultures, responses that appear to be mediated by class 1 hemoglobin (Dordas et al. 2003, 2004).

More than 25 years ago, it was shown that the application of herbicides to soybean plants increased the NO˙ production (Klepper 1979). More recently, several studies have confirmed that treatment with NO˙ donors protect potato and rice leaves against the deleterious effects of the methylviologen herbicides, diquat and paraquat, and that NO˙ acts as a ROS scavenger (Beligni and Lamattina 1999, 2002; Huang et al. 2002)

The analysis of NOS activity in leaves of pea plants exposed to six environmental/abiotic stress conditions (continuous light, high light intensity, continuous dark, low- and high-temperature, and mechanical wounding) showed that NOS activity was significantly induced only by high light intensity, low temperature, and mechanical wounding (F.J. Corpas et al., unpublished results). The same pattern of protein nitrotyrosylation was observed except under the cold treatment where nitrosylation was induced. However, GSNOR activity, which regulates the level of GSNO, was induced in leaves of pea seedlings exposed to all the environmental/abiotic stress conditions except for high light intensity (F.J. Corpas et al., unpublished results).

5 Nitrotyrosine as a Biological Marker of Nitrosative Stress

In plant cells it is necessary to establish parameters to define whether a specific abiotic condition induces nitrosative stress, as established for oxidative stress using lipid peroxidation and protein oxidation as markers (Fig. 1a). Protein tyrosine nitration is a covalent modification of proteins resulting from the addition of a nitro ($-NO_2$) group to one of the two equivalent *ortho* carbons of the aromatic ring of tyrosine residues (Gow et al. 2004). This process can be mediated by peroxynitrite (Radi 2004) and has been routinely used as a marker of pathological disease and oxidative stress in animal cells (Ischiropoulos 2003). However, in plant cells there is little information on nitrosative stress. In suspension cultures of *Taxus cuspidate* an increase of 31% in the free 3-nitrotyrosine content during shear stress has been reported (Gong and Yuan 2005). This result together with that of the increase of tyrosine nitration in leaves of olive plants by salinity (Fig. 1b), indicates that certain stress conditions are associated with the increase of protein nitrotyrosylation. An additional conclusion is that in plant cells the 3-nitrotyrosine content could be used as a marker of nitrosative stress, as proposed for animal cells. However, the possibility cannot be discarded that nitration of proteins could have other important biological functions such as protein turnover or signal transduction (Ischropoulos 2003).

6
Conclusions and Future Prospects

Although much has been learned about the relationship between NO˙ and different stress conditions, our knowledge of the metabolism of endogenous NO˙ and other RNS is still rudimentary. There is a considerable lack of knowledge on the function of NO˙ and other RNS in the biochemistry and physiology of plants under stress conditions. Identification of the specific sources of NO˙ at tissue and subcellular levels, and of its involvement in each physiological process is a basic challenge. Identification and characterization of the RNS targets under physiological and stress conditions will help understand how these molecules participate in the mechanism of response to environmental stress. Protein nitrotyrosylation could be used as a criteria to establish the concept of nitrosative stress in plants, independently of the overproduction of NO˙ or NO-derived molecules. On the other hand, abiotic-induced oxidative stress is not necessarily accompanied by nitrosative stress (Barroso et al. 2006). Further research is necessary to solve the puzzle of relationships between ROS, RNS, and abiotic stresses, and to establish biotechnological strategies against abiotic stresses, which are responsible for important losses in plant yield and crop productivity.

Acknowledgements This work was supported by grants from the DGES, Ministry of Science and Technology (AGL2003-05524), Universidad de Jaén (OA/2/2004), and Junta de Andalucía (groups CVI 0192 and CVI 0286), Spain.

References

Alvarez B, Radi R (2003) Peroxynitrite reactivity with amino acids and proteins. Amino Acids 25:295–311

An LZ, Liu YH, Zhang MX, Chen T, Wang XL (2005) Effects of nitric oxide on growth of maize seedling leaves in the presence or absence of ultraviolet-B radiation. J Plant Physiol 162:317–326

Bartha B, Kolbert Z, Erdei L (2005) Nitric oxide production induced by heavy metals in *Brassica juncea* L. Czern. and *Pisum sativum* L. Acta Biol Szegediensis 49(1–2):9–12

Barroso JB, Corpas FJ, Carreras A, Sandalio LM, Valderrama R, Palma JM, Lupiáñez JA, del Río LA (1999) Localization of nitric oxide synthase in plant peroxisomes. J Biol Chem 274:36729–36733

Barroso JB, Corpas FJ, Carreras A, Rodríguez-Serrano M, Esteban FJ, Fernández-Ocaña A, Chaki M, Romero-Puertas MC, Valderrama R, Sandalio LM, del Río LA (2006) Localization of *S*-nitrosoglutathione and expression of *S*-nitrosoglutathione reductase in pea plants under cadmium stress. J Exp Bot 57:1785–1793

Beligni MV, Lamattina L (1999) Nitric oxide protects against cellular damage produced by methylviologen herbicides in potato plants. Nitric Oxide 3:199–208

Beligni MV, Lamattina L (2002) Nitric oxide interferes with plant photo-oxidative stress by detoxifying reactive oxygen species. Plant Cell Environ 25:737–748

Bright J, Desikan R, Hancock JT, Weir IS, Neill SJ (2006) ABA-induced NO generation and stomatal closure in Arabidopsis are dependent on H_2O_2 synthesis. Plant J 45:113–122

Chung KK, Dawson TM, Dawson VL (2005) Nitric oxide, S-nitrosylation and neurodegeneration. Cell Mol Biol 51:247–254

Cobbett C, Goldsbrough P (2002) Phytochelatins and metallothioneins: roles in heavy metal detoxification and homeostasis. Annu Rev Plant Biol 53:159–182

Corpas FJ, Barroso JB, Carreras A, Quirós M, León AM, Romero-Puertas MC, Esteban FJ, Valderrama R, Palma JM, Sandalio LM, Gómez M, del Río LA (2004a) Cellular and subcellular localization of endogenous nitric oxide in young and senescent pea plants. Plant Physiol 136:2722–2733

Corpas FJ, Barroso JB, Carreras A, Valderrama R, Palma JM, León AM, Sandalio LM, del Río LA (2006) Constitutive arginine-dependent nitric oxide synthase activity in different organs of pea seedlings during plant development. Planta 224:246–254

Corpas FJ, Barroso JB, León AM, Carreras A, Quirós M, Palma JM, Sandalio LM, del Río LA (2004b) Peroxisomes as a source of nitric oxide. In: Magalhaes JR, Singh RP, Passos LP (eds) Nitric oxide signaling in plants. Studium, Houston USA, pp 111–129

Corpas FJ, Barroso JB, del Río LA (2001) Peroxisomes as a source of reactive oxygen species and nitric oxide signal molecules in plant cells. Trends Plant Sci 6:145–150

Corpas FJ, Barroso JB, del Río LA (2004c) Enzymatic sources of nitric oxide in plant cells - beyond one protein-one function. New Phytol 162:246–248

Dalle-Donne I, Scaloni A, Giustarini D, Cavarra E, Tell G, Lungarella G, Colombo R, Rossi R, Milzani A (2005) Proteins as biomarkers of oxidative/nitrosative stress in diseases: the contribution of redox proteomics. Mass Spectrom Rev 24:55–99

Dean JV, Harper JE (1988) The conversion of nitrite to nitrogen oxide(s) by the constitutive NAD(P)H-nitrate reductase enzyme from soybean. Plant Physiol 88:389–395

Desikan R, Griffiths R, Hancock J, Neill S (2002) A new role for an old enzyme: nitrate reductase-mediated nitric oxide generation is required for abscisic acid-induced stomatal closure in *Arabidopsis thaliana*. Proc Natl Acad Sci USA 99:16314–16318

Díaz M, Achkor H, Titarenko E, Martínez MC (2003) The gene encoding glutathione-dependent formaldehyde dehydrogenase/GSNO reductase is responsive to wounding, jasmonic acid and salicylic acid. FEBS Lett 543:136–139

Dordas C, Hasinoff BB, Rivoal J, Hill RD (2004) Class-1 hemoglobins, nitrate and NO levels in anoxic maize cell-suspension cultures. Planta 219:66–72

Dordas C, Hasinoff BB, Igamberdiev AU, Manac'h N, Rivoal J, Hill RD (2003) Expression of a stress-induced hemoglobin affects NO levels produced by alfalfa root cultures under hypoxic stress. Plant J 35:763–770

Durner J, Klessig DF (1999) Nitric oxide as a signal in plants. Curr Opin Plant Biol 2:369–374

Durzan D (2002) Stress-induced nitric oxide and adaptive plasticity in conifers. J Forest Sci 48:281–291

Feechan A, Kwon E, Yun BW, Wang Y, Pallas JA, Loake GJ (2005) A central role for S-nitrosothiols in plant disease resistance. Proc Natl Acad Sci USA 102:8054–8059

Foster MW, McMahon TJ, Stamler JS (2003) S-Nitrosylation in health and disease. Trends Mol Med 9:160–168

Foyer CH (2001) Prospects for enhancement of the soluble antioxidants, ascorbate and glutathione. Biofactors 15:75–78

Garcês H, Durzan D, Pedroso MC (2001) Mechanical stress elicits nitric oxide formation and DNA fragmentation in *Arabidopsis thaliana*. Ann Bot 87:567–574

García-Mata CG, Lamattina L (2001) Nitric oxide induces stomatal closure and enhances the adaptive plant responses against drought stress. Plant Physiol 126:1196–1204

García-Mata C, Lamattina L (2003) Abscisic acid, nitric oxide and stomatal closure – is nitrate reductase one of the missing links? Trends Plant Sci 8:20–26

Gong YW, Yuan YJ (2006) Nitric oxide mediates inactivation of glutathione S-transferase in suspension culture of *Taxus cuspidata* during shear stress. J Biotechnol 123: 185–92

Gould KS, Lamotte O, Klinguer A, Pugin A, Wendehenne D (2003) Nitric oxide production in tobacco leaf cells: a generalized stress response? Plant Cell Environ 26:1851–1862

Gow AJ, Farkouh CR, Munson DA, Posencheg MA, Ischiropoulos H (2004) Biological significance of nitric oxide-mediated protein modifications. Am J Physiol Lung Cell Mol Physiol 287:L262–268

Graziano M, Beligni MV, Lamattina L (2002) Nitric oxide improves internal iron availability in plants. Plant Physiol 130:1852–1859

Graziano M, Lamattina L (2005) Nitric oxide and iron in plants: an emerging and converging story. Trends Plant Sci 10:4–8

Grün S, Lindermayr C, Sell S, Durner J (2006) Nitric oxide and gene regulation in plants. J Exp Bot 57:507–516

Guo FQ, Okamoto M, Crawford NM (2003) Identification of a plant nitric oxide synthase gene involved in hormonal signaling. Science 302:100–103

Guo FQ, Crawford NM (2005) Arabidopsis nitric oxide synthase1 is targeted to mitochondria and protects against oxidative damage and dark-induced senescence. Plant Cell 17:3436–3450

Halliwell B, Gutteridge JMC (2000) Free radicals in biology and medicine, 3rd edn. Oxford University Press, Oxford, UK

Haqqani AS, Do SK, Birnboim HC (2003) The role of a formaldehyde dehydrogenase-glutathione pathway in protein S-nitrosation in mammalian cells. Nitric Oxide 9:172–181

Hasegawa PM, Bressan RA (2000) Plant cellular and molecular responses to high salinity. Annu Rev Plant Physiol Plant Mol Biol 51:463–99

Hausladen A, Stamler JS (1999) Nitrosative stress. Methods Enzymol 300:389–395

He J-M, Xu H, She X-P, Song X-G, Zhao W-M (2005) The role and the interrelationship of hydrogen peroxide and nitric oxide in the UV-B-induced stomatal closure in broad bean. Funct Plant Biol 32:237–247

Hogg N (2000) Biological chemistry and clinical potential of S-nitrosothiols. Free Rad Biol Med 28:1478–1486

Huang X, Stettmaier K, Michel C, Hutzler P, Mueller MJ, Durner J (2004) Nitric oxide is induced by wounding and influences jasmonic acid signaling in *Arabidopsis thaliana*. Planta 218:938–946

Hung KT, Chang CJ, Kao CH (2002) Paraquat toxicity is reduced by nitric oxide in rice leaves. J Plant Physiol 159:159–166

Ischiropoulos H (1998) Biological tyrosine nitration: a pathophysiological function of nitric oxide and reactive oxygen species. Arch Biochem Biophys 356:1–11

Ischiropoulos H (2003) Biological selectivity and functional aspects of protein tyrosine nitration. Biochem Biophys Res Commun 305:776–783

Kaiser WM, Huber SC (2001) Post-translational regulation of nitrate reductase: mechanism, physiological relevance and environmental triggers. J Exp Bot 52:1981–1989

Klatt P, Lamas S (2000) Regulation of protein function by S-glutathiolation in response to oxidative and nitrosative stress. Eur J Biochem 267(16):4928–4944

Klepper L (1979) Nitric-oxide (NO) and nitrogen-dioxide (NO_2) emissions from herbicide-treated soybean plants. Atmos Environ 13:537–542

Kopyra M, Gwóżdz E (2003) Nitric oxide stimulates seed germination and counteracts the inhibitory effect of heavy metals and salinity on root growth of *Lupinus luteus*. Plant Physiol Biochem 41:1011–1017

Lamattina L, García-Mata C, Graziano M, Pagnussat G (2003) Nitric oxide: the versatility of an extensive signal molecule. Annu Rev Plant Biol 54:109–36

Lamotte O, Courtois C, Barnavon L, Pugin A, Wendehenne D (2005) Nitric oxide in plants: the biosynthesis and cell signaling properties of a fascinating molecule. Planta 221:1–4

Laspina VN, Groppas MD, Tomaro ML, Benavides MP (2005) Nitric oxide protects sunflower leaves against Cd-induced oxidative stress. Plant Sci 169:323–330

Leshem YY, Haramaty E (1996) The characterization and contrasting effects of the nitric oxide free radical in vegetative stress and senescence of *Pisum sativum* Linn. foliage. J Plant Physiol 148:258–263

Lindermayr C, Saalbach G, Durner J (2005) Proteomic identification of S-nitrosylated proteins in Arabidopsis. Plant Physiol 137:921–930

Liu L, Hausladen A, Zeng M, Que L, Heitman J, Stamler JS (2001) A metabolic enzyme for *S*-nitrosothiol conserved from bacteria to humans. Nature 410:490–494

Lum HK, Butt YK, Lo SC(2002) Hydrogen peroxide induces a rapid production of nitric oxide in mung bean (*Phaseolus aureus*). Nitric Oxide 6:205–213

Mackerness SAH, John CF, Jordan B, Thomas B (2001) Early signaling components in ultraviolet-B responses: distinct roles for different reactive oxygen species and nitric oxide. FEBS Lett 489:237–242

Martínez MC, Achkor H, Persson B, Fernández MR, Shafqat J, Farrés J, Jörnvall H, Parés X (1996) Arabidopsis formaldehyde dehydrogenase. Molecular properties of plant class III alcohol dehydrogenase provide further insights into the origins, structure and function of plant class P and liver class I dehydrogenases. Eur J Biochem 241:849–857

Martínez-Ruíz A, Lamas S (2004) *S*-Nitrosylation: a potential new paradigm in signal transduction. Cardiovasc Res 62:43–52

Murgia I, de Pinto MC, Delledonne M, Soave C, De Gara L (2004) Comparative effects of various nitric oxide donors on ferritin regulation, programmed cell death, and cell redox state in plant cells. J Plant Physiol 161:777–783

Neill SJ, Desikan R, Hancock JT (2003) Nitric oxide signaling in plants. New Phytol 159:11–35

Orozco-Cárdenas ML, Ryan CA (2002) Nitric oxide negatively modulates wound signaling in tomato plants. Plant Physiol 130:487–493

Pedroso MC, Durzan DJ (2000) Effect of different gravity environments on DNA fragmentation and cell death in *Kalanchoe leaves*. Ann Bot (Lond) 86:983–994

Pedroso MC, Magalhaes JR, Durzan D (2000) Nitric oxide induces cell death in Taxus cells. Plant Sci 157:173–180

Pfeiffer S, Mayer B, Hemmens B (1999) Nitric oxide: chemical puzzles posed by a biological messenger. Angew Chem Int Ed 38:1714–1731

Pryor WA, Squadrito GL (1995) The chemistry of peroxynitrite: a product from the reaction of nitric oxide with superoxide. Am J Physiol 268:L699–722

Puppo A, Groten K, Bastian F, Carzaniga R, Soussi M, Lucas MM, de Felipe MR, Harrison J, Vanacker H, Foyer CH (2005) Legume nodule senescence: roles for redox and hormone signaling in the orchestration of the natural aging process. New Phytol 165:683–701

Radi R (2004) Nitric oxide, oxidants, and protein tyrosine nitration. Proc Natl Acad Sci USA 101:4003–4008

Rao MV, Davis KR (2001) The physiology of ozone induced cell death. Planta 213:682–690

Redondo-Horcajo M, Lamas S (2005) Oxidative and nitrosative stress in kidney disease: a case for cyclosporine A. J Nephrol 18:453–457

Ricciardolo FL, Di Stefano A, Sabatini F, Folkerts G (2006) Reactive nitrogen species in the respiratory tract. Eur J Pharmacol 533:240–252

del Río LA, Corpas FJ, Barroso JB (2004) Nitric oxide and nitric oxide synthase activity in plants. Phytochemistry 65:783–792

Rodríguez-Serrano M, Romero-Puertas MC, Zabalza A, Corpas FJ, Gómez M, del Río LA, Sandalio LM (2006) Cadmium effect on the oxidative metabolism of pea (*Pisum sativum* L.) roots. Imaging of ROS and NO accumulation in vivo. Plant Cell Environ 29:1532–1544

Rockel P, Strube F, Rockel A, Wildt J, Kaiser WM (2002) Regulation of nitric oxide (NO) production by plant nitrate reductase in vivo and in vitro. J Exp Bot 53:103–110

Romero-Puertas MC, McCarthy I, Gómez M, Sandalio LM, Corpas FJ, del Río LA, Palma JM (2004a) Reactive oxygen species-mediated enzymatic systems involved in the oxidative action of 2,4-dichlorophenoxyacetic acid. Plant Cell Environ 27:1135–1148

Romero-Puertas MC, Palma JM, Gómez M, del Río LA, Sandalio LM (2002) Cadmium causes the oxidative modification of proteins in pea plants. Plant Cell Environ 25:677–686

Romero-Puertas MC, Rodríguez-Serrano M, Corpas FJ, Gómez M, del Río LA, Sandalio LM (2004b) Cadmium-induced subcellular accumulation of $O_2^{\bullet-}$ and H_2O_2 in pea leaves. Plant Cell Environ 27:1122–1134

Sakamoto A, Ueda M, Morikawa H (2002) Arabidopsis glutathione-dependent formaldehyde dehydrogenase is an *S*-nitrosoglutathione reductase. FEBS Lett 515:20–24

Sandalio LM, Dalurzo HC, Gómez M, Romero-Puertas MC, del Río LA (2002) Cadmium-induced changes in the growth and oxidative metabolism of pea plants. J Exp Bot 52:2115–2126

Shafqat J, El-Ahmad M, Danielsson O, Martínez MC, Persson B, Parés X, Jörnvall H (1996) Pea formaldehyde-active class III alcohol dehydrogenase: common derivation of the plant and animal forms but not of the corresponding ethanol-active forms (classes I and P). Proc Natl Acad Sci USA 93:5595–5599

Shapiro AD (2005) Nitric oxide signaling in plants. Vitam Horm 72:339–398

Shi S, Wang G, Wang Y, Zhang L, Zhang L (2005) Protective effect of nitric oxide against oxidative stress under ultraviolet-B radiation. Nitric Oxide 13:1–9

Stamler JS, Lamas S, Fang FC (2001) Nitrosylation: the prototypic redox-based signaling mechanism. Cell 106:675–683

Steffens JC (1990) The heavy metal-binding peptide of plants. Annu Rev Plant Physiol Plant Mol Biol 41:553–575

Stevens MJ (2005) Oxidative-nitrosative stress as a contributing factor to cardiovascular disease in subjects with diabetes. Curr Vasc Pharmacol 3:253–266

Uotila L, Koivusalo M (1979) Purification of formaldehyde and formate dehydrogenases from pea seeds by affinity chromatography and *S*-formylglutathione as the intermediate of formaldehyde metabolism. Arch Biochem Biophys 196:33–45

Valderrama R, Corpas FJ, Carreras A, Gómez-Rodríguez MV, Chaki M, Pedrajas JR, Fernández-Ocaña A, del Río LA, Barroso JB (2006) The dehydrogenase-mediated recycling of NADPH is a key antioxidant system against salt-induced oxidative stress in olive plants. Plant Cell Environ 29:1449–1459

Velikova V, Pinelli P, Pasqualini S, Reale L, Ferranti F, Loreto F (2005) Isoprene decreases the concentration of nitric oxide in leaves exposed to elevated ozone. New Phytol 166:419–425

Wang YS, Yang ZM (2005) Nitric oxide reduces aluminum toxicity by preventing oxidative stress in the roots of *Cassia tora* L. Plant Cell Physiol 46:1915–1923

Xing H, Tan L, An L, Zhao Z, Wang S, Zhang C (2004) Evidence for the involvement of nitric oxide and reactive oxygen species in osmotic stress tolerance of wheat seedlings: Inverse correlation between leaf abscisic acid accumulation and leaf water loss. Plant Growth Regul 42:61–68

Yamasaki H, Sakihama Y, Takahashi S (1999) An alternative pathway for nitric oxide production in plants: new features of an old enzyme. Trends Plant Sci 4:128–129

Zhao L, Zhang F, Guo J, Yang Y, Li B, Zhang L (2004) Nitric oxide functions as a signal in salt resistance in the calluses from two ecotypes of reed. Plant Physiol 134:849–857

Zhang Y, Wang L, Liu Y, Zhang Q, Wei Q, Zhang W (2006) Nitric oxide enhances salt tolerance in maize seedlings through increasing activities of proton-pump and Na(+)/H(+) antiport in the tonoplast. Planta 224:545–555

Zhao Z, Chen G, Zhang C (2001) Interaction between reactive oxygen species and nitric oxide in drought-induced abscisic acid synthesis in root tips of wheat seedlings Aust J Plant Physiol 28:1055–1061

Plant Cell Monogr (6)
L. Lamattina · J. C. Polacco: Nitric Oxide in Plant Growth
DOI 10.1007/7089_2006_093/Published online: 11 October 2006

Nitric Oxide-Mediated Signaling Functions During the Plant Hypersensitive Response

Matteo De Stefano · Elodie Vandelle · Annalisa Polverari · Alberto Ferrarini · Massimo Delledonne (✉)

Dipartimento Scientifico e Tecnologico, Università degli Studi di Verona,
Strada le Grazie 15, 37134 Verona, Italy
massimo.delledonne@univr.it

Abstract Growing evidence suggests that nitric oxide (NO) is a central molecule in several physiological functions, ranging from plant development to defence responses. Plants use NO as a signaling molecule in pathways comparable to those of mammals, suggesting the existence of many commonalities between the action of NO in plants and animals.

In this chapter, we examine the mechanisms through which plants respond to pathogen challenge and focus on inherent NO signaling functions. In particular, we describe the major NO-generating systems and their involvement in the response to pathogen attack, the role of NO in the activation of the hypersensitive response (HR), and the participation in the establishment of systemic acquired resistance (SAR). Next we describe the evidence for the involvement of Ca^{2+}, cyclic GMP (cGMP), cyclic ADP-ribose (cADPR), and mitogen-activated protein kinases (MAPK) in NO-mediated signal cascades, the role of NO in posttransational protein modifications, and its participation in modulating gene expression.

1 Introduction

NO is a bioactive molecule that regulates an ever-growing list of biological processes in phylogenetically distant species (Beligni and Lamattina 2001). It is a gaseous free radical with a relatively short half-life, estimated to be less than 6 s (Bethke et al. 2004). This extremely short half-life reflects the highly reactive nature of NO, due to its molecular structure comprising an unshared electron. Its broad chemistry involves an interplay between three species differing in their physical properties and chemical reactivity: the nitrosonium cation (NO^+), the radical (NO), and the nitroxyl anion (NO^-) (Neill et al. 2003). It also begets a number of chemical offspring, including NO_2, N_2O_3, and $ONOO^-$. These NO species, collectively termed reactive nitrogen species (RNS), are generated during the oxidative metabolism of NO and possess their own unique biological effects. Typically, NO rapidly reacts with reactive oxygen species (ROS), proteins (especially those containing reactive amino acids such as cysteine and tyrosine), as well as with various receptors and transcription factors (Romero-Puertas et al. 2004). Since its discovery as a biological messenger molecule, NO is now well recognized for its involvement in diverse animal biological processes, including vasodilation, bronchodilation, neurotransmission, tumor surveillance, antimicrobial defense, and regulation of inflammatory–immune processes. By the late 1990s, NO has also emerged as an important messenger in plant signaling, playing a crucial role in the regulation of normal physiological processes in plants, including stomatal closure, growth and development (Neill et al. 2002; Guo et al. 2003; Pagnussat et al. 2003), and during plant–pathogen interactions (Delledonne et al. 1998; Durner et al. 1998). The role of NO in the response of plants to pathogens is the object of the present chapter and will be discussed in detail.

Plants protect themselves against pathogen attacks using constitutive resistance provided by both physical and chemical barriers, such as the cuticle, cell wall, and preformed chemical compounds, but also by active resistance mechanisms, which are induced when a pathogen overcomes these constitutive barriers and is recognized as a potential invader. Active resistance is characterized by a complex set of events that can eventually stop pathogen spreading and can work at the species level or at the race/cultivar level of interaction. At the species level, plants can exhibit an active non-host resistance against some pathogens, which means that all genotypes within the species are equally able to react and stop infection by all races of a given pathogen. This reaction is due to the perception of pathogen-associated molecular patterns (PAMPs), also termed general elicitors, which are characteristic of microbial organisms (Nurnberger et al. 2004). At the race/cultivar level, the ability of the otherwise susceptible plant species to recognize the pathogen and to activate these responses is regulated in a gene-for-gene specific manner by the direct or indirect interaction between the products of a plant resistance (R) gene and a pathogen avirulence (Avr) gene (McDowell and Woffenden 2003). When either the plant or the pathogen lacks its cognate gene, activation of the plant's defense responses either fails or is delayed sufficiently so that pathogen colonization ensues.

Both kinds of resistance are characterized by the activation of a largely overlapping set of defense responses. Pathogen recognition (PAMP or R/Avr) results in the activation of signal transduction pathways, involving ion fluxes (Atkinson et al. 1996; Heath 2000; Wendehenne et al. 2002), protein phosphorylation/dephosphorylation, production of ROS (Blumwald 1997; Lamb and Dixon 1997; Nurnberger and Scheel 2001) and NO (Delledonne et al. 2001; Lamotte et al. 2004), up to the expression of defense-related genes, encoding for example, pathogenesis-related (PR) proteins (Fritig et al. 1998) and enzymes responsible for antimicrobial compound synthesis such as phytoalexins (Somssich 1998).

Plant defence responses are frequently associated with rapid and localized cell death, known as the hypersensitive response (HR) (Hammond-Kosack and Jones 1996), which contributes to limiting pathogen development. Based on several arguments, the HR is considered by some authors to be programmed cell death (PCD): this active process requires protein synthesis for its induction and presents some apoptotic features, such as chromatin condensation, formation of apoptotic-like bodies (Levine et al. 1996), changes in mitochondrial morphology (Wakabayashi and Karbowski 2001), membrane dysfunction, endonucleolytic cleavage of DNA, and protease activities (Heath 2000) such as caspase-like activities (Elbaz et al. 2002; Lincoln et al. 2002; del Pozo and Lam 2003).

Moreover, the hypersensitive reaction determines the activation of a salicylic acid-dependent systemic response in the whole plant, known as systemic acquired resistance (SAR) (Ryals et al. 1996). This reduces subsequent infection of healthy tissues by a broad range of pathogens over several weeks.

2
NO Production During HR

An ever growing amount of evidence supports the notion that NO production, temporarily coordinated with the oxidative burst, is associated with the establishment of a hypersensitive reaction in different model systems. In plants or cell cultures of *Arabidopsis thaliana*, soybean and tobacco, NO production rapidly increases following elicitation with avirulent strains of *Pseudomonas syringae* and is maintained for several hours (Delledonne et al. 1998; Clarke et al. 2000; Mur et al. 2003; Zhang et al. 2003). Transient NO production is also observed prior to the onset of the HR in resistant barley epidermal cells challenged with the powdery mildew fungus *Blumeria graminis* f. sp. *hordei* (Prats et al. 2005). Finally, plants produce NO in response to general elicitors of plant defences, notably in tobacco leaves and cells challenged with cryptogein, a necrotizing elicitor produced by the oomycete *Phytophthora cryptogea* (Foissner et al. 2000; Lamotte et al. 2004), or in *A. thaliana* leaves treated with lipopolysaccharides (LPS), cell-surface components of Gram-negative bacteria (Zeidler et al. 2004).

In animals, NO is produced mainly by NO synthases (NOSs), calmodulin-dependent hemoproteins that oxidize L-arginine to form NO and L-citrulline (Nathan and Xie 1994). In plants, a NO synthase has eluded detection for a long period, but finally AtNOS1 protein was discovered in *A. thaliana* (Guo et al. 2003). This enzyme has no obvious homology to known mammalian NOS, although it has an unexpected sequence homology with proteins involved in NO synthesis in the snail *Helix pomatia*.

The protein belongs to a novel family of putative NOS that is conserved in both eukaryotes and bacteria (Zemojtel et al. 2004) and is involved in NO accumulation in roots and leaves (Guo et al. 2003). The biochemical analysis on AtNOS1 shows that its activity is stimulated by Ca^{2+} and calmodulin (Guo et al. 2003). In addition, two papers show that Ca^{2+} influx from the extracellular space is required for the initiation and maintenance of NOS-dependent NO production in tobacco and grape cells when stimulated by elicitors of defence responses (Lamotte et al. 2004, Vandelle et al. 2006).

Recently, AtNOS1 has been shown to participate in NO production during plant–pathogen interaction. When *Atnos1* mutant plants are treated with LPS, the NO burst is reduced by 80% in comparison to wild type (Zeidler et al. 2004). Moreover this mutant shows enhanced susceptibility to the virulent pathogen *P. syringae* pv. *tomato* DC3000 and the expression of *PR* genes is inhibited.

Plants can synthesize NO by other enzymatic and non-enzymatic routes (Bethke et al. 2004; Corpas et al. 2004). Among the enzymatic sources of NO, nitrate reductase (NR) was the first to be identified (Yamasaki and Sakihama 2000). Plant NR-dependent NO production has been demonstrated both in vitro and in vivo (Yamasaki and Sakihama 2000; Morot-Gaudry-Talarmain et al. 2002; Sakihama et al. 2003). During anoxia, high rate of NO emission correlates with high nitrite levels and NR activation (Yamasaki and Sakihama 2000; Rockel et al. 2002). Nevertheless, the involvement of NR in NO production during plant–pathogen interactions remains questionable. Indeed, NR is not required for NO production in tobacco challenged with cryptogein (Kaiser et al. 2002; Lamotte et al. 2004). By contrast, the involvement of NR in NO production was shown using NR-deficient *A. thaliana* plants (*nia1*, *nia2*), challenged with an avirulent strain of *P. syringae* (Modolo et al. 2005). This apparent discrepancy could be explained by the existence of different NO sources that cooperate in plants to produce NO during different processes. For instance, at least three enzymatic activities appear to be involved in producing NO during *A. thaliana/P. syringae pv. maculicola* interaction (Modolo et al. 2005), NOS activity, NR activity, and nitrite reduction to NO by the mitochondrial electron transport system (Modolo et al. 2005). Two other systems also use nitrite as a substrate to generate NO: a plasma membrane-bound nitrite:NO-reductase (NiNOR) identified in tobacco roots (Stöhr et al. 2001) and the xanthine oxidoreductase that converts nitrite to NO under anoxic conditions (Harrison 2002). However, the involvement of these enzymes in NO production during plant–pathogen interaction has not been demonstrated.

Finally, NO synthesis can also occur through the non-enzymatic reduction of nitrite to NO under acidic conditions (Yamasaki and Sakihama 2000). This has been observed in barley aleurone layer apoplast, acidified by abscissic acid (ABA) or by gibberellic acid (GA)-treatment, when nitrite is added to the incubation medium (Bethke et al. 2004). However, this source of NO production is unlikely to be significant in response to pathogens or elicitors, which normally trigger alkalinization of the extracellular medium, incompatible with the chemical reaction (Bolwell et al. 2002).

3 NO and Cell Death

In plants, hypersensitive cell death is characterized by the rapid accumulation of ROS and NO. A functional implication of NO in cell death has been thoroughly demonstrated (Delledonne et al. 1998, 2001; Durner et al. 1998; Delledonne 2005). In *A. thaliana* suspension cells, NO donors, induce cell death at concentrations similar to those generated in cells challenged by avirulent bacteria (Clarke et al. 2000). In soybean cell suspensions, cell death appears to be mediated by balanced levels of NO and H_2O_2 (Delledonne et al. 2001). The same results emerged from experiments with cultured tobacco cells, in which separate administration of either NO donors or H_2O_2 generators did not have any effect on cell viability, while if NO and H_2O_2 were added simultaneously, cells died in a concentration-dependent manner (de Pinto et al. 2002). The mechanisms by which NO kills cells depend on its rate of production and diffusion, on the redox state of the cell, and on its interaction with ROS (Delledonne 2005).

Exactly how NO cooperates with ROS to trigger hypersensitive cell death is still the object of extensive investigation. In many biological systems, the cytotoxic effects of NO and ROS derive from the diffusion-limited reaction of NO with O_2^- to form the peroxynitrite anion ($ONOO^-$) (Koppenol et al. 1992), a highly reactive oxidant molecule that interacts with many molecular components and may modulate downstream signaling (Beckman et al. 1990). In mammals, $ONOO^-$ induces apoptosis and is also cytotoxic due to oxidative tissue damage (Lin et al. 1995). However, $ONOO^-$ does not seem to be an essential mediator of NO/ROS-induced cell death in plants: exposure of soybean cells to exogenous $ONOO^-$ does not cause cell death, despite a strong increase in nitrated proteins (Delledonne et al. 2001). Currently, the relative concentration of H_2O_2 and NO appears to be at the core of cell death initiation (Delledonne et al. 2001). In vitro studies have demonstrated that reaction of NO with H_2O_2 can generate different highly reactive oxygen species with cytotoxic potential, such as hydroxyl radicals (Kanner et al. 1991) and singlet oxygen (Noronha-Dutra et al. 1993). Moreover, both NO and H_2O_2 can provoke lipid peroxidation and DNA breakage. In animal cells, one of the hallmarks of apoptosis is the release of cytochrome C from mitochondria, which in turn activates a cascade of caspases that selectively cleave and activate crucial substrates in the cell, including the nucleases responsible for DNA degradation (Ueda et al. 2002). In plants, both H_2O_2 and NO cause the release of cytochrome C from mitochondria and a caspase-like signaling cascade has also been identified during the HR (Chichkova et al. 2004; Mur et al. 2006). Furthermore, the initiation of NO-mediated cell death in plants can be suppressed by an irreversible inhibitor of mammalian caspase-1 (Clarke et al. 2000) and expression of a cysteine (cystatin-class) protease inhibitor (AtCYS1) in transgenic tobacco. *A. thaliana* cells could suppress the cell death program triggered by NO or avirulent bacteria (Belenghi et al. 2003). Taken together, these suggest that NO and H_2O_2 might cooperate in a signal cascade leading to cell death, which is intriguingly similar to apoptosis initiation and execution in animal cells.

4 NO and SAR

The HR and some examples of necrogenic pathogen infections lead to the biological activation of SAR (Durrant and Dong 2004). This is the induction of plant defence responses in uninfected parts of the plant, which then become more resistant to secondary infections by a broad range of pathogens. The crucial player of SAR establishment is salicylic acid (SA): SA is a well-known plant hormone with an important role in resistance, both locally and systemically, in a number of incompatible interactions. Exogenous application of SA has been found to mimic SAR and induce the transcription of *PR* genes. Transgenic (*NahG*) plants, expressing a SA hydroxylase enzyme and unable to accumulate SA, are often more susceptible to a range of pathogens (Shirasu et al. 1997) and are also unable to mount a SAR response following appropriate stimuli (Seskar et al. 1998).

Several lines of evidence highlight the role of NO in the modulation of signaling leading to SAR, although its activity is fully dependent on the functions of SA. *PR-1* induction by NO is mediated by SA, as it is blocked in *NahG* transgenic plants (Durner et al. 1998). However, NO activity in this context is not irrelevant as, in tobacco, NO treatment has been shown to stimulate the accumulation of SA and its conjugates (Durner et al. 1998) and, accordingly, treatment of tobacco plants with NOS inhibitors or NO scavengers partially inhibits SA-induced SAR (Song and Goodman 2001).

Although SA is essential to SAR and can move through the phloem to the entire plant, it is not the systemic signal responsible for SAR activation (Mauch-Mani and Métraux 1998). Several molecules, such as short peptides and selected lipids and lipid derivatives, have been suggested to be putative short- or long-distance signals mediating the development of a variety of defense mechanisms (Van Bel and Gaupels 2004). For example, a putative lipid transfer protein (DIR1) has been shown to be essential for loading systemic SAR signals in *A. thaliana* (Maldonado et al. 2002).

Yet another candidate that could serve as a mobile signal is *S*-nitrosoglutathione (GSNO) (Durner and Klessig 1999). GSNO is a low molecular-weight compound responsible for NO mobility in the phloem. Treatments with GSNO have been shown to induce systemic resistance against TMV infection (Song and Goodman 2001) and *PAL* expression in tobacco through a SA-dependent mechanism (Durner et al. 1998). It is possible that GSNO produced during pathogen attack is loaded into the phloem, systemically dispersed, and unloaded to initiate systemic SA synthesis. Recently, a GSNO reductase (GSNOR) that metabolizes GSNO with high specificity has been identified in *A. thaliana* (Diaz et al. 2003). Mutations in the gene encoding GSNOR lead to increased *S*-nitrosothiol (SNO) levels and reduction of plant defense responses. Disease susceptibility in *AtGSNOR1* mutant plants has been associated with a reduction of SA levels and reduced or delayed expression of SA-dependent genes, such as *PR-1*. These data support the hypothesis that AtGSNOR1, and hence SNO levels, could control the SA signaling network (Feechan et al. 2005) and suppress plant defence responses (Mur et al. 2006).

5 NO Signaling in Plant Defense Responses

Ca^{2+}, as a ubiquitous intracellular second messenger, can regulate diverse cellular processes in plants. Ca^{2+} fluxes are thought to play key roles in plant defense responses against pathogens. In animals, NO is known to regulate Ca^{2+} channels located in the plasma membrane or in endomembranes, directly by S-nitrosylation and tyrosine-nitration or indirectly via second messengers, notably cGMP (Clementi et al. 1998; Xu et al. 1998; Eu et al. 1999). Many studies have found that Ca^{2+}-permeable channels, including plasma membrane voltage-dependent Ca^{2+} channels, cyclic nucleotide-gated Ca^{2+} channels, and the intracellular channels inositol-triphosphate receptor and ryanodine receptor (RYR) are direct targets of NO. NO can also interact with the heme prosthetic group of guanylate cyclase, leading to its activation and generation of intracellular cGMP. In turn, cGMP regulates different targets, allowing NO signal transduction. cGMP can activate protein kinases, which lead to RYR activation by phosphorylation (Suko et al. 1993), or can activate the ADPR cyclase, responsible for cADPR synthesis, a second messenger also able to activate RYR channels (Clementi et al. 1996; Willmott et al. 1996).

Many studies over the last few years have demonstrated that NO may also regulate Ca^{2+} homeostasis in plants (Wendehenne et al. 2001). In tobacco cells challenged with cryptogein, NO participates in the increase in free cytosolic Ca^{2+} concentration ($[Ca^{2+}]$cyt), by activating Ca^{2+} release from internal stores (Lamotte et al. 2004). Moreover, NO donors induce $[Ca^{2+}]$cyt elevation in *Vicia faba* guard cells and tobacco cells (Garcia-Mata et al. 2003; Lamotte et al. 2004). These events, as in animals, seem to be mediated by cGMP and cADPR, although neither NO-sensitive guanylate cyclase (GC) nor ADPR cyclase have been identified in plant cells. Indeed, NO donor treatment of tobacco leaves or cells induces a transient increase in intracellular cGMP levels (Durner et al. 1998). GC inhibitors, cADPR antagonists, and RYR inhibitors suppress the Ca^{2+}-mobilizing actions of NO, suggesting that NO might activate RYR through cGMP and cADPR (Fig. 1) (Garcia-Mata et al. 2003; Lamotte et al. 2004). In addition, a GC inhibitor also blocks NO-induced cell death in *A. thaliana* cells (Clarke et al. 2000). Finally, using GC inhibitors and cGMP analogs, it was shown that cGMP is required for phenylalanine ammonia lyase (*PAL*) and *PR-1* gene expression induced by a NO donor in tobacco (Durner et al. 1998). However, GC inhibitors produce only a partial suppression of *PAL* transcript accumulation, suggesting the existence of both cGMP-dependent and independent pathways (Durner et al. 1998).

PAL and *PR-1* gene expression in tobacco can also be induced by cADPR treatment and inhibited by cADPR-gated Ca^{2+} channel inhibitor (Durner and Klessig 1999). Similar to that observed for GC, NO-dependent induction of *PR-1* transcript accumulation can be suppressed by a cADPR antagonist, such as 8-Br-cADPR, but only partially, suggesting once again that NO activation of defense responses may occur through more than one pathway (Klessig et al. 2000) (Fig. 2).

Other important signaling functions are achieved in plants through NO activity on major protein kinases, such as mitogen-activated protein kinases (MAPK), which play important roles in plant responses to multiple stresses and seem to be one of the converging points in the defense-signaling network

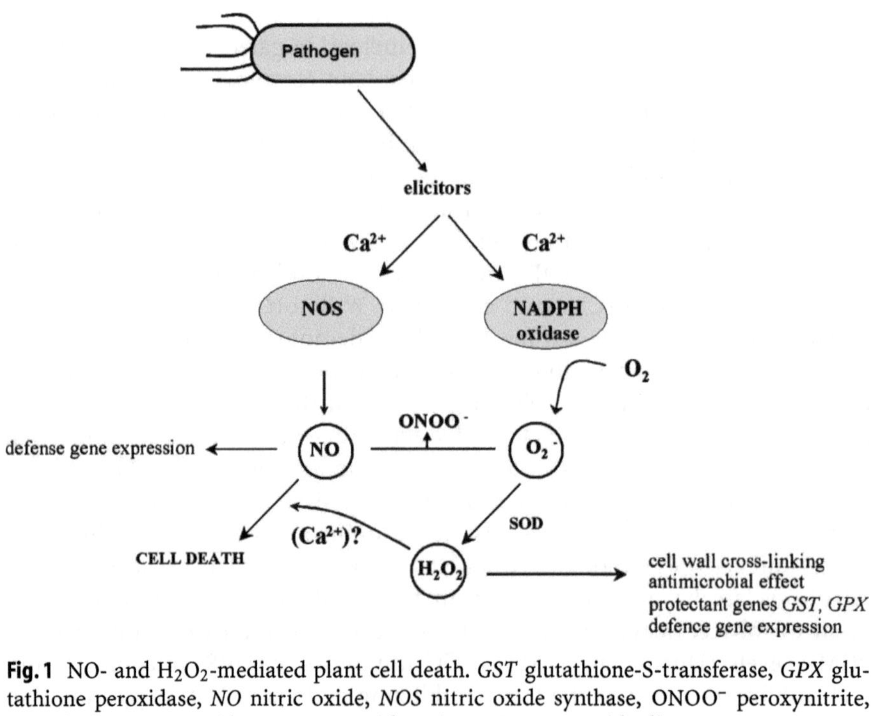

Fig. 1 NO- and H_2O_2-mediated plant cell death. *GST* glutathione-S-transferase, *GPX* glutathione peroxidase, *NO* nitric oxide, *NOS* nitric oxide synthase, $ONOO^-$ peroxynitrite, H_2O_2 hydrogen peroxide, O_2^- superoxide anion, *SOD* superoxide dismutase

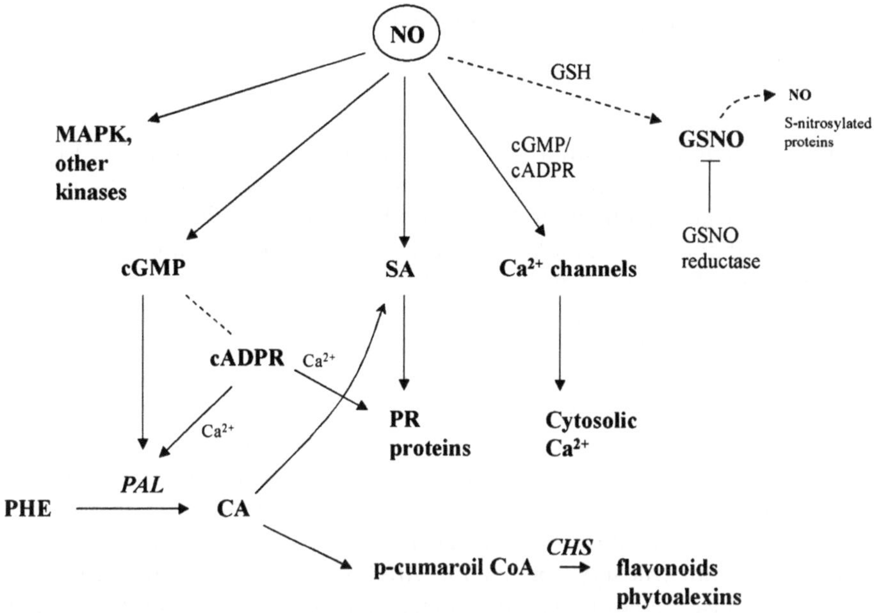

Fig. 2 Representation of NO signaling functions. *CHS* chalcone synthase, *CA* cinnamic acid, Ca^{2+} calcium influx, *cADPR* cyclic cADP ribose, *cGMP* cyclic GMP, *GSH* glutathione, *GSNO* *S*-nitrosoglutathione, *MAPK* mitogen-activated protein kinase, *PAL* phenylalanine ammonia lyase, *PHE* phenylalanine, *PR proteins* pathogenesis-related proteins, *SA* salicylic acid. *Dotted lines* represent potential NO functions

of plants (Neill et al. 2002). In both *A. thaliana* and tobacco leaves, artificially generated NO stimulates MAPK activation (Clarke et al. 2000; Kumar and Klessig 2000; Capone et al. 2004) and it was shown that NO can also exert its modulating function on MAPK activity systemically through the entire plant, as exposure of *A. thaliana* roots to oxidative and nitrosative stress can induce a rapid activation of protein kinases with MAPK-like properties in shoots. In tobacco, the SA-induced MAPK SIPK is typically induced by SA and H_2O_2 (Yang et al. 2001) but can also be induced by NO in a SA-dependent manner (Kumar and Klessig 2000).

6 NO-Dependent Gene Induction

NO is known to affect the expression of numerous genes in different plant species such as *A. thaliana*, tobacco and soybean. The NO-dependent intracellular signaling pathways that lead to the activation or repression of gene transcription have not yet been defined. Given the diverse nature of NO functions, large-scale gene expression studies would appear to be a more informative approach towards the understanding of the signaling networks affected (Grun et al. 2006).

Polverari et al. (2003) investigated the changes of expression profiles in *A. thaliana* plants after infiltration with the NO donor sodium nitroprusside (SNP) by cDNA-amplification fragment length polymorphism (AFLP) transcript profiling. Expression profiling by microarrays was assessed by Parani et al. (2004), who examined the NO-induced alterations in the expression profile of *A. thaliana* in leaves following administration of SNP in roots. In both these analyses NO was found to modulate transcription of genes related to signal transduction, such as transcription factors (WRKYs and zinc finger proteins), receptor-like protein kinases and MAPKs, genes related to plant defence response and related to protection against oxidative stress such as glutathione-S-transferase or glutaredoxin (Polverari et al. 2003; Parani et al. 2004).

Another large-scale transcriptional analysis was carried out by Zeidler et al. (2004) and revealed the role of NO synthase-generated NO on gene expression. LPS treatment was used to generate a nitrosative burst in *A. thaliana* leaves from both wild-type plants and *AtNOS1* insertion mutants impaired in NO synthase activity. Transcriptional changes were investigated using a custom-built microarray containing about 700 defense-related genes. The results of this analysis showed that in wild-type plants LPS-mediated NO burst induces an array of defense or stress-associated genes including GST, cytochrome P450 and PR proteins and, moreover, that this induction was abolished in *Atnos1* mutant plants (Zeidler et al. 2004).

7
A Novel Mechanism of NO Action: Posttranslational Regulation of Proteins

It is now recognized that NO and its related species can introduce posttranslational modifications of proteins using different routes. In addition to its capability to bind metal ions of heme groups, as reported in animals for GC activation (Brandish et al. 1998; Russwurm and Koesling 2004), NO can perform important posttranslational protein modifications through S-nitrosylation and nitration.

S-Nitrosylation is the formation of *S*-nitrosothiols by covalent addition to cysteine residues of a NO moiety. Protein cysteinyl thiol groups can be nitrosylated by low molecular weight *S*-nitrosothiols (SNOs) or by transfer of a NO group from another S-nitrosylated protein (Fig. 2) (Delledonne 2005). Cysteine nitrosylation is both reversible and specific. S-Nitrosylated proteins can be easily denitrosylated because the S – NO bond is labile in a reductive microenvironment (Delledonne 2005), allowing cells to flexibly and precisely adapt protein function in response to environmental signals (Mannick and Schonhoff 2002). The remarkable specificity of S-nitrosylation is conferred by different factors that are the subcellular compartmentalization of NO sources, the target protein, and the presence of consensus motifs flanking cysteine residues (Hess et al. 2005). In this framework, the formation of SNOs may serve to both stabilize and diversify NO-related signals, while GSNO could function as a mobile reservoir of NO bioactivity (Feechan et al. 2005). Recently, some potential targets for S-nitrosylation have been identified in plants (Lindermayr et al. 2005). Among them, one isoform of *S*-methionine adenosyl transferase, an enzyme involved in ethylene biosynthesis, has been shown to be inhibited by S-nitrosylation in *A. thaliana* (Lindermayr et al. 2006).

Nitration is the process by which a nitrite group is added to the ortho position of tyrosine residues forming 3-nitrotyrosine. Tyrosine nitration is mediated by reactive nitrogen species such as $ONOO^-$ and nitrogen dioxide (NO_2), formed as secondary products of NO metabolism in the presence of oxidants including O_2^-, H_2O_2 and transition metal centers (Radi 2004). A number of reports on mammals have shown protein tyrosine nitration during stress responses (Giasson et al. 2000). Recent works indicated that protein nitration operates also in plants: nitrotyrosine-containing proteins as a product of $ONOO^-$ direct reaction were detected using anti-nitrotyrosine antibodies in soybean (Delledonne et al. 2001) and tobacco (Saito et al. 2006) cell suspensions. In addition, increased protein nitration has been observed in antisense NR tobacco accumulating higher nitrate and NO levels (Morot-Gaudry-Talarmain et al. 2002). In contrast to S-nitrosylation, tyrosine nitration can introduce irreversible modifications that may alter protein conformation and structure, and also catalytic activity (Souza et al. 2000). In particular, tyrosine nitration may interfere with tyrosine phosphorylation of key proteins involved in plant cell signaling.

8 Conclusions

Plant resistance is a complex phenomenon in which plants recognize the invading pathogen and activate a rapid and effective set of defenses responses, often characterized by a localized cell death at the infection site, that limits the growth and spread of the invading pathogen. In this framework, although several hypotheses still await experimental demonstration, it is now clear that NO plays a key role during plant–pathogen interaction modulating the HR, SAR and defense gene induction. Thus, new insights on NO signaling functions at the molecular and subcellular levels will soon make it possible to characterize several important physiological and pathological processes in plants, as has already been demonstrated in mammals. Future efforts will be focused on the identification of targets for NO using proteomic and transcriptomic approaches.

References

Atkinson MM, Midland SL, Sims JJ, Keen NT (1996) Syringolide 1 triggers Ca^{2+} influx, K^{+} efflux, and extracellular alkalization in soybean cells carrying the disease-resistance gene Rpg4. Plant Physiol 112:297–302

Beckman JS, Beckman TW, Chen J, Marshall PA, Freeman BA (1990) Apparent hydroxyl radical production by peroxynitrite: implications for endothelial injury from nitric oxide and superoxide. Proc Natl Acad Sci USA 87:1620–1624

Belenghi B, Acconcia F, Trovato M, Perazzolli M, Bocedi A, Polticelli F, Ascenzi P, Delledonne M (2003) AtCYS1, a cystatin from Arabidopsis thaliana, suppresses hypersensitive cell death. Eur J Biochem 270:2593–2604

Beligni MV, Lamattina L (2001) Nitric oxide in plants: the history is just beginning. Plant Cell Environm 24:267–278

Bethke PC, Badger MR, Jones RL (2004) Apoplastic synthesis of nitric oxide by plant tissues. Plant Cell 16:332–341

Blumwald E (1997) Signal transduction in plant–pathogen interactions. Plant Physiol 114:50001–50001

Bolwell GP, Bindschedler LV, Blee KA, Butt VS, Davies DR, Gardner SL, Gerrish C, Minibayeva F (2002) The apoplastic oxidative burst in response to biotic stress in plants: a three-component system. J Exp Bot 53:1367–1376

Brandish PE, Buechler W, Marletta MA (1998) Regeneration of the ferrous heme of soluble guanylate cyclase from the nitric oxide complex: acceleration by thiols and oxyhemoglobin. Biochem 37:16898–16907

Capone R, Tiwari BS, Levine A (2004) Rapid transmission of oxidative and nitrosative stress signals from roots to shoots in Arabidopsis. Plant Physiol Biochem 42:425–428

Chichkova NV, Kim SH, Titova ES, Kalkum M, Morozov VS, Rubtsov YP, Kalinina NO, Taliansky ME, Vartapetian AB (2004) A plant caspase-like protease activated during the hypersensitive response. Plant Cell 16:157–171

Clarke A, Desikan R, Hurst RD, Hancock JT, Neill SJ (2000) NO way back: nitric oxide and programmed cell death in Arabidopsis thaliana suspension cultures. Plant J 24:667–677

Clementi E, Brown GC, Feelisch M, Moncada S (1998) Persistent inhibition of cell respiration by nitric oxide: crucial role of S-nitrosylation of mitochondrial complex I and protective action of glutathione. Proc Natl Acad Sci USA 95:7631–7636

Clementi E, Riccio M, Sciorati C, Nistico G, Meldolesi J (1996) The type 2 ryanodine receptor of neurosecretory PC12 cells is activated by cyclic ADP-ribose. Role of the nitric oxide/cGMP pathway. J Biol Chem 271:17739–17745

Corpas FJ, Barroso JB, del Rio LA (2004) Enzymatic sources of nitric oxide in plant cells – beyond one protein-one function. New Phytol 162:246–248

de Pinto MC, Tommasi F, De Gara L (2002) Changes in the antioxidant systems as part of the signaling pathway responsible for the programmed cell death activated by nitric oxide and reactive oxygen species in tobacco Bright-Yellow 2 cells. Plant Physiol 130:698–708

del Pozo O, Lam E (2003) Expression of the baculovirus p35 protein in tobacco affects cell death progression and compromises N gene-mediated disease resistance response to Tobacco mosaic virus. Mol Plant Microbe Interact 16:485–494

Delledonne M (2005) NO news is good news for plants. Curr Opin Plant Biol 8:390–396

Delledonne M, Xia Y, Dixon RA, Lamb C (1998) Nitric oxide functions as a signal in plant disease resistance. Nature 394:585–588

Delledonne M, Zeier J, Marocco A, Lamb C (2001) Signal interactions between nitric oxide and reactive oxygen intermediates in the plant hypersensitive disease resistance response. Proc Natl Acad Sci USA 98:13454–13459

Diaz M, Achkor H, Titarenko E, Martinez MC (2003) The gene encoding glutathione-dependent formaldehyde dehydrogenase/GSNO reductase is responsive to wounding, jasmonic acid and salicylic acid. FEBS Lett 543:136–139

Durner J, Klessig DF (1999) Nitric oxide as a signal in plants. Curr Opin Plant Biol 2:369–374

Durner J, Wendehenne D, Klessig DF (1998) Defense gene induction in tobacco by nitric oxide, cyclic GMP, and cyclic ADP-ribose. Proc Natl Acad Sci USA 95:10328–10333

Durrant WE, Dong X (2004) Systemic acquired resistance. Annu Rev Phytopathol 42:185–209

Elbaz M, Avni A, Weil M (2002) Constitutive caspase-like machinery executes programmed cell death in plant cells. Cell Death Differ 9:726–733

Eu JP, Xu L, Stamler JS, Meissner G (1999) Regulation of ryanodine receptors by reactive nitrogen species. Biochem Pharmacol 57:1079–1084

Feechan A, Kwon E, Yun BW, Wang Y, Pallas JA, Loake GJ (2005) A central role for S-nitrosothiols in plant disease resistance. Proc Natl Acad Sci USA 102:8054–8059

Foissner I, Wendehenne D, Langebartels C, Durner J (2000) In vivo imaging of an elicitor-induced nitric oxide burst in tobacco. Plant J 23:817–824

Fritig B, Heitz T, Legrand M (1998) Antimicrobial proteins in induced plant defense. Curr Opin Immunol 10:16–22

Garcia-Mata C, Gay R, Sokolovski S, Hills A, Lamattina L, Blatt MR (2003) Nitric oxide regulates K^+ and Cl^- channels in guard cells through a subset of abscisic acid-evoked signaling pathways. Proc Natl Acad Sci USA 100:11116–11121

Giasson BI, Duda JE, Murray IV, Chen Q, Souza JM, Hurtig HI, Ischiropoulos H, Trojanowski JQ, Lee VM (2000) Oxidative damage linked to neurodegeneration by selective alpha-synuclein nitration in synucleinopathy lesions. Science 290:985–989

Grun S, Lindermayr C, Sell S, Durner J (2006) Nitric oxide and gene regulation in plants. J Exp Bot 57:507–516

Guo FQ, Okamoto M, Crawford NM (2003) Identification of a plant nitric oxide synthase gene involved in hormonal signaling. Science 302:100–103

Hammond-Kosack KE, Jones JD (1996) Resistance gene-dependent plant defense responses. Plant Cell 8:1773–1791

Harrison R (2002) Structure and function of xanthine oxidoreductase: where are we now? Free Radic Biol Med 33:774–797

Heath MC (2000) Hypersensitive response-related death. Plant Mol Biol 44:321–334

Hess DT, Matsumoto A, Kim SO, Marshall HE, Stamler JS (2005) Protein S-nitrosylation: purview and parameters. Nat Rev Mol Cell Biol 6:150–166

Kaiser WM, Weiner H, Kandlbinder A, Tsai CB, Rockel P, Sonoda M, Planchet E (2002) Modulation of nitrate reductase: some new insights, an unusual case and a potentially important side reaction. J Exp Bot 53:875–882

Kanner J, Harel S, Granit R (1991) Nitric oxide as an antioxidant. Arch Biochem Biophys 289:130–136

Klessig DF, Durner J, Noad R, Navarre DA, Wendehenne D, Kumar D, Zhou JM, Shah J, Zhang S, Kachroo P, Trifa Y, Pontier D, Lam E, Silva H (2000) Nitric oxide and salicylic acid signaling in plant defense. Proc Natl Acad Sci USA 97:8849–8855

Koppenol WH, Moreno JJ, Pryor WA, Ischiropoulos H, Beckman JS (1992) Peroxynitrite, a cloaked oxidant formed by nitric oxide and superoxide. Chem Res Toxicol 5:834–842

Kumar D, Klessig DF (2000) Differential induction of tobacco MAP kinases by the defense signals nitric oxide, salicylic acid, ethylene, and jasmonic acid. Mol Plant Microbe Interact 13:347–351

Lamb C, Dixon RA (1997) The oxidative burst in plant disease resistance. Annu Rev Plant Physiol Plant Mol Biol 48:251–275

Lamotte O, Gould K, Lecourieux D, Sequeira-Legrand A, Lebrun-Garcia A, Durner J, Pugin A, Wendehenne D (2004) Analysis of nitric oxide signaling functions in tobacco cells challenged by the elicitor cryptogein. Plant Physiol 135:516–529

Levine A, Pennell RI, Alvarez ME, Palmer R, Lamb C (1996) Calcium-mediated apoptosis in a plant hypersensitive disease resistance response. Curr Biol 6:427–437

Lin KI, Lee SH, Narayanan R, Baraban JM, Hardwick JM, Ratan RR (1995) Thiol agents and Bcl-2 identify an alphavirus-induced apoptotic pathway that requires activation of the transcription factor NF-kappa B. J Cell Biol 131:1149–1161

Lincoln JE, Richael C, Overduin B, Smith K, Bostock R, Gilchrist DG (2002) Expression of the antiapoptotic baculovirus p35 gene in tomato blocks programmed cell death and provides broad-spectrum resistance to disease. Proc Natl Acad Sci USA 99:15217–15221

Lindermayr C, Saalbach G, Bahnweg G, Durner J (2006) Differential inhibition of Arabidopsis methionine adenosyltransferases by protein S-nitrosylation. J Biol Chem 281:4285–4291

Lindermayr C, Saalbach G, Durner J (2005) Proteomic identification of S-nitrosylated proteins in Arabidopsis. Plant Physiol 137:921–930

Maldonado AM, Doerner P, Dixon RA, Lamb CJ, Cameron RK (2002) A putative lipid transfer protein involved in systemic resistance signaling in Arabidopsis. Nature 419:399–403

Mannick JB, Schonhoff CM (2002) Nitrosylation: the next phosphorylation? Arch Biochem Biophys 408:1–6

Mauch-Mani B, Métraux JP (1998) Salicylic acid and systemic acquired resistance to pathogen attack. Annal Bot 82:535–540

McDowell JM, Woffenden BJ (2003) Plant disease resistance genes: recent insights and potential applications. Trends Biotechnol 21:178–183

Modolo LV, Augusto O, Almeida IM, Magalhaes JR, Salgado I (2005) Nitrite as the major source of nitric oxide production by Arabidopsis thaliana in response to Pseudomonas syringae. FEBS Lett 579:3814–3820

Morot-Gaudry-Talarmain Y, Rockel P, Moureaux T, Quillere I, Leydecker MT, Kaiser WM, Morot-Gaudry JF (2002) Nitrite accumulation and nitric oxide emission in relation to cellular signaling in nitrite reductase antisense tobacco. Planta 215:708–715

Mur LA, Carver TL, Prats E (2006) NO way to live; the various roles of nitric oxide in plant–pathogen interactions. J Exp Bot 57:489–505

Mur LAJ, Santosa IE, Laarhoven L-JJ, Harren F, Smith AR (2003) A new partner in the *danse macabre*: the role of nitric oxide in the hypersensitive response. Bulg J Plant Physiol (special issue):110–123

Nathan C, Xie QW (1994) Nitric oxide synthases: roles, tolls, and controls. Cell 78:915–918

Neill SJ, Desikan R, Clarke A, Hancock JT (2002) Nitric oxide is a novel component of abscisic acid signaling in stomatal guard cells. Plant Physiol 128:13–16

Neill SJ, Desikan R, Clarke A, Hurst RD, Hancock JT (2002) Hydrogen peroxide and nitric oxide as signaling molecules in plants. J Exp Bot 53:1237–1247

Neill SJ, Desikan R, Hancock JT (2003) Nitric oxide signaling in plants. New Phytol 159:11–35

Noronha-Dutra AA, Epperlein MM, Woolf N (1993) Reaction of nitric oxide with hydrogen peroxide to produce potentially cytotoxic singlet oxygen as a model for nitric oxide-mediated killing. FEBS Lett 321:59–62

Nurnberger T, Brunner F, Kemmerling B, Piater L (2004) Innate immunity in plants and animals: striking similarities and obvious differences. Immunol Rev 198:249–266

Nurnberger T, Scheel D (2001) Signal transmission in the plant immune response. Trends Plant Sci 6:372–379

Pagnussat GC, Lanteri ML, Lamattina L (2003) Nitric oxide and cyclic GMP are messengers in the indole acetic acid-induced adventitious rooting process. Plant Physiol 132:1241–1248

Parani M, Rudrabhatla S, Myers R, Weirich H, Smith B, Leaman DW, Goldman SL (2004) Microarray analysis of nitric oxide responsive transcripts in *Arabidopsis*. Plant Biotechnol J 2:359–366

Polverari A, Molesini B, Pezzotti M, Buonaurio R, Marte M, Delledonne M (2003) Nitric oxide-mediated transcriptional changes in Arabidopsis thaliana. Mol Plant Microbe Interact 16:1094–1105

Prats E, Mur LAJ, Sanderson R, Carver TLW (2005) Nitric oxide contributes both to papilla-based resistance and the hypersensitive response in barley attacked by Blumeria graminis f. sp. hordei. Molecular Plant Pathol 6:65–78

Radi R (2004) Nitric oxide, oxidants, and protein tyrosine nitration. Proc Natl Acad Sci USA 101:4003–4008

Rockel P, Strube F, Rockel A, Wildt J, Kaiser WM (2002) Regulation of nitric oxide (NO) production by plant nitrate reductase in vivo and in vitro. J Exp Bot 53:103–110

Romero-Puertas MC, Perazzolli M, Zago ED, Delledonne M (2004) Nitric oxide signaling functions in plant–pathogen interactions. Cell Microbiol 6:795–803

Russwurm M, Koesling D (2004) Guanylyl cyclase: NO hits its target. Biochem Soc Symp:51–63

Ryals JA, Neuenschwander UH, Willits MG, Molina A, Steiner HY, Hunt MD (1996) Systemic acquired resistance. Plant Cell 8:1809–1819

Saito S, Yamamoto-Katou A, Yoshioka H, Doke N, Kawakita K (2006) Peroxynitrite generation and tyrosine nitration in defense responses in Tobacco BY-2 cells. Plant Cell Physiol

Sakihama Y, Tamaki R, Shimoji H, Ichiba T, Fukushi Y, Tahara S, Yamasaki H (2003) Enzymatic nitration of phytophenolics: evidence for peroxynitrite-independent nitration of plant secondary metabolites. FEBS Lett 553:377–380

Seskar M, Shulaev V, Raskin I (1998) Endogenous methyl salicylate in pathogen-inoculated tobacco plants. Plant Physiol 116:387–392

Shirasu K, Nakajima H, Rajasekhar VK, Dixon RA, Lamb C (1997) Salicylic acid potentiates an agonist-dependent gain control that amplifies pathogen signals in the activation of defense mechanisms. Plant Cell 9:261–270

Somssich IE, Hahlbrock K(1998) Pathogen defence in plants – a paradigm of biological complexity. Trends Plant Sci 3:86–90

Song F, Goodman RM (2001) Activity of nitric oxide is dependent on, but is partially required for function of, salicylic acid in the signaling pathway in tobacco systemic acquired resistance. Mol Plant Microbe Interact 14:1458–1462

Souza JM, Choi I, Chen Q, Weisse M, Daikhin E, Yudkoff M, Obin M, Ara J, Horwitz J, Ischiropoulos H (2000) Proteolytic degradation of tyrosine nitrated proteins. Arch Biochem Biophys 380:360–366

Stöhr C, Strube F, Marx G, Ullrich WR, Rockel P (2001) A plasma membrane-bound enzyme of tobacco roots catalyses the formation of nitric oxide from nitrite. Planta 212:835–841

Suko J, Maurer-Fogy I, Plank B, Bertel O, Wyskovsky W, Hohenegger M, Hellmann G (1993) Phosphorylation of serine 2843 in ryandodine receptor-calcium release channel of skeletal muscle by cAMP-, cGMP- and CaM-dependent protein kinase. Biochem Biophys Acta 1175:193–206

Ueda S, Masutani H, Nakamura H, Tanaka T, Ueno M, Yodoi J (2002) Redox control of cell death. Antioxid Redox Signal 4:405–414

Vandelle E, Poinssot B, Wendehenne D, Bentéjac M, Pugin A (2006) Integrated signaling network involving calcium, nitric oxide, and active oxygen species but not mitogen-activated protein kinases in BcPG1-elicited grapevine defenses. Mol Plant Microbe Interact 19:429–440

Van Bel AJE, Gaupels F (2004) Pathogen-induced resistance and alarm signals in the phloem. Mol Plant Pathol 5:495–504

Wakabayashi T, Karbowski M (2001) Structural changes of mitochondria related to apoptosis. Biol Signals Recept 10:26–56

Wendehenne D, Lamotte O, Frachisse JM, Barbier-Brygoo H, Pugin A (2002) Nitrate efflux is an essential component of the cryptogein signaling pathway leading to defense responses and hypersensitive cell death in tobacco. Plant Cell 14:1937–1951

Wendehenne D, Pugin A, Klessig DF, Durner J (2001) Nitric oxide: comparative synthesis and signaling in animal and plant cells. Trends Plant Sci 6:177–183

Willmott N, Sethi JK, Walseth TF, Lee HC, White AM, Galione A (1996) Nitric oxide-induced mobilization of intracellular calcium via the cyclic ADP-ribose signaling pathway. J Biol Chem 271:3699–3705

Xu L, Eu JP, Meissner G, Stamler JS (1998) Activation of the cardiac calcium release channel (ryanodine receptor) by poly-S-nitrosylation. Science 279:234–237

Yamasaki H, Sakihama Y (2000) Simultaneous production of nitric oxide and peroxynitrite by plant nitrate reductase: in vitro evidence for the NR-dependent formation of active nitrogen species. FEBS Lett 468:89–92

Yang KY, Liu Y, Zhang S (2001) Activation of a mitogen-activated protein kinase pathway is involved in disease resistance in tobacco. Proc Natl Acad Sci USA 98:741–746

Zeidler D, Zahringer U, Gerber I, Dubery I, Hartung T, Bors W, Hutzler P, Durner J (2004) Innate immunity in *Arabidopsis thaliana*: lipopolysaccharides activate nitric oxide synthase (NOS) and induce defense genes. Proc Natl Acad Sci USA 101:15811–15816

Zemojtel T, Penzkofer T, Dandekar T, Schultz J (2004) A novel conserved family of nitric oxide synthase? Trends Biochem Sci 29:224–226

Zhang C, Czymmek KJ, Shapiro AD (2003) Nitric oxide does not trigger early programmed cell death events but may contribute to cell-to-cell signaling governing progression of the Arabidopsis hypersensitive response. Mol Plant Microbe Interact 16:962–972

Plant Cell Monogr (6)
L. Lamattina · J. C. Polacco: Nitric Oxide in Plant Growth
DOI 10.1007/7089_2006_094/Published online: 14 October 2006

Nitric Oxide in Cell-to-Cell Communication Coordinating the Plant Hypersensitive Response

Allan D. Shapiro

Department of Biological Sciences and Center for Biomolecular Science and Engineering
Florida Gulf Coast University, 10501 FGCU Boulevard South,
Fort Myers, Florida 33965-6565, USA
ashapiro@fgcu.edu

Abstract The primary and probably only important role that NO˙ plays in the hypersensitive response is communication between dying cells and neighboring cells. NO˙ accumulates extracellularly immediately prior to programmed cell death of infected cells and inhibits extracellular H_2O_2-degrading enzymes, leading to H_2O_2 accumulation. NO˙ and/or H_2O_2 travel to adjacent cells, where H_2O_2 accumulation induces salicylic acid biosynthesis. Salicylic acid mediates the observed NO˙-dependent potentiation of programmed cell death triggering. These effects appear to depend upon augmentation of plasma membrane depolarization by direct effects of salicylic acid; whereas delayed negative feedback on programmed cell death is gene expression-dependent.

1
Introduction

NO˙ first captured the attention of mammalian biologists as an intercellular messenger. It was implicated as the endothelium-derived relaxing factor that crossed into smooth muscle cells surrounding blood vessels and caused muscle relaxation that dilated the vessels (Rees et al. 1989). Subsequently, NO˙ was implicated as a signal sent from postsynaptic cells back to presynaptic cells in long-term potentiation processes implicated in memory (Arancio et al. 1996). NO˙ has long been known to be emitted from plants under certain conditions (Klepper 1979) and has physical properties that indicate cell-to-cell movement in plant tissues (Shapiro 2005). As such, recent evidence that its primary role in plant defense responses is in cell-to-cell communication processes should not have been surprising.

However, this realization has been obscured by a tendency to view the hypersensitive response (HR) to avirulent (gene-for-gene disease resistance-eliciting) plant pathogens as a property of individual infected cells. This tendency has led to serious misinterpretations of signaling data. Herein, the argument will be made that the HR is better viewed as a progressive response that spreads within leaf tissue. This "progression" is subject both to potentiation and to negative feedback that alter its kinetics and extent of spread. NO˙ has emerged as an important player in these multicellular coordination processes. An integrated picture of control over the HR and the cell-to-cell communication role played by NO˙ is developed and presented in this chapter.

2
Watching the HR Unfold in Time with NO•-Dependent Kinetics

The classic assay for the HR is easy to perform. A high titer of avirulent pathogen is inoculated into a leaf and a day or so later, either the leaf tissue has collapsed or it hasn't. Collapse in this assay is correlated with disease resistance in a lower titer infection. Collapse results from programmed cell death (PCD) of plant cells. PCD leads to ion leakage through compromised plasma membranes, followed by osmosis-driven water loss. Excess water is eliminated largely via transpiration. The dried tissue collapses.

This simple assay was the basis of seminal experiments implicating NO˙ in the HR. Work in mammalian systems had led to development of chemical inhibitors of mammalian NO˙ biosynthesis enzymes (NO˙ synthases, NOSs). These chemical inhibitors were shown to partially block or delay leaf collapse associated with the HR of Arabidopsis plants to avirulent *Pseudomonas syringae* (Delledonne et al. 1998) and of tobacco plants to avirulent *Ralstonia solanacearum* (Huang and Knopp 1998). A NOS-like activity was indeed shown to be induced in tobacco by *R. solanacearum* (Huang and Knopp 1998) or by Tobacco Mosaic Virus (Durner et al. 1998).

The ease of performance of HR assays (and of disease assays) also facilitated genetic analysis (Shapiro 2000). Mutant pathogen strains (and later plant mutant alleles) were isolated via screening for gain or loss of the HR, disease resistance, or correlated physiology. These studies implicated numerous gene products in the HR. They also provided mutant pathogen strains and plant lines that have been invaluable in piecing together mechanistic details of molecular recognition of pathogen and downstream plant signaling. However, these studies revealed a baroque complexity in the signaling network (Glazebrook 2005). As such, the simple assays that launched the field are no longer adequate.

Complicated networks involving cross-talk and feedback regulation require time course data to sort out signaling circuitry. Although it is possible to follow the kinetics of the leaf collapse reaction, the rate-limiting step for leaf collapse is transpirational water loss. As such, these measurements are not comparable between laboratories, as they mainly reflect fan speed in the growth chamber (Zhang et al. 2004). Instead, ion leakage can be followed as a proxy for PCD (Keppler and Novacky 1986). Leakage of ions from edges of cut leaf disks immersed in distilled water can be measured using a conductivity meter. This procedure is often used semiquantitatively to compare different plant lines or treatments by merely floating a leaf disk on water for a lengthy incubation period. Alternatively, this assay can be performed in a reproducible and quantitative fashion that much better approximates conditions in the intact leaves by taking leaf samples over a time course postinfection (Bestwick et al. 1995; Zhang et al. 2004).

In initial experiments using these methods, the author's laboratory could not document reproducible effects with NOS inhibitors or $NO^{\bullet}$ scavengers (Zhang C, Worley CK, Shapiro AD, unpublished data). Variability was also documented in inhibitor/scavenger effects on leaf collapse (Delledonne et al. 1998; Worley CK, Shapiro AD, unpublished data). As such, a more sensitive assay was employed. PCD was followed with single cell resolution using fluorescence microscopy (Zhang et al. 2003). Leaves were infected as with the conventional HR assay. Ten minutes prior to a desired time point, a leaf was infiltrated with propidium iodide. This dye enters cells with compromised plasma membranes and subsequently fluoresces a bright red color. In most experiments, a dye that fluoresced green following reaction with $NO^{\bullet}$ was introduced simultaneously. In this fashion, the relative kinetics of $NO^{\bullet}$ accumulation and PCD were established. Although the absolute timing depended upon which pathogen strain was used (in response to *P. syringae* bacteria carrying *avrB*, the first PCD events were seen ~ 3 h postinfection, and in response to bacteria carrying *avrRpt2*, ~ 5.5 h), the relative timing of $NO^{\bullet}$ production and PCD was identical. Surprisingly, $NO^{\bullet}$ production was not an early event in the process leading to PCD; in fact, initial $NO^{\bullet}$ production preceded the first PCD events by only approximately 10 min. Absolute timing of onset of these events showed some variability (varying by approximately 30 min between experiments), but relative timing was highly reproducible.

A second major surprise was that $NO^{\bullet}$ accumulation was not required for the HR. When either NOS inhibitors or a $NO^{\bullet}$ scavenger were employed, PCD, as assessed by propidium iodide internalization, still occurred. However, the first PCD events were delayed by ~ 1 h when $NO^{\bullet}$ accumulation was prevented. The subtlety of these results explained earlier inconsistencies in experiments implicating $NO^{\bullet}$ in the HR. One additional caveat to these earlier results was that it had not been proven that enzyme inhibitors and scavengers remained effective throughout the experiments. In this experiment, reagent efficacy was assessed using the $NO^{\bullet}$ reporter dye. Indeed, these reagents were only fully effective for 4 h postinfiltration, with partial efficacy following that point. However, in response to bacteria carrying *avrB*, PCD occurred prior to this point. As such, the conclusion was made that $NO^{\bullet}$ is not essential for the HR; however, it does facilitate it by potentiating the events triggering PCD (Zhang et al. 2003). This conclusion probably also applies to fungal HRs, as a $NO^{\bullet}$ scavenger only slightly delayed the barley HR to avirulent powdery mildew (Prats et al. 2005).

These results were very surprising because of the tendency in the field to view the HR as a property of individual infected cells. In fact, many early experiments on the role of NO˙ in the HR used plant cell culture systems. These experiments were guided by knowledge of NO˙ function in mammalian systems. In these systems, when NO˙ is supplied at high levels, it will damage cells sufficiently to kill them. In many cases, the "killer" is actually peroxynitrite, which is a strong oxidant formed by reaction of NO˙ with superoxide. As the central signaling event in the HR had long been assumed to be an "oxidative burst", a complementary role for NO˙ seemed logical. However, experiments with plant cell cultures indicated that although NO˙ could kill cells, it required H_2O_2 as a partner, rather than superoxide, to do so (Delledonne et al. 2001). NO˙ does not react with H_2O_2 at an appreciable rate (Brunelli et al. 1995). As such, these puzzling results were merely described as a need for "balance" between reactive oxygen and NO˙ (Delledonne et al. 2002). Although it was not universally appreciated at the time, an initial clue as to why this interaction was necessary to kill cultured cells was provided by the demonstration that NO˙ inhibits the major H_2O_2 scavenging enzymes, catalase and ascorbate peroxidase (Clark et al. 2000). In hindsight, it is clear that NO˙-dependent, persistent, high-level accumulation of H_2O_2 played a role in the cell death seen in these experiments. However, understanding this role and that of NO˙ requires abandoning the deeply entrenched dogma that an "oxidative burst" mediates hypersensitive cell death of individual infected cells.

3
The Oxidative Burst Hypothesis is Wrong

The term "oxidative burst" was borrowed from mammalian immunology to describe a rapid accumulation of H_2O_2 by plant cell cultures following application of "elicitor" molecules derived from plant pathogens (Apostol et al. 1989). Shortly thereafter, the term was also applied to describe reactive oxygen accumulation during the HR (Devlin and Gustine 1992). Although this particular paper argued, based on negative evidence, that the oxidative burst was probably not required for the HR, the opposite conclusion was reached by other researchers in a study showing the HR could be delayed by antioxidants (Ádám et al. 1989). The latter view prevailed in the literature in subsequent years, to the point of being repeated as a truism. A consensus "oxidative burst hypothesis" could be said to have the following tenets:

1. The interaction between a pathogen avirulence (*avr*) gene and a plant resistance (*R*) gene leads to a markedly increased rate of production of superoxide ion, which rapidly undergoes dismutation to H_2O_2.
2. In most systems, superoxide production is catalyzed by a plasma-membrane localized NADPH oxidase (Desikan et al. 1996; van Gestelen et al. 1997). Other enzymes have been proposed in specific systems (Best-

wick et al. 1998; Bolwell et al. 2002; Martinez et al. 1998). Nonetheless, relevant superoxide production is extracellular, not intracellular.
3. In some systems, the oxidative burst is biphasic, with only the second phase being dependent on the *avr* gene/*R* gene interaction. However, both phases precede hypersensitive cell death.
4. At least initially, this superoxide production is from cells receiving the avirulence gene-specified "signals" from the pathogen.
5. The oxidative burst is an essential signaling component leading to PCD in the HR.

The preponderance of evidence indicates that much of the above-stated hypothesis (tenets #1, #3, #4 and #5) is wrong. These issues have been most thoroughly investigated with the HR shown by Arabidopsis leaves to avirulent *Pseudomonas syringae*. Detailed time course measurements of reactive oxygen accumulation during the HR have been published. Reactive oxygen measurements depended upon a dye that fluoresced following oxidation by H_2O_2. Arabidopsis leaves were infected with avirulent *Pseudomonas syringae*, and fluorescence was quantified from leaves taken at a series of time points postinfection. Use of a version of the dye that did not cross membranes verified that measurements were of extracellular H_2O_2. A gradual increase in accumulation of H_2O_2 following an initial $\sim$4–6 h (depending on *avr* gene) lag phase was documented in wild-type Columbia plants (Shapiro and Zhang 2001; Zhang et al. 2004). In response to bacteria carrying two different *avr* genes and with wild type plants as well as several mutant plant lines, initiation of PCD always preceded H_2O_2 accumulation. If reactive oxygen accumulation does not precede PCD, it logically cannot be the cause of it, nor can it occur in the first set of cells responding to the avirulence signal.

An important caveat to these experiments was the sensitivity of the assay for H_2O_2. This caveat was addressed in experiments using the same system. These experiments used an exquisitely sensitive electron microscopy-based assay to detect H_2O_2 accumulation. Very low levels of H_2O_2 were indeed detected at early time points. However, abolishing this early H_2O_2 accumulation pharmacologically did not appear to affect the HR (Grant et al. 2000). Moreover, genetic disruption of the *AtrbohD* gene, which codes for the NADPH oxidase primarily responsible for H_2O_2 accumulation during the HR, largely eliminated H_2O_2 accumulation with minimal effects on the HR (Torres et al. 2002).

HRs of other plants to avirulent bacteria differ in some respects; however, no evidence has been obtained to date proving a requirement for an oxidative burst. In the tobacco nonhost HR to *Pseudomonas syringae* pv. *syringae*, H_2O_2 accumulates in a two-phase oxidative burst that does precede cell death. However, coinfiltration of catalase with the bacteria led only to a minor delay in cell death kinetics while largely abolishing H_2O_2 accumulation (Mur et al. 2005a). In an earlier study using the same pathosystem, superoxide production was documented and found to precede hypersensitive cell death. However, coinfiltration with superoxide dismutase (SOD) had only minor effects on the extent of PCD (Ádám et al. 1989). A two-phase oxidative burst that appeared to depend on apoplastic peroxidases rather than NADPH oxidase did precede PCD in the HR of lettuce to *Pseudomonas syringae* pv. *phaseolicola* (Bestwick et al. 1998). However, no SOD/catalase-based loss-of-function experiments were reported. The oxidative burst was rapid (and peroxidase-dependent) also in cotton cotyledons responding to avirulent *Xanthomonas campestris* pv. *malvacearum* (Martinez et al. 1998, 2000). However, neither loss-of-function experiments nor cell death kinetics were reported. Importantly, although similar in its requirement for peroxidases, the oxidative burst induced by avirulent *Xanthomonas* on pepper is subsequent to the appearance of visible HR symptoms (Do et al. 2003).

Several cell culture-based model systems showed no role for reactive oxygen in the HR (Dorey et al. 1999; Naton et al. 1996; Tenhaken and Rübel 1997). However, a cell-to-cell signaling role was suggested by experiments with the HR of soybean culture cells to avirulent *Pseudomonas syringae* bacteria. In these experiments, populations of cells were separated using dialysis membrane. Inclusion of catalase was used to prove that a signal sent from an infected population of soybean cells to an uninfected population to induce gene expression (albeit with no discernible effects on PCD in the uninfected cell population) was H_2O_2 (Levine et al. 1994).

The HR to oomycete, fungal, and viral pathogens has also been studied. The original report of the oxidative burst of potato tubers to avirulent *Phytophthora infestans* (an oomycete pathogen) (Doke 1983) was later shown to have been misinterpreted, with the observed reaction due not to superoxide, but to other reducing agents (Moreau and Osman 1989). These issues were revisited using *Phytophthora parasitica* and cultured tobacco cells. A two-phase oxidative burst was reported, with the second phase immediately preceding cell death. Exogenous SOD reduced the extent of cell death. However, when the tobacco cells were filtered to remove aggregates prior to infection, superoxide production and hypersensitive cell death were nearly abolished. These results were interpreted to indicate a requirement for cell contact/cell-to-cell signaling in this model of HR (Able et al. 1998).

One report in a true fungus did indeed show reactive oxygen accumulation prior to hypersensitive cell death. H_2O_2 was shown to accumulate in the plant cell wall underneath a germ tube of powdery mildew in infected barley 9 h prior to whole cell epifluorescence and 12 h prior to whole cell collapse (Thordal-Christensen et al. 1997). However, these have been argued to be very late signs of cell death. Earlier markers of cell death were employed (e.g., cytoplasmic disorganization) in studies of the HR of cowpea to avirulent cowpea rust fungus. In these studies, PCD preceded reactive oxygen accumulation. Moreover, neither SOD nor catalase affected any cell death-related phenotype assessed (Heath 1998). Initial PCD events clearly preceded H_2O_2 (and $NO^{\bullet}$) accumulation in oat responding to avirulent crown rust (Tada et al. 2004).

The role of the oxidative burst in a viral HR has been addressed mainly with avirulent Tobacco Mosaic Virus on tobacco carrying the *N* resistance gene. This HR is temperature-sensitive. As such, the virus was allowed to spread systemically at the higher, nonpermissive temperature, prior to shifting temperature to allow a confluent, whole-plant HR. In these experiments, superoxide production clearly preceded the HR and was necessary for PCD to occur (Doke and Ohashi 1988). As these results are strikingly different from those in all other systems studied, some consideration as to the implications of the unusual methods employed is merited. The artificially synchronized HR had the potential to amplify normally minor responses. The high-level reactive oxygen species accumulation seen with the synchronized HR might reflect amplification of what would otherwise be minor accumulation. As H_2O_2 can travel between cells, the "amplified" response is undoubtedly qualitatively different from a more typical HR. Assay of the HR in directly inoculated leaves would also be problematic, because mechanical infection involves wounding. Wounding also leads to H_2O_2 accumulation (Orozco-Cardenas et al. 2001).

A further problem with the oxidative burst hypothesis is that one key tenet, that the rate of production of superoxide ion increases markedly, cannot be investigated. When the oxidative burst is measured in vivo, usually H_2O_2, rather than superoxide is quantified. However, as reactive oxygen is continuously produced and consumed, increased H_2O_2 need not be reflective of increased superoxide production. This author's laboratory attempted to measure extracellular superoxide production semiquantitatively using an in vivo, dye-based assay. This task proved impossible because basal superoxide production in chloroplasts was vastly greater. Infiltration of the dye rapidly produced a colored precipitate that made measurement of the smaller extracellular pool impossible (Zhang C and Shapiro AD, unpublished data). Assaying the oxidative burst in the dark would be nonphysiological and thus irrelevant. Hypersensitive cell death is markedly inhibited in the absence of light (Genoud et al. 2002; Zeier et al. 2004b), and HR assays attempted in the dark instead showed bacterial growth and development of disease symptoms (Zeier et al. 2004b). Interestingly, H_2O_2 still accumulated under these conditions. However, the observation of jasmonate production indicated that this signaling was nonphysiological.

In summary, reactive oxygen accumulation during the HR does not generally appear as a burst, in many systems does not precede PCD, and in most cases is not necessary for PCD. Nonetheless, a facilitatory role in the HR for cell-to-cell transport of H_2O_2 does seem plausible in some cases.

4 NO˙ and Salicylic Acid as Cell-to-Cell Signals

NO˙ and H_2O_2 thus appeared to play related roles in the HR. Both can facilitate the HR in a nonessential fashion, and they can cooperate to kill cultured cells. A further clue towards the function of NO˙ came with the realization that, like H_2O_2, it functions in cell-to-cell signaling. The key experiments involved confocal microscopy (Zhang et al. 2003). Using the NO˙-reporter dye, the initial sites of NO˙ accumulation were shown to be punctate foci on the cell surface. As such, the enzymatic or non-enzymatic activities responsible for NO˙ production must be localized either in the plasma membrane or in the cell wall. Time course studies established that these cells underwent PCD soon after NO˙ accumulation was first observed. Later, NO˙ accumulation was seen near cells directly adjacent to the original ones, and then these cells underwent PCD. The hypothesis was made that NO˙ was acting in a cell-to-cell signaling process that facilitated PCD events in neighboring cells.

Like NO˙ and H_2O_2, salicylic acid (SA) had been shown to be a nonessential facilitator of the HR. This role was first documented with cultured soybean cells. Exogenous SA was shown to potentiate PCD such that levels of avirulent bacteria that would not otherwise have elicited PCD did so (Shirasu et al. 1997). This "gain control" role for SA was later confirmed in intact plants. These studies relied on *ndr1* mutants. In *ndr1* plants, activation of SA biosynthesis in response to avirulent bacteria or exogenous generators of reactive oxygen was impaired (Shapiro and Zhang 2001). This impairment blocked the HR to bacteria carrying *avrRpt2* but not to bacteria carrying other *avr* genes (Century et al. 1995, 1997). Similar results were obtained using either engineered SA catabolism (Rate et al. 1999) or the *sid2* mutation in SA biosynthesis (Zhang and Shapiro 2002). As the *avrRpt2*-conditioned HR is relatively slow and gradual (Grant et al. 2000; Ritter and Dangl 1996; Shapiro and Zhang 2001), the conclusion was made that HR triggering is subject to a threshold that is lowered by increasing SA levels. The ability of *ndr1* and *sid2* plants to show a HR to very high levels of *avrRpt2* (Shapiro 2000; Zhang and Shapiro 2002) further supported this interpretation.

As with NO˙, this facilitatory role depends at least in part on cell-to-cell communication. Experiments using *npr1* mutant plants infected with *P. syringae* carrying *avrRpt2* established that most of the SA that functions in PCD facilitation is synthesized in the last 2 h prior to death of the first wave of cells that will undergo PCD. Parallel quantification of PCD and SA accumulation led to the conclusion that more rapid PCD in *npr1* mutants, caused by interrupting a negative feedback circuit (Rate and Greenberg 2001; Zhang et al. 2004), led to cell death before these cells could synthesize SA (Agrawal et al. 2004; Zhang et al. 2004). As SA diffuses across cell membranes (Yalpani et al. 1991), in wild type plants, this synthesis raises the pool of SA in neighboring cells. These striking similarities suggested that understanding PCD facilitation required an explanation of how NO•, H_2O_2, and SA work together.

5
A NO•-Based, Alternative Hypothesis to Explain H_2O_2 and SA Accumulation during the HR

Insight as to how NO•, H_2O_2, and SA work together was acquired recently by expressing an enzyme known to degrade NO• in Arabidopsis (Zeier et al. 2004a). In this study, reducing NO• accumulation led to greater H_2O_2 degradation, decreased induction of PAL (phenylalanine ammonia lyase, the first enzyme in one of the SA biosynthesis pathways) transcription, and decreased SA accumulation. The HR was also reduced in severity. As the links between H_2O_2 accumulation, SA accumulation, and HR severity had already been made (see above), this new evidence combined with the earlier demonstration that NO• inhibits catalase and ascorbate peroxidase indicated a primary role in suppressing H_2O_2 degradation. The cell-to-cell signaling role for NO• and the extracellular location of its activity in the HR were further supported by the observation that expressing the NO• degrading enzyme in the bacteria rather than the plant led to similar results.

However, these results must be interpreted with caution because of potential pleiotropic effects of the transgene. The *E. coli* flavohemoglobin used as an NO•-degrading enzyme is also known to be an NADH-dependent superoxide synthase (Membrillo-Hernández et al. 1996). This enzyme also has efficient alkylhydroperoxide reductase activity that appears to function in *E. coli* in the repair of oxidatively damaged membrane phospholipids (Bonamore et al. 2003). Expression of other globins has been documented to perturb NADPH/NADP ratios (Igamberdiev et al. 2004) or to lead to peroxidase activity with diverse substrates (Sakamoto et al. 2004). Any of these activities might affect the HR and related physiology. Furthermore, the bacteria expressing the NO•-degrading enzyme showed increased doubling time in culture and other evidence that the enzyme was toxic to these cells (Boccara et al. 2005). These caveats notwithstanding, assuming inhibition of extracellular H_2O_2 scavenging to be the principal direct role of NO• in the HR explains all extant data. This author knows of no other idea that does.

6
An Integrated Model for PCD Control During the HR

Given the proposed direct role of NO•, a testable, working model for how NO• and other signals act together to control hypersensitive cell death was developed and is presented in Fig. 1. The close coincidence in time between apoplastic NO˙ accumulation and loss of plasma membrane integrity (Zhang et al. 2003) suggested that the two events are coupled. One possibility for this coupling would be that small molecules required for the NOS reaction leak from damaged plasma membranes. Of the known cofactors, all are larger than the dye used to track PCD events and would thus not leak prior to PCD. However, the substrate, arginine, is much smaller than the dye. This hypothesis was tested by infiltrating up to 10 mM arginine. No NO˙ accumulated (Shapiro 2005), ruling out this hypothesis. As such, it is instead assumed in Fig. 1 that the activator of NOS is a membrane lipid breakdown product. Upon accumulation of this signal from the dying cells, NO˙ would accumulate and inhibit apoplastic catalase and ascorbate peroxidase activities, leading to H_2O_2 accumulation. NO˙ and/or H_2O_2 would travel intercellularly to carry signals to surrounding cells. H_2O_2 accumulation would then induce transcription of PAL in cells surrounding the dying cell, leading to SA accumulation. SA accumulation would also result from transport from infected cells prior to their death.

With all intercellular signaling converging upon increases in SA levels, it is now clear why this phytohormone is so prominent in PCD control. SA not only potentiates PCD (see above) but also triggers delayed negative feedback on PCD. Negative feedback appears to be transcriptional, with one proposed pathway dependent upon *NPR1* (and associated transcription factors) and the other pathway upon induction of SOD genes (Agrawal et al. 2004; Zhang

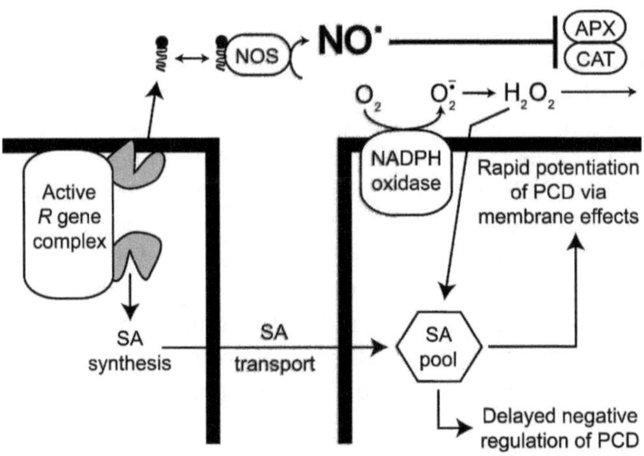

Fig. 1 Nitric oxide influences hypersensitive cell death via intercellular signaling that serves to modulate salicylic acid levels. *SA* salicylic acid, *PCD* programmed cell death, *NOS* nitric oxide synthase, *APX* ascorbate peroxidase, *CAT* catalase. The putative membrane lipid-derived activator of the NOS is depicted as a *filled circle* attached to a *wavy line*. The *upper "pacman"* represents the putative enzyme responsible for generation of this signal. The *lower "pacman"* represents a putative signaling component that, perhaps indirectly, causes increases in SA biosynthesis

et al. 2004). Time required for gene induction would explain the observed delays.

In Fig. 1, the potentiation of PCD triggering is shown as taking place at the plasma membrane. This placement is a result of preliminary results from the author's laboratory that indicated SA levels early in the HR were at least in the tens of millimolar, perhaps hundreds of millimolar, range. Anatomy experiments performed on Arabidopsis leaves (Shapiro AD and Czymmek KJ, unpublished data) provided conversion factors allowing expression of previous measurements of SA levels (Shapiro and Gutsche 2003; Shapiro and Zhang 2001) as concentrations. Levels of SA $\leq$ 1 mM have been shown to result in membrane depolarization (Glass and Dunlop 1974; Gordon et al. 2002; Macrì et al. 1986). Membrane depolarization in these experiments was caused by movement of SA from a more acidic compartment (apoplast) to a less acidic compartment (cytoplasm), driven by electrogenic proton transport. SA import in Arabidopsis cells was shown to occur simultaneously with SA export fluxes (Clarke et al. 2005). As such, salicylic acid accumulation will cause net influx of protons, depolarizing the membrane. Since membrane depolarization is a signaling intermediate connecting avr gene product-dependent molecular recognition of pathogen to PCD (Pavlovkin et al. 1986; Pike et al. 2005), this author proposes that the cellular rheostat responsible for "gain control" of hypersensitive cell death is the voltage across the plasma membrane. Perhaps membrane depolarization serves to facilitate activity of enzymes that catalyze plasma membrane breakdown.

This explanation would account for why potentiation of PCD triggering is necessary only with "weak" avr gene signals. It would also explain why phenotypes of mutants showing altered SA levels as well as experiments using exogenous application of SA are often confusing. Timing of changes in SA levels relative to other signaling events would dictate whether one saw HR suppression or potentiation. Interpretation of experiments with type III effector proteins shown to suppress the HR (Abramovitch and Martin 2004) are similarly confounded. Are these proteins interfering with molecular recognition, interfering with potentiation of PCD triggering, interfering with the death process itself, or enhancing negative feedback regulation? Careful time course studies will be essential for further dissection of this signaling network.

7
Is That All There Is?

At present, the case for $NO^{\cdot}$ accumulating and acting intercellularly with the ultimate effect of raising SA levels and thus manipulating PCD control is compelling. However, a sufficient explanation of $NO^{\cdot}$ function in the HR is not necessarily a complete explanation. Recently, application of laser photoacoustic detection has revealed low level emission of $NO^{\cdot}$ much earlier in the HR than the accumulation seen with confocal microscopy approaches (Mur et al. 2005b). As the limit of detection of the dye used with confocal microscopy is 3 nM (Kojima et al. 1999), any proposed response to this earlier accumulation must be extremely sensitive. Although additional direct targets of NO˙ in plants are known (e.g., cytochrome oxidase, photosynthetic electron transport), none appear to be this sensitive (Caro and Puntarulo 1999; Shapiro 2005; Takahashi and Yamasaki 2002). As such, the relevance of this early NO˙ accumulation is as-yet unestablished.

It is tempting to wield Occam's razor and declare that the role of NO˙ in cell-to-cell communication is the only important role it plays in the HR. It may be more prudent, however, to keep an open mind to additional possible roles.

Acknowledgements This publication is contribution #0013 for the Center for Biomolecular Science and Engineering, Florida Gulf Coast University.

References

Able AJ, Guest DI, Sutherland MW (1998) Use of a new tetrazolium-based assay to study the production of superoxide radicals by tobacco cell cultures challenged with avirulent zoospores of *Phytophthora parasitica* var *nicotianae*. Plant Physiol 117:491–499

Abramovitch RB, Martin GB (2004) Strategies used by bacterial pathogens to suppress plant defenses. Curr Opin Plant Biol 7:356–364

Ádám A, Farkas T, Somlyai G, Hevesi M, Király Z (1989) Consequence of $O_2^{\cdot -}$ Generation during a bacterially induced hypersensitive reaction in tobacco: deterioration of membrane lipids. Physiol Mol Plant Path 34:13–26

Agrawal V, Zhang C, Shapiro AD, Dhurjati PS (2004) A dynamic mathematical model to clarify signaling circuitry underlying programmed cell death control in Arabidopsis disease resistance. Biotechnol Prog 20:426–442

Apostol I, Heinstein PF, Low PS (1989) Rapid stimulation of an oxidative burst during elicitation of cultured plant cells. Plant Physiol 90:109–116

Arancio O, Kiebler M, Lee CJ, Lev-Ram V, Tsien RY, Kandel ER, Hawkins RD (1996) Nitric oxide acts directly in the presynaptic neuron to produce long-term potentiation in cultured hippocampal neurons. Cell 87:1025–1035

Bestwick CS, Bennett MH, Mansfield JW (1995) Hrp mutant of *Pseudomonas syringae* pv. *phaseolicola* induces cell wall alterations but not membrane damage leading to the hypersensitive reaction in lettuce. Plant Physiol 108:503–516

Bestwick CS, Brown IR, Mansfield JW (1998) Localized changes in peroxidase activity accompany hydrogen peroxide generation during the development of a nonhost hypersensitive reaction in lettuce. Plant Physiol 118:1067–1078

Boccara M, Mills CE, Zeier J, Anzi C, Lamb C, Poole RK, Delledonne M (2005) Flavohaemoglobin HmpX from *Erwinia chrysanthemi* confers nitrosative stress tolerance and affects the plant hypersensitive reaction by intercepting nitric oxide produced by the host. Plant J 43:226–237

Bolwell GP, Bindschedler LV, Blee KA, Butt VS, Davies DR, Gardner SL, Gerrish C, Minibayeva F (2002) The apoplastic oxidative burst in response to biotic stress in plants: a three-component system. J Exp Bot 53:1367–1376

Bonamore A, Gentili P, Ilari A, Schinina ME, Boffi A (2003) *Escherichia coli* flavohemoglobin is an efficient alkylhydroperoxide reductase. J Biol Chem 278:22272–22277.

Brunelli L, Crow JP, Beckman JS (1995) The comparative toxicity of nitric oxide and peroxynitrite to *Escherichia coli*. Arch Biochem Biophys 316:327–334

Caro A, Puntarulo S (1999) Nitric oxide generation by soybean embryonic axes. Possible effect on mitochondrial function. Free Radic Res 31:S205-S212

Century KS, Holub EB, Staskawicz BJ (1995) *NDR1*, a locus of *Arabidopsis thaliana* that is required for disease resistance to both a bacterial and a fungal pathogen. Proc Natl Acad Sci USA 92:6597–6601

Century KS, Shapiro AD, Repetti PP, Dahlbeck D, Holub E, Staskawicz BJ (1997) *NDR1*: a pathogen-induced component required for Arabidopsis disease resistance. Science 278:1963–1965.

Clark D, Durner J, Navarre DA, Klessig DF (2000) Nitric oxide inhibition of tobacco catalase and ascorbate peroxidase. Molec Plant-Microbe Interact 13:1380–1384

Clarke A, Mur LAJ, Darby RM, Kenton P (2005) Harpin modulates the accumulation of salicylic acid by *Arabidopsis* cells via apoplastic alkalinization. J Exp Bot 56:3129–3136

Delledonne M, Murgia I, Ederle D, Sbicego PF, Biondani A, Polverari A, Lamb C (2002) Reactive oxygen intermediates modulate nitric oxide signaling in the plant hypersensitive response. Plant Physiol Biochem 40:605–610

Delledonne M, Xia Y, Dixon RA, Lamb C (1998) Nitric oxide functions as a signal in plant disease resistance. Nature 394:585–588

Delledonne M, Zeier J, Marocco A, Lamb C (2001) Signal interactions between nitric oxide and reactive oxygen intermediates in the plant hypersensitive disease resistance response. Proc Natl Acad Sci USA 98:13454–13459

Desikan R, Hancock JT, Coffey MJ, Neill SJ (1996) Generation of active oxygen in elicited cells of *Arabidopsis thaliana* is mediated by a NADPH oxidase-like enzyme. FEBS Letts 382:213–217

Devlin WS, Gustine DL (1992) Involvement of the oxidative burst in phytoalexin accumulation and the hypersensitive reaction. Plant Physiol 100:1189–1195

Do HM, Hong JK, Jung HW, Kim SH, Ham JH, Hwang BK (2003) Expression of peroxidase-like genes, H_2O_2 production, and peroxidase activity during the hypersensitive response to *Xanthomonas campestris* pv. *vesicatoria* in *Capsicum annuum*. Molec Plant-Microbe Interact 16:196–205

Doke N (1983) Involvement of superoxide anion generation in the hypersensitive response of potato tuber tissues to infection with an incompatible race of *Phytophthora infestans* and to hyphal wall components. Physiol Plant Path 23:345–357

Doke N, Ohashi Y (1988) Involvement of an O_2^- generating system in the induction of necrotic lesions on tobacco leaves infected with Tobacco Mosaic Virus. Physiol Molec Plant Pathol 32:163–175

Dorey S, Kopp M, Geoffroy P, Fritig B, Kauffmann S (1999) Hydrogen peroxide from the oxidative burst is neither necessary nor sufficient for hypersensitive cell death induction, phenylalanine ammonia lyase stimulation, salicylic acid accumulation, or scopoletin consumption in cultured tobacco cells treated with elicitin. Plant Physiol 121:163–171

Durner J, Wendehenne D, Klessig DF (1998) Defense gene induction in tobacco by nitric oxide, cyclic GMP, and cyclic ADP-ribose. Proc Natl Acad Sci USA 95:10328–10333

Genoud T, Buchala AJ, Chua N-H, Métraux J-P (2002) Phytochrome signaling modulates the SA-perceptive pathway in Arabidopsis. Plant J 31:87–95

Glass ADM, Dunlop J (1974) Influence of phenolic acids on ion uptake: IV. depolarization of membrane potentials. Plant Physiol 54:855–858

Glazebrook J (2005) Contrasting mechanisms of defense against biotrophic and necrotrophic pathogens. Annu Rev Phytopathol 43:205–227

Gordon LK, Minibayeva FV, Ogorodnikova TI, Rakhmatullina DF, Tzentzevitzky AN, Kolesnikov OP, Maksyutin DA, Valitova JN (2002) Salicylic acid induces dissipation of the proton gradient on the plant cell plasma membrane. Doklady Biol Sci 387:581–583

Grant M, Brown I, Adams S, Knight M, Ainslie A, Mansfield J (2000) The *RPM1* plant disease resistance gene facilitates a rapid and sustained increase in cytosolic calcium that is necessary for the oxidative burst and hypersensitive cell death. Plant J 23:441–450

Heath MC (1998) Involvement of reactive oxygen species in the response of resistant (hypersensitive) or susceptible cowpeas to the cowpea rust fungus. New Phytol 138:251–263

Huang J, Knopp JA (1998) Involvement of nitric oxide in *Ralstonia solanacearum*-induced hypersensitive reaction in tobacco. In: Prior P, Allen C, Elphinstone J (eds) Bacterial wilt disease: molecular and ecological aspects. INRA and Springer Editions, Berlin, Germany, pp 218–224

Igamberdiev AU, Seregélyes C, Manac'h N, Hill RD (2004) NADH-dependent metabolism of nitric oxide in alfalfa root cultures expressing barley hemoglobin. Planta 219:95–102

Keppler LD, Novacky A (1986) Involvement of membrane lipid peroxidation in the development of a bacterially induced hypersensitive reaction. Phytopathol 76:104–108

Klepper L (1979) Nitric oxide (NO) and nitrogen dioxide (NO_2) emissions from herbicide-treated soybean plants. Atmos Environ 13:537–542

Kojima H, Urano Y, Kikuchi K, Higuchi T, Hirata Y, Nagano T (1999) Fluorescent indicators for imaging nitric oxide production. Angew Chem Int Ed 38:3209–3212

Levine A, Tenhaken R, Dixon R, Lamb C (1994) H_2O_2 from the oxidative burst orchestrates the plant hypersensitive disease resistance response. Cell 79:583–593

Macrì F, Vianello A, Pennazio S (1986) Salicylate-collapsed membrane potential in pea stem mitochondria. Physiol Plant 67:136–140

Martinez C, Baccou J-C, Bresson E, Baissac Y, Daniel J-F, Jalloul A, Montillet J-L, Geiger J-P, Assigbetsé K, Nicole M (2000) Salicylic acid mediated by the oxidative burst is a key molecule in local and systemic responses of cotton challenged by an avirulent race of *Xanthomonas campestris* pv. *malvacearum*. Plant Physiol 122:757–766

Martinez C, Montillet JL, Bresson E, Agnel JP, Dai GH, Daniel JF, Geiger JP, Nicole M (1998) Apoplastic peroxidase generates superoxide anions in cells of cotton cotyledons undergoing the hypersensitive reaction to *Xanthomonas campestris* pv. *malvacearum* Race 18. Molec. Plant-Microbe Interact 11:1038–1047

Membrillo-Hernández J, Ioannidis N, Poole RK (1996) The flavohaemoglobin (HMP) of *Escherichia coli* generates superoxide in vitro and causes oxidative stress in vivo. FEBS Letts 382:141–144

Moreau RA, Osman SF (1989) The properties of reducing agents released by treatment of *Solanum tuberosum* with elicitors from *Phytophthora infestans*. Physiol Mol Plant Path 35:1–10

Mur LAJ, Kenton P, Draper J (2005a) *In Planta* measurements of oxidative bursts elicited by avirulent and virulent bacterial pathogens suggests that H_2O_2 is insufficient to elicit cell death in tobacco. Plant Cell Environ 28:548–561

Mur LAJ, Santosa IE, Laarhoven LJJ, Holton NJ, Harren FJM, Smith AR (2005b) Laser photoacoustic detection allows in planta detection of nitric oxide in tobacco following challenge with avirulent and virulent *Pseudomonas syringae* pathovars. Plant Physiol 138:1247–1258

Naton B, Hahlbrock K, Schmelzer E (1996) Correlation of rapid cell death with metabolic changes in fungus-infected, cultured parsley cells. Plant Physiol 112:433–444

Orozco-Cardenas ML, Narvaez-Vasquez J, Ryan CA (2001) Hydrogen peroxide acts as a second messenger for the induction of defense genes in tomato plants in response to wounding, systemin, and methyl jasmonate. Plant Cell 13:179–191

Pavlovkin J, Novacky A, Ullrich-Eberius CI (1986) Membrane potential changes during bacteria-induced hypersensitive reaction. Physiol Mol Plant Pathol 28:125–135

Pike SM, Zhang X-C, Gassmann W (2005) Electrophysiological characterization of the Arabidopsis avrRpt2-specific hypersensitive response in the absence of other bacterial signals. Plant Physiol 138:1009–1017

Prats E, Mur LAJ, Sanderson R, Carver TLW (2005) Nitric oxide contributes both to papilla-based resistance and the hypersensitive response in barley attacked by *Blumeria graminis* f. sp. *hordei*. Molec Plant Pathol 6:65–78

Rate DN, Cuenca JV, Bowman GR, Guttman DS, Greenberg JT (1999) The gain-of-function Arabidopsis *acd6* mutant reveals novel regulation and function of the salicylic acid signaling pathway in controlling cell death, defenses, and cell growth. Plant Cell 11:1695–1708

Rate DN, Greenberg JT (2001) The Arabidopsis *aberrant growth and death2* mutant shows resistance to *Pseudomonas syringae* and reveals a role for NPR1 in suppressing hypersensitive cell death. Plant J 27:203–211

Rees DD, Palmer RMJ, Moncada S (1989) Role of Endothelium-derived nitric oxide in the regulation of blood pressure. Proc Natl Acad Sci USA 86:3375–3378

Ritter C, Dangl JL (1996) Interference between Two specific pathogen recognition events mediated by distinct plant disease resistance genes. Plant Cell 8:251–257

Sakamoto A, Sakurao S-H, Fukunaga K, Matsubara T, Ueda-Hashimoto M, Tsukamoto S, Takahashi M, Morikawa H (2004) Three distinct Arabidopsis hemoglobins exhibit peroxidase-like activity and differentially mediate nitrite-dependent protein nitration. FEBS Letts 572:27–32

Shapiro AD (2000) Using Arabidopsis mutants to delineate disease resistance signaling pathways. Can J Plant Pathol 22:199–216

Shapiro AD (2005) Nitric oxide signaling in plants. Vitamins and Hormones 72:339–398

Shapiro AD, Gutsche AT (2003) Capillary electrophoresis-based profiling and quantitation of total salicylic acid and related phenolics for analysis of early signaling in Arabidopsis disease resistance. Anal Biochem 320:223–233

Shapiro AD, Zhang C (2001) The role of *NDR1* in avirulence gene-directed signaling and control of programmed cell death in Arabidopsis. Plant Physiol 127:1089–1101

Shirasu K, Nakajima H, Rajasekhar VK, Dixon RA, Lamb C (1997) Salicylic acid potentiates an agonist-dependent gain control that amplifies pathogen signals in the activation of defense mechanisms. Plant Cell 9:261–270

Tada Y, Mori T, Shinogi T, Yao N, Takahashi S, Betsuyaku S, Sakamoto M, Park P, Nakayashiki H, Tosa Y, Mayama S (2004) Nitric oxide and reactive oxygen species do not elicit hypersensitive cell death but induce apoptosis in the adjacent cells during the defense response of oat. Molec Plant-Microbe Interact 17:245–253

Takahashi S, Yamasaki H (2002) Reversible inhibition of photophosphorylation in chloroplasts by nitric oxide. FEBS Letts 512:145–148

Tenhaken R, Rübel C (1997) Salicylic acid is needed in hypersensitive cell death in soybean but does not act as a catalase inhibitor. Plant Physiol 115:291–298

Thordal-Christensen H, Zhang Z, Wei Y, Collinge DB (1997) Subcellular localization of H_2O_2 in plants. H_2O_2 Accumulation in papillae and hypersensitive response during the barley-powdery mildew interaction. Plant J 11:1187–1194

Torres MA, Dangl JL, Jones JDG (2002) Arabidopsis gp91^{phox} homologues *AtrbohD* and *AtrBohF* are required for accumulation of reactive oxygen intermediates in the plant defense response. Proc Natl Acad Sci USA 99:523–528

van Gestelen P, Asard H, Caubergs RJ (1997) Solubilization and separation of a plant plasma membrane NADPH – O_2^- synthase from other NAD(P)H oxidoreductases. Plant Physiol 115:543–550

Yalpani N, Silverman P, Wilson TMA, Kleier DA, Raskin I (1991) Salicylic acid is a systemic signal and an inducer of pathogenesis-related proteins in virus-infected tobacco. Plant Cell 3:809–818

Zeier J, Delledonne M, Mishina T, Severi E, Sonoda M, Lamb C (2004a) Genetic elucidation of nitric oxide signaling in incompatible plant–pathogen interactions. Plant Physiol 136:2875–2886

Zeier J, Pink B, Mueller MJ, Berger S (2004b) Light conditions influence specific defence responses in incompatible plant–pathogen interactions: uncoupling systemic resistance from salicylic acid and PR-1 accumulation. Planta 219:673–683

Zhang C, Czymmek KJ, Shapiro AD (2003) Nitric oxide does not trigger early programmed cell death events but may contribute to cell-to-cell signaling governing progression of the Arabidopsis hypersensitive response. Mol Plant-Microbe Interact 16:962–972

Zhang C, Gutsche AT, Shapiro AD (2004) Feedback control of the Arabidopsis hypersensitive response. Molec Plant-Microbe Interact 17:357–365

Zhang C, Shapiro AD (2002) Two Pathways act in an additive rather than obligatorily synergistic fashion to induce systemic acquired resistance and *PR* gene expression. BMC Plant Biology 2:9

Plant Cell Monogr (6)
L. Lamattina · J. C. Polacco: Nitric Oxide in Plant Growth
DOI 10.1007/7089_2006_095/Published online: 14 October 2006

Mitochondrial Nitric Oxide Synthesis During Plant–Pathogen Interactions: Role of Nitrate Reductase in Providing Substrates

Ione Salgado[1] (✉) · Luzia V. Modolo[1] · Ohara Augusto[2] · Márcia R. Braga[3] · Halley C. Oliveira[1]

[1]Departamento de Bioquímica, Instituto de Biologia, Universidade Estadual de Campinas (UNICAMP), CP 6109, 13083-970 Campinas, Brazil
ionesm@unicamp.br

[2]Departamento de Bioquímica, Instituto de Química, Universidade de São Paulo (USP), CP 26077, 05513-970 São Paulo, Brazil

[3]Seção de Fisiologia e Bioquímica de Plantas, Instituto de Botânica, CP 4005, 01061-970 São Paulo, Brazil

Abstract Nitric oxide (•NO) is an important signaling molecule that regulates plant metabolism and mediates defense responses against biotic and abiotic stresses. Although the cellular mechanisms by which •NO is generated in plants have been intensively investigated, they still remain controversial, particularly those implicated in plant resistance to pathogens. •NO can be synthesized in plants via the oxidation of L-arginine into L-citrulline by a nitric oxide synthase (NOS) that has no homology with the animal NOS family. In addition to L-arginine, nitrite is also an important source for •NO in plants. Since nitrate reductase (NR) activity is required for •NO production, reduction of nitrite catalyzed by the enzyme has been considered a major •NO source in plants. Recent experimental data, however, indicate that NR-defective mutant plants can synthesize •NO from exogenous nitrite. Nevertheless, NR-deficient plants lack enough endogenous substrates (L-arginine and nitrite) for adequate •NO synthesis, resulting in an impaired hypersensitive response (HR). These findings indicate that NR activity is an important source of substrates for •NO production. The main pathways for •NO production from L-arginine and nitrite in plants are located in mitochondria, suggesting that these organelles play a central role in •NO signaling.

1 Introduction

Nitric oxide (•NO) is a major effector in biological systems and can influence a broad range of physiological and pathophysiological processes. Although uncharged, •NO contains an unpaired electron that makes it reactive with oxygen, superoxide anion, and transition metals to generate other reactive nitrogen and oxygen species (peroxynitrite, nitrogen oxides, etc.) and finally *S*-nitrosothiols, *N*-nitrosyl compounds, nitro-tyrosine, metal-nitrosyl complexes and other compounds (Stamler et al. 1992; Augusto et al. 2004). The chemical properties of $^{\bullet}$NO and its derivatives that enable them to react with diverse molecular targets initially suggested that $^{\bullet}$NO was a promiscuous molecule. However, more recent findings, obtained mainly in animal systems, reveal that $^{\bullet}$NO acts basically by modifying heme and sulfhydryl sites (Hare and Stamler 2005). In addition, the production of $^{\bullet}$NO and its signaling components are compartmentalized, and this provides a basis for understanding the specificity of $^{\bullet}$NO action in different metabolic conditions (Hare and Stamler 2005).

$^{\bullet}$NO was first identified as the factor responsible for the endothelium-dependent relaxation of mammalian blood vessels (Palmer et al. 1987; Ignarro et al. 1987). Since then, $^{\bullet}$NO has been implicated in numerous physiological processes, including the control of blood pressure, platelet aggregation, neurotransmission, and the immune response in animal cells (Mayer and Hemmens 1997). Moreover, uncontrolled $^{\bullet}$NO production has been implicated in a number of pathophysiologies, e.g., excessive $^{\bullet}$NO formation has been associated with various neurodegenerative diseases (Guix et al. 2005) whereas an attenuated production has been associated with diabetes, hypertension and other cardiovascular diseases (Naseem 2005).

The ability of plants to produce $^{\bullet}$NO was initially observed in herbicide-treated soybean plants (Klepper 1979). However, only after the discovery of its broad range of bioactivities in mammals was it realized that this radical could also be an important signaling molecule in plants. Indeed, $^{\bullet}$NO influences several plant metabolic and developmental processes, including seed germination, root growth, leaf expansion, stomata movement, cell wall lignification, flowering, growth of the pollen tubes, and senescence (Lamattina et al. 2003; del Rio et al. 2004; Magalhães et al. 2005; Salgado et al. 2006).

Besides its involvement in plant signal transduction pathways related to growth and development, $^{\bullet}$NO has also been implicated in plant responses to biotic and abiotic signals (Neill et al. 2003; del Rio et al. 2004). For instance, wound signaling (Orozco-Cárdenas and Ryan 2002; Huang et al. 2004), protection against oxidative stress under UV light (Shi et al. 2005), and lateral root formation in tomato/*Azospirillum brasilense* interactions (Creus et al. 2005) are all influenced by this radical.

$^{\bullet}$NO has also been implicated in plant defense against pathogen attack. Thus, increased $^{\bullet}$NO production occurs during various plant–pathogen interactions (Salgado et al. 2004; Delledonne et al. 2005). Additionally, specific and general elicitors of plant defense induce $^{\bullet}$NO production in plant tissues and cells in culture (Foissner et al. 2000; Modolo et al. 2002; Lamotte et al. 2004; Yamamoto et al. 2004; Zeidler et al. 2004). $^{\bullet}$NO, together with reactive oxygen species, is also involved in localized cell death in host tissue (hypersensitive response, HR) during plant–pathogen incompatible interactions (Delledonne et al. 2001; Zhang et al. 2003; Tada et al. 2004; Murgia et al. 2004). The expression of genes such as those encoding phenylalanine ammonia-lyase (*PAL*), chalcone synthase (*CHS*), and pathogenesis-related protein-1 (*PR-1*) is enhanced by ˙NO, with the consequent activation of biosynthetic pathways that lead to the accumulation of antimicrobial compounds (Delledonne et al. 1998; Durner et al. 1998; Modolo et al. 2002). Moreover, ˙NO stimulates the accumulation of salicylic acid, a signaling molecule that plays a central role in systemic acquired resistance (Durner et al. 1998).

Currently, ˙NO is considered a multifunctional effector in plants. However, the mechanisms and subcellular location of its formation in plants are only beginning to be understood. In this review, we discuss recent advances in our knowledge of the mechanisms responsible for ˙NO production in plants, with special emphasis on the main sources of ˙NO during plant–pathogen interactions and their compartmentalization.

2 Sources of ˙NO in Plants

2.1 ˙NO Synthesis from L-arginine

In mammalian cells, a family of nitric oxide synthases (NOS, EC.1.14.13.39) composed of three highly homologous isoforms (Lamas et al. 1992) is the primary source of ˙NO. All NOS enzymes catalyze the formation of ˙NO and L-citrulline from L-arginine (Tayeh and Marletta 1989) by a reaction that requires oxygen and NADPH as cosubstrates and enzyme-bound heme, tetrahydrobiopterin (H_4B), calmodulin, FAD, and FMN as cofactors (Stuehr 1997). NOS have been found in various subcellular compartments, including the plasma membrane, endoplasmic reticulum, mitochondria, and cytosol. This compartmentalization is one of the important aspects in determining their specificity (Barouch et al. 2002).

Interestingly, several plant tissues produce L-citrulline from L-arginine via a process that is dependent on mammalian NOS cofactors and sensitive to its inhibitors (Delledonne et al. 1998; Durner et al. 1998; Ribeiro et al. 1999; Barroso et al. 1999; Modolo et al. 2002). Additionally, antibodies produced against mammalian NOS isoforms bind to proteins in the cytosolic and nuclear compartments of maize root tips (Ribeiro et al. 1999) and in the peroxisomes and chloroplasts of pea leaves (Barroso et al. 1999; Corpas et al. 2001). However, no gene or protein with homology to mammalian NOS has been found in plants, in particular in *Arabidopsis thaliana* (The Arabidopsis Genome Initiative 2000). These apparently contradictory findings were resolved when Guo et al. (2003) identified the gene *AtNOS1* in *A. thaliana* that encodes a protein that converts L-arginine into L-citrulline and ˙NO. The activity is stimulated by NADPH, Ca^{2+}, and calmodulin, indicating that this protein shares some of the properties described for animal NOS. AtNOS1 has no sequence similarity to any mammalian NOS isoform but is homologous to a protein identified in the snail *Helix pomatia* that cross-reacts with human neuronal NOS antibodies and produces ˙NO from L-arginine by an unknown mechanism (Huang et al. 1997). Whereas all mammalian NOS isoforms are homodimers of subunits with molecular masses ranging from 130 to 160 kDa (Stuehr 1997), AtNOS1 is a monomeric protein (∼ 60 kDa) that lacks the NADPH-binding site typical of NOS (Crawford 2006).

AtNOS1-deficient mutants show a considerable decrease in ˙NO production, with a consequent reduction in root and shoot growth, seed fertility, and abscisic acid-induced stomatal movements (Guo et al. 2003). Such mutants also have high levels of reactive oxygen species and accelerated dark-induced senescence (Guo and Crawford 2005). Compared to wild-type plants, these mutants show early flowering, which suggests that ˙NO produced by AtNOS1 controls this process (He et al. 2004). Recently, N-terminal peptide targeting showed that AtNOS1 is located in mitochondria. In agreement, mitochondria isolated from wild-type *A. thaliana* leaves produce ˙NO in response to L-arginine whereas those from AtNOS1-deficient mutants do not (Guo and Crawford 2005). Sequence alignment shows that this family of proteins is conserved in all eukaryotes and many bacteria (Zemojtel et al. 2004). Recently, a mouse ortholog of AtNOS1 was cloned and also localized in mitochondria (Zemojtel et al. 2006).

The discovery of this novel AtNOS1 partly explains the effects attributed to ˙NO in plants. However, the production of this radical is not restricted to NOS activity since NOS inhibitors do not affect certain ˙NO-mediated events (Neill et al. 2003).

2.2
˙NO Synthesis from Nitrite

In addition to L-arginine, nitrite has emerged as a potential source of ˙NO in plants in which the existence of a specialized system for nitrogen assimilation reinforces the importance of this pathway. The production of ˙NO from nitrite may occur through various enzymatic and non-enzymatic mechanisms. The latter may occur under acidic conditions, a process that is accelerated by reducing agents such as ascorbate (Yamasaki et al. 1999). The reduction of nitrite to ˙NO was shown to occur in the acidic apoplastic space of barley aleurone layers, whose physiological relevance was related to coordination of different tissues activities during seed germination (Bethke et al. 2004).

Potential enzymatic mechanisms for nitrite reduction in plants include xanthine oxidase (XO) (Li et al. 2004), a plasma membrane-bound nitrite:NO reductase (NI-NOR) (Stöhr et al. 2001), nitrite reductase (NiR) (Kuznetsova et al. 2004) and nitrate reductase (NR) (Klepper 1990; Yamasaki and Sakihama 2000; Rockel et al. 2002).

Xanthine oxidase (XO) from animal sources catalyzes the conversion of nitrite to ˙NO (Zhang et al. 1998; Li et al. 2004). Recently, a XO gene was identified in *A. thaliana* and showed high homology with its animal counterparts (Hesberg et al. 2004). However, to date there is no evidence that XO has a relevant role in the generation of ˙NO in plants and, at least in *A. thaliana* leaves, allopurinol (a potent XO inhibitor) has no effect on ˙NO production (Modolo et al. 2005).

Nitrite : NO reductase (NI : NOR) is a plasma membrane-bound enzyme with nitrite-reducing activity that has been identified in root tissues of tobacco plants (Stöhr et al. 2001). This enzyme was named NI : NOR and, together with bacterial nitrification and denitrification in the rhizosphere, may constitute an important source for ˙NO formation in roots (Stöhr and Ullrich 2002). The molecular identity of the NI : NOR remains unknown.

Nitrite reductase (NiR) present in chloroplasts reduces nitrite to ammonium using ferredoxin as electron donor. This enzyme contains one siroheme and one [4Fe – 4S] cluster as prosthetic groups. Spectroscopic evidence was provided for an intermediate state in which NO-ferrous siroheme is formed during the catalytic cycle of NiR purified from spinach chloroplasts (Kuznetsova et al. 2004). Therefore, the possibility that ˙NO is produced from NiR reaction has been established. However, NiR anti-sense tobacco plants (Morot-Gaudry-Talarmain et al. 2004) and NiR-deficient green alga *Chlorella sorokiniana* (Tischner et al. 2004) accumulate nitrite and release increased amounts of ˙NO. In addition, in various higher non-leguminous plants, conditions that decrease NiR activity result in increased ˙NO production (Xu and Zhao 2003). Hence, it seems unlikely that NiR has important ˙NO-producing activity in plants.

Nitrate reductase (NR) is a key enzyme for nitrate assimilation in plants since it catalyzes the NAD(P)H-dependent reduction of nitrate to nitrite that is further reduced to ammonium by NiR (see Lea 1990). In addition to the main nitrate-reducing activity of NR, results obtained in vivo and in vitro have suggested that this enzyme may also reduce nitrite to ˙NO (Dean and Harper 1986; Yamasaki and Sakihama 2000; Rockel et al. 2002; Kaiser et al. 2002). Early studies detected an enzymatic emission of $NO_{(x)}$ (˙NO and NO_2) in leaf tissues and this alternative fate for nitrite was attributed to the activity of the constitutive NR (Klepper 1987). This $NO_{(x)}$ emission was first suggested to be limited to the tribe Phaseoleae of the Leguminosae (Klepper 1987) but more recently it was shown to occur in plants belonging to other families such as the Poaceae, Asteraceae, and Brassicaceae (Yamasaki et al. 1999; Rockel et al. 2002). The nitrite-reducing activity of NR seems to be favored under anaerobic conditions, when the nitrite concentration increases (Yamasaki et al. 1999; Yamasaki 2000; Rockel et al. 2002). In *Arabidopsis* leaves, electron paramagnetic resonance (EPR) studies revealed that ˙NO production from nitrite was reduced by high concentrations of nitrate (Vanin et al. 2004). This effect was attributed to competitive inhibition of the nitrite-reducing activity of NR by nitrate (Vanin et al. 2004). NR purified from maize reduces nitrite to ˙NO using NADH as electron donor in a reaction suggested to be catalyzed by the same domain in which nitrate is reduced (Yamasaki et al. 1999).

Studies with NR-deficient mutants of *Arapidopsis* have suggested that the ˙NO production induced by abscisic acid (ABA) and stomata closing are dependent on NR (Desikan et al. 2002; Garcia-Mata and Lamattina 2003). *Arabidopsis* guard cells of NR-deficient mutants neither synthesize ˙NO nor show stomatal closure in response to nitrite or ABA (Desikan et al. 2002). A role for NR in ˙NO production during sorghum seed germination has also been suggested (Simontacchi et al. 2004).

Overall, these data provide evidence that NR can reduce nitrite to ˙NO and that various physiological processes are modulated by the NR-dependent ˙NO generation. However, ˙NO production from nitrite may also occur via a mitochondrial-dependent process that we have recently shown to be important during plant–pathogen interactions (Modolo et al. 2005, 2006).

3
Synthesis of •NO During Plant–Pathogen Interaction

In some plant–pathogen incompatible interactions, such as those between *A. thaliana-Pseudomonas syringae* (Delledonne et al. 1998; Zeidler et al. 2004), tobacco–tobacco mosaic virus (Durner et al. 1998), and soybean-*Diaporthe phaseolorum* (Modolo et al. 2002), an increased production of L-citrulline from L-arginine has been detected, indicating participation of an NOS-like enzyme in plant defensive responses to pathogen attack. Recent findings suggest that AtNOS1 or its homologs are involved in ˙NO production in these pathosystems. Indeed, *AtNOS1* silencing dramatically increases the susceptibility of *A. thaliana* to *P. syringae* pv. *tomato* by preventing the induction of local and systemic defensive genes (Zeidler et al. 2004).

Although NOS activity increases in *A. thaliana* after inoculation with *P. syringae* pv. *maculicola*, this is not the main source of ˙NO under these conditions (Modolo et al. 2005). Indeed, electron paramagnetic resonance (EPR) spin-trapping experiments using the ferrous-*N*-methyl-D-glucamine complex [$(MGD)_2Fe(II)$] have shown that *A. thaliana* plants have an endogenous production of ˙NO in the range of 50–57 pmol/min/mg that is not derived from L-arginine. This was confirmed by isotopic labeling of L-arginine with ^{15}N. This substrate would produce labeled $^{\cdot 15}NO$ which, when complexed with $(MGD)_2Fe(II)$, should yield a two-line EPR spectrum instead of the normal three-line one. Since the typical $(MGD)_2Fe(II)^{15}NO$ complex was not detected (Fig. 1), L-arginine is not the main ˙NO source. Indeed, parallel experiments showed that ˙NO production from NOS activity was much lower, in the range of 12 pmol/min/mg (Modolo et al. 2005), and hence difficult to discriminate by EPR experiments (Giorgio et al. 1998).

Additionally, the EPR-detectable $^{\cdot}$NO was barely observed in homogenates of *A. thaliana* bearing double mutations for the NR enzyme (*nia1 nia2*) (Fig. 1). This result, together with the claim that NR is an important source of $^{\cdot}$NO in plants (Rockel et al. 2002; Lea et al. 2004), suggested that nitrate could be the source of $^{\cdot}$NO during the *A. thaliana–P. syringae* interaction. However, the incubation of homogenates from the leaf tissues of wild-type and NR-deficient mutant plants (controls or *P. syringae*-inoculated) with the spin trap, $^{15}NO_3^-$ and NADH (an important cofactor for NR activity), discarded this possibility. The results were not significantly different from those obtained in experiments with L-^{15}N-arginine (Fig. 1), and indicated that the

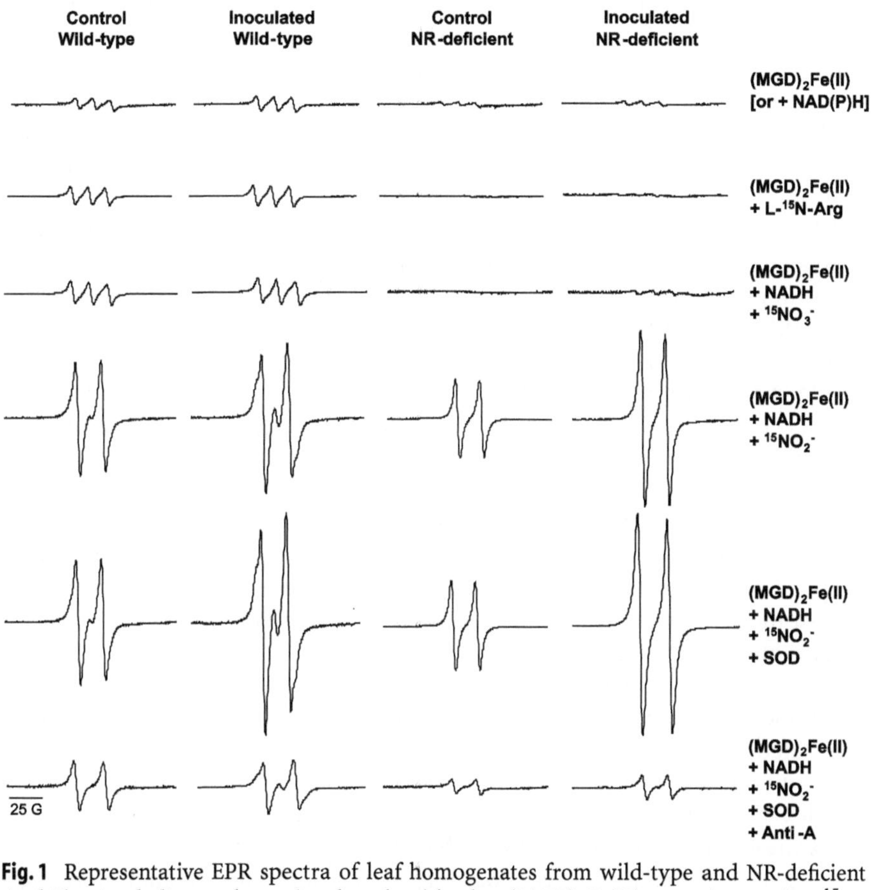

Fig. 1 Representative EPR spectra of leaf homogenates from wild-type and NR-deficient *Arabidopsis thaliana* plants incubated with the $(MGD)_2Fe(II)$ complex and L-^{15}N-arginine, $^{15}NO_3^-$, or $^{15}NO_2^-$. NAD(P)H, superoxide dismutase (SOD) and antimycin-A (Anti-A) were added as indicated. The spectra were obtained using homogenates of control and 6 h *Pseudomonas syringae*-inoculated leaves. [Reprinted by permission of Federation of the European Biochemical Societies from Modolo et al. (2005)]

EPR-detectable ˙NO from an endogenous source in wild-type plants was also not derived from nitrate. Dordas et al. (2004) were similarly unable to detect ˙NO production from nitrate by EPR in *Zea mays* cells cultured under aerobic conditions, although nitrate-derived ˙NO was produced under hypoxia.

In contrast to the results obtained with L-^{15}N-arginine and $^{15}NO_3^-$, the presence of $^{15}NO_2^-$ elicited a substantial increase in the levels of EPR-detectable ˙^{15}NO in leaf homogenates of control and *P. syringae*-inoculated plants (Fig. 1). The detection of ˙^{15}NO was even more accentuated in the presence of superoxide dismutase (Fig. 1), which prevents the reaction of ˙NO with superoxide anion. Furthermore, most of this nitrite-derived ˙NO production was not attributable to NR activity since plants lacking the two structural genes *NIA1* and *NIA2* were still able to produce ˙NO as efficiently as the wild-type plants (Fig. 1). These observations led to the suggestion that the limiting factor for ˙NO synthesis in *nia1 nia2* plants was the low level of endogenous nitrite because the plants were NR-deficient and were cultivated in nitrate-free medium. Accordingly, endogenous nitrite was undetectable in *nia1 nia2 A. thaliana* plants (Modolo et al. 2006) and basal ˙NO production was detected only upon nitrite supplementation of leaf homogenates (Fig. 1).

These data suggest that the main role of NR for ˙NO generation in plants is related to the production of nitrite, the substrate for such production. Accordingly, various reports have shown that ˙NO production in plants correlates with endogenous nitrite levels. In leaves of NiR antisense tobacco plants that present very low NiR activity, the nitrite concentration is tenfold higher and ˙NO emission rates are 100-fold higher than in wild-type leaves (Morot-Gaudry-Talarmain et al. 2002). In *Nicotiana plumbaginifolia*, the expression of a tobacco NR with a mutation in the regulatory phosphorylation site that produces a permanently active NR resulted in higher levels of nitrite and ˙NO emission (Lea et al. 2004). Tobacco NR-free plants or cell suspensions supplied with nitrate for short periods never emitted ˙NO. However, when supplied with nitrite, virtually NR-free cells emitted ˙NO (under anoxia) at almost the same rates as NR-induced cells (Planchet et al. 2005).

The observation that significant nitrite-derived ˙NO synthesis can occur in NR-deficient plants (Modolo et al. 2005; Planchet et al. 2005) raised questions about the mechanism responsible for the synthesis that could not be attributed to NR. Mammalian mitochondria can reduce nitrite to ˙NO, an activity that is prevented by inhibitors of mitochondrial electron transport (Kozlov et al. 1999). Similar results were reported for the algae *Chlorella sorokiniana* (Tischner et al. 2004), tobacco cell suspension cultures (Planchet et al. 2005), *A. thaliana* leaves (Modolo et al. 2005) and isolated mitochondria from the roots of several plant species (Gupta et al. 2005). These findings suggested that mitochondrial-dependent nitrite reduction may be a general mechanism for ˙NO generation in eukaryotic organisms (Chen et al. 2005). In plants, NR is fundamental for this mode of ˙NO generation from nitrite.

NR is not only important to provide nitrite but it also has a relevant role in L-arginine biosynthesis. The total content of amino acids, the end products of nitrate assimilation, is markedly lower in transgenic tobacco plants with reduced NR activity (Hänsch et al. 2001). Analysis of the amino acid contents of *nia1 nia2 A. thaliana* leaves revealed reduced levels, particularly of L-arginine, whose content was ten times lower than that of wild-type plants (Modolo et al. 2006). A low content of L-arginine, in addition to a nitrite deficiency, may adversely affect ˙NO production during defensive responses.

In situ analysis using the indicator 4,5-diaminofluorescein (DAF-2DA) (Modolo et al. 2006) showed that the leaves of wild-type *A. thaliana* plants have a basal ˙NO production (Fig. 2a), which increased significantly upon *P. syringae* inoculation (Fig. 2b). In contrast, the leaves of *nia1 nia2* plants showed marginal ˙NO production (Fig. 2c), even after the inoculation of

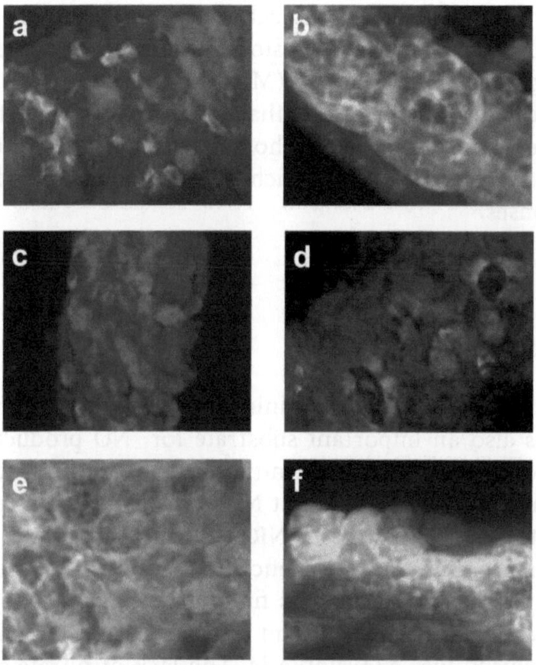

Fig. 2 ˙NO production in situ in wild-type and NR-deficient *A. thaliana* leaves treated with L-arginine, NO_2^-, or *Pseudomonas syringae*. The leaves were infiltrated with water (control) or *P. syringae* (inoculated) and maintained in the dark for 6 h after which they were stained with 4,5-diaminofluorescein diacetate (DAF-2DA) to visualize ˙NO: **a** wild-type (control), **b** wild-type (inoculated), **c** NR-deficient (control), **d** NR-deficient (inoculated), **e** NR-deficient infiltrated with L-arginine, **f** NR-deficient infiltrated with NO_2^-. [Reprinted from Plant Science, Modolo LV et al. (2006), with permission from Elsevier]

P. syringae (Fig. 2d). However, when non-inoculated *nia1 nia2* leaves were infiltrated with L-arginine or with nitrite, a fluorescent emission indicative of ˙NO production was observed (Fig. 2e,f). These results showed that NR-deficient plants were unable to increase ˙NO production in response to pathogen attack because they were deficient in both nitrite and L-arginine. The importance of endogenous levels of L-arginine in regulating ˙NO synthesis is supported by studies with ˙NO-overproducing (*nox*) mutants of *A. thaliana*. These *nox* mutant plants possess larger amounts of L-arginine because of disruption of a chloroplast phosphoenolpyruvate/phosphate translocator and, consequently, they produce more ˙NO than wild-type plants (He et al. 2004).

Relevantly, NR-deficient *A. thaliana* plants lose their ability to develop a normal hypersensitive response (HR) at the site of *P. syringae* inoculation, a response clearly seen in wild-type plants (Modolo et al. 2006). Since ˙NO is essential for the induction of defense-associated genes, the inability of *nia1 nia2* plants to develop a normal HR to *P. syringae* may reflect their incapacity to promptly produce this free radical because they lack the substrates for ˙NO synthesis. In agreement, HR is restored if nitrite is provided to *nia1 nia2* mutants during bacterial inoculation (Modolo et al. 2006).

Overall, the available data indicate that NR activity is important for proper plant resistance against invading pathogens by providing nitrogenous substrates for the synthesis of ˙NO, which is an important mediator of plant defensive responses.

4 Conclusions and Future Scope

In contrast to animals, in which L-arginine is the primary source for ˙NO synthesis, nitrite is also an important substrate for ˙NO production in plants. The reduction of nitrite to ˙NO was initially attributed to the activity of NR. However, recent findings indicate that NR is instead responsible for providing the substrates for ˙NO synthesis. NR-deficient plants show reduced levels of nitrite and L-arginine and this deficiency accounts for the impaired response against pathogen attack. This finding reveals a connection between primary nitrogen metabolism and plant disease resistance, a still obscure and controversial subject, via ˙NO synthesis. The lack of nitrate assimilation in NR-deficient mutants also affects the overall synthesis of amino acids and the compounds produced from them. Hence, it will be necessary to investigate the extent to which altered metabolic conditions can influence plant resistance to pathogens, independent of a deficiency in the substrates for ˙NO synthesis.

Mitochondria are important subcellular compartments for ˙NO generation from nitrite and L-arginine in plants. Our understanding about the mitochondrial nitrite-reducing activity of plants is still limited and requires further investigation. The acidic milieu created across the inner mitochondrial membrane as a result of the electrochemical gradient generated from coupled electron transport via the respiratory chain may favor nitrite reduction to ˙NO. The respiratory chain can also provide the required reducing equivalents as proposed for mammalian mitochondria (Kozlov et al. 1999). Thus, functional mitochondria are required for nitrite reduction to ˙NO. The question of how this process is enhanced during plant–pathogen interactions remains unanswered.

The discovery that the synthesis of ˙NO from L-arginine is mediated by a novel NOS and that it is located in mitochondria partially explained the well-documented L-arginine-dependent ˙NO production in plants. Since the NADPH-binding site is absent in AtNOS1, Crawford (2006) suggested that a mitochondrial redox cofactor may be necessary for the enzyme to function or, alternatively, that the enzyme interacts with other proteins to form a ˙NO synthesizing complex. Plant mitochondria can express alternative NAD(P)H dehydrogenases in the inner mitochondrial membrane, and their expression levels are increased by stress (Rasmusson et al. 2004). These NAD(P)H dehydrogenases may cooperate with AtNOS1 to produce ˙NO.

The observation that mitochondria are important subcellular compartments for ˙NO generation in plants has many implications, because these organelles are important targets of ˙NO action. A distinguishing feature of plant mitochondria is that they can express the alternative oxidase (AOX) under conditions of stress, such as pathogen attack (Lennon et al. 1997; Simons et al. 1999). In contrast to cytochrome *c* oxidase, AOX is resistant to ˙NO (Millar and Day 1996). Thus, this alternative fate for oxygen reduction may ameliorate the extent of the deleterious effects of ˙NO on mitochondrial functionality during plant defense responses. Mitochondria also participate in ˙NO-induced programmed cell death by formation of the permeability transition pore (Saviani et al. 2002) through which pro-apoptotic mitochondrial factors are released into the cytosol. Consequently, mitochondrial ˙NO production may play an important role in the programmed cell death seen during the hypersensitive response.

In conclusion, the available data on ˙NO production in plants during plant–pathogen interactions suggest that more than one mechanism operates to generate the radical, with mitochondria probably playing a central role. There is no doubt that an NOS-like enzyme, possibly AtNOS1, and/or its homologs, participates in the process. However, such activity is insufficient to explain the ˙NO burst observed during the acute response of plants to pathogens. Experiments designed to elucidate the mitochondrial reduction of nitrite to ˙NO in response to pathogen attack are therefore necessary. In addition, there is a need to investigate how these different sources of ˙NO may influence plant physiology.

Acknowledgements Work in our laboratories is funded by the Fundação de Amparo à Pesquisa do Estado de São Paulo (FAPESP) and Conselho Nacional de Desenvolvimento Científico e Tecnológico (CNPq) (Projeto Milênio: Redoxoma).

References

Augusto O, Bonini MG, Trindade D (2004) Spin trapping of glutathiyl and protein radicals produced from nitric oxide-derived oxidants. Free Radic Biol Med 15:1224–1232

Barouch LA, Harrison RW, Skaf MW, Rosas GO, Cappola TP, Kobeissi ZA, Hobai IA, Lemmon CA, Burnett AL, O'Rourke B, Rodriguez ER, Huang PL, Lima JA, Berkowitz DE, Hare JM (2002) Nitric oxide regulates the heart by spatial confinement of nitric oxide synthase isoforms. Nature 416:337–339

Barroso JB, Corpas FJ, Carreras A, Sandalio LM, Valderrama R, Palma JM, Lupianez JA, del Rio LA (1999) Localization of nitric-oxide synthase in plant peroxisomes. J Biol Chem 274:36729–36733

Bethke PC, Badger MR, Jones RL (2004) Apoplastic synthesis of nitric oxide by plant tissues. Plant Cell 16:332–341

Chen Z, Foster MW, Zhang J, Mao L, Rockman HA, Kawamoto T, Kitagawa K, Nakayama KI, Hess DT, Stamler JS (2005) An essential role for mitochondrial aldehyde dehydrogenase in nitroglycerin bioactivation. Proc Natl Acad Sci USA 102:12159–12164

Corpas FJ, Barroso JB, del Rio LA (2001) Peroxisomes as a source of reactive oxygen species and nitric oxide signal molecules in plant cells. Trends Plant Sci 6:145–150

Crawford NM (2006) Mechanisms for nitric oxide synthesis in plants. J Exp Bot 57:471–478

Creus CM, Graziano M, Casanovas EM, Pereyra MA, Simontacchi M, Puntarulo S, Barassi CA, Lamattina L (2005) Nitric oxide is involved in the *Azospirillum brasilense*-induced lateral root formation in tomato. Planta 221:297–303

Dean JV, Harper JE (1986) Nitric oxide and nitrous oxide production by soybean and winged bean during in vivo nitrate reductase assay. Plant Physiol 82:718–723

del Rio LA, Corpas FJ, Barroso JB (2004) Nitric oxide and nitric oxide synthase activity in plants. Phytochemistry 65:783–792

Delledonne M (2005) NO news is good news for plants. Curr Opin Plant Biol 8:390–396

Delledonne M, Xia Y, Dixon RA, Lamb C (1998) Nitric oxide functions as a signal in plant disease resistance. Nature 394:585–588

Delledonne M, Zeier J, Marocco A, Lamb C (2001) Signal interactions between nitric oxide and reactive oxygen intermediates in the plant hypersensitive disease resistance response. Proc Natl Acad Sci USA 98:13454–13459

Desikan R, Griffiths R, Hancock J, Neill S (2002) A new role for an old enzyme: nitrate reductase-mediated nitric oxide generation is required for abscisic acid-induced stomatal closure in *Arabidopsis thaliana*. Proc Natl Acad Sci USA 99:16314–16318

Dordas C, Hasinoff BB, Rivoal J, Hill RD (2004) Class-1 hemoglobins, nitrate and NO levels in anoxic maize cell-suspension cultures. Planta 219:66–72

Durner J, Wendehenne D, Klessig DF (1998) Defense gene induction in tobacco by nitric oxide, cyclic GMP, and cyclic ADP-ribose. Proc Natl Acad Sci USA 95:10328–10333

Foissner I, Wendehenne D, Langebartels C, Durner J (2000) In vivo imaging of an elicitor-induced nitric oxide burst in tobacco. Plant J 23:817–824

Garcia-Mata C, Lamattina L (2003) Abscisic acid, nitric oxide and stomatal closure – is nitrate reductase one of the missing links? Trends Plant Sci 8:20–26

Giorgio S, Linares E, Ischiropoulos H, Von Zuben FJ, Yamada A, Augusto O (1998) In vivo formation of electron paramagnetic resonance-detectable nitric oxide and of nitrotyrosine is not impaired during murine leishmaniasis. Infect Immun 66:807–814

Graziano M, Lamattina L (2005) Nitric oxide and iron in plants: an emerging and converging story. Trends Plant Sci 10:4–8

Guix FX, Uribesalgo I, Coma M, Munoz FJ (2005) The physiology and pathophysiology of nitric oxide in the brain. Prog Neurobiol 76:126–152

Guo FQ, Crawford NM (2005) *Arabidopsis* nitric oxide synthase1 is targeted to mitochondria and protects against oxidative damage and dark-induced senescence. Plant Cell 17:3436–3450

Guo FQ, Okamoto M, Crawford NM (2003) Identification of a plant nitric oxide synthase gene involved in hormonal signaling. Science 302:100–103

Gupta KJ, Stoimenova M, Kaiser WM (2005) In higher plants, only root mitochondria, but not leaf mitochondria reduce nitrite to NO, in vitro and in situ. J Exp Bot 56:2601–2609

Hänsch R, Fessel DG, Witt C, Hesberg C, Hoffmann G, Walch-Liu P, Engels C, Kruse J, Rennenberg H, Kaiser WM, Mendel RR (2001) Tobacco plants that lack expression of functional nitrate reductase in roots show changes in growth rates and metabolite accumulation. J Exp Bot 52:1251–1258

Hare JM, Stamler JS (2005) NO/redox disequilibrium in the failing heart and cardiovascular system. J Clin Invest 115:509–517

He Y, Tang RH, Hao Y, Stevens RD, Cook CW, Ahn SM, Jing L, Yang Z, Chen L, Guo F, Fiorani F, Jackson RB, Crawford NM, Pei ZM (2004) Nitric oxide represses the *Arabidopsis* floral transition. Science 305:1968–1971.

Hesberg C, Hansch R, Mendel RR, Bittner F (2004) Tandem orientation of duplicated xanthine dehydrogenase genes from *Arabidopsis thaliana*: differential gene expression and enzyme activities. J Biol Chem 279:13547–13554

Huang S, Kerschbaum HH, Engel E, Hermann A (1997) Biochemical characterization and histochemical localization of nitric oxide synthase in the nervous system of the snail, *Helix pomatia*. J Neurochem 69:2516–2528

Huang X, Stettmaier K, Michel C, Hutzler P, Mueller MJ, Durner J (2004) Nitric oxide is induced by wounding and influences jasmonic acid signaling in *Arabidopsis thaliana*. Planta 218:938–946

Ignarro LJ, Buga GM, Wood KS, Byrns RE, Chaudhuri G (1987) Endothelium-derived relaxing factor produced and released from artery and vein is nitric oxide. Proc Natl Acad Sci USA 84:9265–9269

Kaiser WM, Weiner H, Kandlbinder A, Tsai CB, Rockel P, Sonoda M, Planchet E (2002) Modulation of nitrate reductase: some new insights, an unusual case and a potentially important side reaction. J Exp Bot 53:875–882

Klepper LA (1979) Nitric oxide (NO) and nitrogen oxide (NO_2) emissions from herbicide-treated soybean plants. Atmos Environ 13:537–542

Klepper LA (1987) Nitric oxide emissions from soybean leaves during in vivo nitrate reductase assays. Plant Physiol 85:96–99

Klepper L (1990) Comparison between NO_x evolution mechanisms of wild type and nr_1 mutant soybean leaves. Plant Physiol 93:26–36

Kozlov AV, Staniek K, Nohl H (1999) Nitrite reductase activity is a novel function of mammalian mitochondria. FEBS Lett 454:127–130

Kuznetsova S, Knaff DB, Hirasawa M, Setif P, Mattioli TA (2004) Reactions of spinach nitrite reductase with its substrate, nitrite, and a putative intermediate, hydroxylamine. Biochemistry 43:10765–10774

Lamas S, Marsden PA, Li GK, Tempst P, Michel T (1992) Endothelial nitric oxide synthase: molecular cloning and characterization of a distinct constitutive enzyme isoform. Proc Natl Acad Sci USA 89:6348–6352

Lamattina L, Garcia-Mata C, Graziano M, Pagnussat G (2003) Nitric oxide: the versatility of an extensive signal molecule. Annu Rev Plant Biol 54:109–136

Lamotte O, Gould K, Lecourieux D, Sequeira-Legrand A, Lebrun-Garcia A, Durner J, Pugin A, Wendehenne D (2004) Analysis of nitric oxide signaling functions in tobacco cells challenged by the elicitor cryptogein. Plant Physiol 135:516–529

Lea PJ, Robinson SA, Stewart GR (1990) Intermediate nitrogen metabolism. In: Miflin BJ, Lea PJ (eds) The biochemistry of plants. Academic, New York, pp 121–158

Lea US, Ten Hoopen F, Provan F, Kaiser WM, Meyer C, Lillo C (2004) Mutation of the regulatory phosphorylation site of tobacco nitrate reductase results in high nitrite excretion and NO emission from leaf and root tissue. Planta 219:59–65

Lennon AM, Neuenschwander UH, Ribas-Carbo M, Giles L, Ryals JA, Siedow JN (1997) The effects of salicylic acid and tobacco mosaic virus infection on the alternative oxidase of tobacco. Plant Physiol 115:783–791

Li H, Samouilov A, Liu X, Zweier JL (2004) Characterization of the effects of oxygen on xanthine oxidase-mediated nitric oxide formation. J Biol Chem 279:16939–16946

Magalhães JR, Singh RP, Passos LP (eds) (2005) Nitric oxide signaling in higher plants. Studium, Houston

Mayer B, Hemmens B (1997) Biosynthesis and action of nitric oxide in mammalian cells. Trends Biochem Sci 22:477–481

Millar AH, Day DA (1996) Nitric oxide inhibits the cytochrome oxidase but not the alternative oxidase of plant mitochondria. FEBS Lett 398:155–158

Modolo LV, Cunha FQ, Braga MR, Salgado I (2002) Nitric oxide synthase-mediated phytoalexin accumulation in soybean cotyledons in response to the *Diaporthe phaseolorum* f. sp. *meridionalis* elicitor. Plant Physiol 130:1288–1297

Modolo LV, Augusto O, Almeida IM, Magalhaes JR, Salgado I (2005) Nitrite as the major source of nitric oxide production by *Arabidopsis thaliana* in response to *Pseudomonas syringae*. FEBS Lett 579:3814–3820

Modolo LV, Augusto O, Almeida IMG, Pinto-Maglio CAF, Oliveira HC, Seligman K, Salgado I (2006) Decreased arginine and nitrite levels in nitrate reductase-deficient *Arabidopsis thaliana* plants impair nitric oxide synthesis and the hypersensitive response to *Pseudomonas syringae*. Plant Sci 171:34–40

Morot-Gaudry-Talarmain Y, Rockel P, Moureaux T, Quillere I, Leydecker MT, Kaiser WM, Morot-Gaudry JF (2004) Nitrite accumulation and nitric oxide emission in relation to cellular signaling in nitrite reductase antisense tobacco. Planta 215:708–715

Murgia I, Tarantino D, Vannini C, Bracale M, Carravieri S, Soave C (2004) *Arabidopsis thaliana* plants overexpressing thylakoidal ascorbate peroxidase show increased resistance to Paraquat-induced photooxidative stress and to nitric oxide-induced cell death. Plant J 38:940–953

Naseem KM (2005) The role of nitric oxide in cardiovascular diseases. Mol Aspects Med 26:33–65

Neill SJ, Desikan R, Hancock JT (2003) Nitric oxide signalling in plants. New Phytologist 159:11–35

Orozco-Cárdenas ML, Ryan CA (2002) Nitric oxide negatively modulates wound signaling in tomato plants. Plant Physiol 130:487–493

Palmer RM, Ferrige AG, Moncada S (1987) Nitric oxide release accounts for the biological activity of endothelium-derived relaxing factor. Nature 327:524–526

Planchet E, Jagadis Gupta K, Sonoda M, Kaiser WM (2005) Nitric oxide emission from tobacco leaves and cell suspensions: rate limiting factors and evidence for the involvement of mitochondrial electron transport. Plant J 41:732–743

Rasmusson AG, Soole KL, Elthon TE (2004) Alternative NAD(P)H dehydrogenases of plant mitochondria. Annu Rev Plant Biol 55:23–39

Ribeiro EA Jr, Cunha FQ, Tamashiro WM, Martins IS (1999) Growth phase-dependent subcellular localization of nitric oxide synthase in maize cells. FEBS Lett 445:283–286

Rockel P, Strube F, Rockel A, Wildt J, Kaiser WM (2002) Regulation of nitric oxide (NO) production by plant nitrate reductase in vivo and in vitro. J Exp Bot 53:103–110

Salgado I, Saviani EE, Modolo LV, Braga MR (2004) Nitric oxide signaling in plant defence responses to pathogen attack. In: Hemantaranjan A (ed) Advances in plant physiology, vol 7. Scientific, Jodhpur, India, pp 117–137

Salgado I, Modolo LV, Magalhães JR, Tamashiro WMSC, Saviani EE (2006) Production of nitric oxide in plants by nitric oxide synthase enzymes. In: Singh RP, Jaiwal PK (eds) Biotechnological approaches to improve nitrogen use efficiency in plants. Studium, Houston, pp 87–102

Saviani EE, Orsi CH, Oliveira JF, Pinto-Maglio CA, Salgado I (2002) Participation of the mitochondrial permeability transition pore in nitric oxide-induced plant cell death. FEBS Lett 510:136–140

Shi S, Wang G, Wang Y, Zhang L, Zhang L (2005) Protective effect of nitric oxide against oxidative stress under ultraviolet-B radiation. Nitric Oxide 13:1–9

Simons BH, Millenaar FF, Mulder L, Van Loon LC, Lambers H (1999) Enhanced expression and activation of the alternative oxidase during infection of *Arabidopsis* with *Pseudomonas syringae* pv *tomato*. Plant Physiol 120:529–538

Simontacchi M, Jasid S, Puntarulo S (2004) Nitric oxide generation during early germination of sorghum seeds. Plant Sci 167:839–847

Stamler JS, Singel DJ, Loscalzo J (1992) Biochemistry of nitric oxide and its redox-activated forms. Science 258:1898–1902

Stöhr C, Ullrich WR (2002) Generation and possible roles of NO in plant roots and their apoplastic space. J Exp Bot 53:2293–2303

Stöhr C, Strube F, Marx G, Ullrich WR, Rockel P (2001) A plasma membrane-bound enzyme of tobacco roots catalyses the formation of nitric oxide from nitrite. Planta 212:835–841

Stuehr DJ (1997) Structure–function aspects in the nitric oxide synthases. Annu Rev Pharmacol Toxicol 37:339–359

Tada Y, Mori T, Shinogi T, Yao N, Takahashi S, Betsuyaku S, Sakamoto M, Park P, Nakayashiki H, Tosa Y, Mayama S (2004) Nitric oxide and reactive oxygen species do not elicit hypersensitive cell death but induce apoptosis in the adjacent cells during the defense response of oat. Mol Plant Microbe Interact 17:245–253

Tayeh MA, Marletta MA (1989) Macrophage oxidation of L-arginine to nitric oxide, nitrite, and nitrate: tetrahydrobiopterin is required as a cofactor. J Biol Chem 264:19654–19658

The Arabidopsis Genome Initiative (2000) Analysis of the genome sequence of the flowering plant *Arabidopsis thaliana*. Nature 408:796–815

Tischner R, Planchet E, Kaiser WM (2004) Mitochondrial electron transport as a source for nitric oxide in the unicellular green alga *Chlorella sorokiniana*. FEBS Lett 576:151–155

Tsuchiya K, Yoshizumi M, Houchi H, Mason RP (2000) Nitric oxide-forming reaction between the iron-*N*-methyl-D-glucamine dithiocarbamate complex and nitrite. J Biol Chem 275:1551–1556

Vanin AF, Svistunenko DA, Mikoyan VD, Serezhenkov VA, Fryer MJ, Baker NR, Cooper CE (2004) Endogenous superoxide production and the nitrite/nitrate ratio control the concentration of bioavailable free nitric oxide in leaves. J Biol Chem 279:24100–24107

Yamamoto A, Katou S, Yoshioka H, Doke N, Kawakita K (2004) Involvement of nitric oxide generation in hypersensitive cell death induced by elicitin in tobacco cell suspension culture. J Gen Plant Pathol 70:85–92

Yamasaki H (2000) Nitrite-dependent nitric oxide production pathway: implications for involvement of active nitrogen species in photoinhibition in vivo. Philos Trans R Soc Lond B Biol Sci 355:1477–1488

Yamasaki H, Sakihama Y (2000) Simultaneous production of nitric oxide and peroxynitrite by plant nitrate reductase: in vitro evidence for the NR-dependent formation of active nitrogen species. FEBS Lett 468:89–92

Yamasaki H, Sakihama Y, Takahashi S (1999) An alternative pathway for nitric oxide production in plants: new features of an old enzyme. Trends Plant Sci 4:128–129

Zeidler D, Zahringer U, Gerber I, Dubery I, Hartung T, Bors W, Hutzler P, Durner J (2004) Innate immunity in *Arabidopsis thaliana*: lipopolysaccharides activate nitric oxide synthase (NOS) and induce defense genes. Proc Natl Acad Sci USA 101:15811–15816

Zemojtel T, Penzkofer T, Dandekar T, Schultz J (2004) A novel conserved family of nitric oxide synthase? Trends Biochem Sci 29:224–226

Zemojtel T, Kolanczyk M, Kossler N, Stricker S, Lurz R, Mikula I, Duchniewicz M, Schuelke M, Ghafourifar P, Martasek P, Vingron M, Mundlos S (2006) Mammalian mitochondrial nitric oxide synthase: characterization of a novel candidate. FEBS Lett 580:455–462

Zhang C, Czymmek KJ, Shapiro AD (2003) Nitric oxide does not trigger early programmed cell death events but may contribute to cell-to-cell signaling governing progression of the *Arabidopsis* hypersensitive response. Mol Plant Microbe Interact 16:962–972

Zhang Z, Naughton D, Winyard PG, Benjamin N, Blake DR, Symons MC (1998) Generation of nitric oxide by a nitrite reductase activity of xanthine oxidase: a potential pathway for nitric oxide formation in the absence of nitric oxide synthase activity. Biochem Biophys Res Commun 249:767–772

Plant Cell Monogr (6)
L. Lamattina · J. C. Polacco: Nitric Oxide in Plant Growth
DOI 10.1007/7089_2006_096/Published online: 7 October 2006

Nitric Oxide as an Alternative Electron Carrier During Oxygen Deprivation

Abir U. Igamberdiev · Kevin N. Baron · Robert D. Hill (✉)

Department of Plant Science, University of Manitoba, Winnipeg, MB R3T 2N2, Canada
rob_hill@umanitoba.ca

Abstract Plant cells exposed to anaerobic stress generate copious amounts of the gaseous free radical nitric oxide (NO). At this time, the concomitant expression of the ubiquitous class 1 plant hemoglobins establishes one component of a soluble terminal NO dioxygenase system, which yields nitrate ions via reaction of oxyhemoglobin with NO. Class 1 hemoglobin expression also enhances the cellular energy status, redox status, and NO metabolism of plant cells exposed to hypoxic stress. The ability of class 1 hemoglobins to ligate oxygen at concentrations two orders of magnitude lower than cytochrome *c* oxidase suggests that hemoglobin and NO may serve as components of an alternative type of respiration that is operative during conditions that impair the operation of mitochondrial terminal oxidases. We suggest that, under hypoxic conditions, NO can be formed by anaerobic reduction of nitrite by a portion of the mitochondrial electron transport chain. NADH and NADPH, accumulated due to glycolytic fermentation and lipid breakdown, contribute electrons to the process, generating a chemiosmotic potential capable of generating ATP. The overall anaerobic reaction sequence is referred to as the Hb/NO cycle.

1 Introduction

There is a wealth of evidence to support the role of nitric oxide (NO) as a signaling molecule in biological systems. There is also an equal abundance of literature dealing with the production of NO and other nitrogen oxides by soil microbes existing under, or exposed to, oxygen deprivation. Nitrogen oxide emission is frequently associated with soil waterlogging where nitrogenous compounds, existing in higher oxidation states, act as alternative electron acceptors to oxygen, producing gaseous nitrogen oxides that are mainly released into the atmosphere (McKenney and Drury 1997).

In this chapter, we review the recent progress demonstrating that NO is formed during hypoxia in plants and that NO production assists in maintaining the redox and energy status of the hypoxic plant cell. In addition, we present a schematic model of the components constituting the hemoglobin-assisted NO turnover (defined as Hb/NO cycle). We also discuss the role and physiological relevance of individual components of the Hb/NO cycle with respect to cellular NO levels and plant energy and/or redox status.

2
Nitric Oxide Formation During Hypoxia

Nitric oxide is produced in significant quantities in plants exposed to oxygen deprivation (Rockel et al. 2002; Dordas et al. 2003a, 2004). Rates of NO formation under hypoxia are in the order of 10–50 nmol g^{-1} FW h^{-1} and are approximately 50-fold higher than the levels generated under aerobic conditions. A significant amount of NO produced is immediately scavenged, so real rates of NO production under hypoxia may potentially differ by at least one order of magnitude (Vanin et al. 2004). The various pathways by which NO may be formed in plants have been discussed in other portions of the monograph. In this chapter, our discussion will be limited to those pathways by which the compound is likely to be produced under hypoxic conditions.

As nitrate has been a known alleviator of plant flooding (hypoxic) stress for many years (Arnon 1937), and nitrate metabolism is linked to both aerobic and anaerobic plant NO generation, it is worthwhile summarizing the beneficial aspects of nitrate to hypoxic plant metabolism as a prelude to NO generation in hypoxic plants. From existing studies, the beneficial effects of nitrate, thus far, have been linked to pH regulation (Fan et al. 1997), carbohydrate utilization (Reggiani et al. 1986; Saglio et al. 1988; Fan et al. 1997), and NAD^+ regeneration (Reggiani et al. 1985). Ultrastructural studies also suggest that the integrity of hypoxically stressed mitochondria are extended in the presence of exogenous nitrate. More recent studies utilizing transgenic plants deficient in nitrate reductase (Stoimenova et al. 2003a,b) implicate products of nitrate metabolism to positively contribute towards flooding tolerance, energy metabolism, and pH regulation of plant during hypoxia.

Cytosolic nitrate reductase (cNR) serves as an attractive enzymatic catalyst of plant NO generation during hypoxic stress. In phosphorylated form, cNR exhibits low activity due to interactions with 14 – 3 – 3 proteins (Huber et al. 2002). However, the dephosphorylation of cNR via cNR phosphatases has been shown to correlate with the activation and stability of cNR and bears particular relevance for hypoxic NO generation (Huber et al. 2002). Degradation and half-life of cNR is also affected by cNR phosphorylation and 14-3-3 binding, with cNR activation being positively correlated with protein stability (Kaiser and Huber 2001). The drop in cytosolic pH under anoxic conditions triggers cNR dephosphorylation leading to a 2.5-fold activation of the enzyme (Botrel and Kaiser 1997). Moreover, when cNR is activated, a shift away from lactate and ethanol formation probably occurs, which is supported by the observation that nitrate reductase-lacking tobacco plants produce substantially more ethanol and lactate during anaerobiosis (Stoimenova et al. 2003b). In Arabidopsis root cultures, two nitrate reductase genes are induced under low-oxygen (5%) pressure. One (NR1) shows moderate induction after 0.5–4 h of hypoxia and strong induction after 20 h, while the other (NR2) is strongly activated in 2–4 h and even more after 20 h (Klok et al. 2002). The potential maximum activity of activated nitrate reductase, although lower than alcohol dehydrogenase, exceeds the rate of hypoxic ethanol formation by more than threefold (Botrel and Kaiser 1997).

Although nitrate reductase is activated under hypoxia, nitrite reduction is limited at the nitrite reductase step (Botrel et al. 1996). This limitation is explained both by a suppressive effect of cellular acidification and by increased flux through nitrate reductase (Botrel and Kaiser 1997; Botrel et al. 1996). When nitrite accumulates, it can also be used by cNR as a substrate to produce NO. The cNR reaction rate with nitrite to produce NO is probably only 1–2% of the maximal cNR reaction in leaves (Yamasaki et al. 2001; Rockel et al. 2002; Sakihama et al. 2002), so its physiological importance is still questionable.

In addition to cNR, the plasma membrane-bound nitrate reductase (PM-NR) and nitrite-NO reductase (Ni-NOR), have also been implicated in hypoxic NO generation (Ward et al. 1988, 1989; Meyerhoff et al. 1994). In the case of the latter, Ni-NOR, it appears as though atmospheric oxygen reversibly inhibits enzyme activity and NO production (Stöhr and Stremlau 2006). Such observations implicate Ni-NOR in hypoxic NO generation. However, it must be noted that the pH optimum, electron donor preference, and response to nitrate are markedly different for Ni-NOR when compared to that of cNR (Stöhr and Stremlau 2006, see also the contribution by Stöhr in this volume).

NO synthase (AtNOS1) from Arabidopsis is the only arginine-dependent nitric oxide synthase identified thus far in plants, and appears to generate the NO required for pathogen response, hormonal signaling, flowering, and regulation of reactive oxygen species (Guo et al. 2003). It is an unlikely candidate for plant NO synthesis under hypoxia for two reasons: first, NO synthase consumes oxygen and thus competes with other processes requiring oxygen in the hypoxic cell, and second, the nitrogen of NO is provided by nitrate during hypoxic NO production in maize cells (Dordas et al. 2004).

In addition to enzymatic sources of NO generation in plants, there is also evidence implying that non-enzymatic sources contribute to NO generation. For instance, in acidic and reducing environments, non-enzymatic reduction of nitrite produces nitrous acid, which can then react with ascorbate to produce dehydroascorbate and NO (Weitzberg and Lundberg 1998). This reaction has been demonstrated in cereal aleurone layers (Bethke et al. 2004).

While all of the above reactions for NO formation can contribute to the hypoxic NO pool, recent studies indicate that NO can be formed in a mitochondrial nitrite:NO reductase reaction associated with the electron transport chain (Tischner et al. 2004; Gupta et al. 2005; Planchet et al. 2005), a reaction possible in anaerobic plant mitochondria (Igamberdiev et al. 2005, for details see the chapter by Kaiser in this volume). Such a reaction may be conserved among species as both fungal (Kobayashi et al. 1996) and protist mitochondria (Finlay et al. 1983) were capable of operating during anaerobiosis in concert with associated nitrite reductase(s). The reaction can involve electrons from either the cytochrome *c* (Tielens et al. 2002) or the ubiquinone pool (Zumft 1997). In higher plants, the capacity of nitrite reduction to NO is associated with root mitochondria, while in leaf mitochondria it is likely absent (Gupta et al. 2005).

Although nitrite reductase is generally not associated with either plant or animal mitochondria, considerable amounts of nitrite are reduced to NO by hypoxic animal or plant mitochondria (Kozlov et al. 1999, 2005). This reaction can be catalyzed by anaerobic cytochrome *c* oxidase in its reduced state at the heme iron site (Brudvig et al. 1980; Paitian 1985; Mason et al. 2006). Six additional plant-specific subunits of cytochrome *c* oxidase (Millar et al. 2004) have also been implicated in this reaction. There is also the possibility of a transfer of electrons to nitrite from ubiquinol at the level of complex III (Kozlov et al. 1999; Nohl et al. 2001; Lacsa et al. 2006). In animal tissues, the major source of hypoxic NO production is arginine-independent and linked to the mitochondrial electron transport chain (Lacsa et al. 2006). Recent studies of plant and algal mitochondria suggest that they can readily convert nitrite to NO under anaerobic conditions at a higher rate than other previously mentioned reactions (Tischner et al. 2004; Planchet et al. 2005).

3 Hypoxia-Induced Hemoglobin Synthesis in Plants

The expression of a hemoglobin gene accompanying hypoxia was first demonstrated in barley (Taylor et al. 1994), following work on the existence of hypoxia-induced hemoglobin in non-nodulating plant species (Bogusz et al. 1988). The properties of the Hb protein (Duff et al. 1997) indicated that it probably did not function as a carrier, store or sensor of O_2 (Hill 1998). Barley class 1 hemoglobin is a homodimer with a monomeric molecular weight of 18 kDa (Duff et al. 1997). Its O_2 dissociation constant (2–3 nM) indicates that it remains oxygenated at extremely low O_2 tensions. The K_m for oxygen of cytochrome *c* oxidase is 140 nM. At concentrations where oxyhemoglobin dissociates, therefore, cytochrome *c* oxidase is effectively non-functional in utilizing oxygen. Other class 1 hemoglobins possess similar properties (Arredondo-Peter et al. 1997; Hargrove et al. 1997; Kundu et al. 2003). The unique features of class 1 hemoglobins result from the hexacoordination of the heme moiety during oxygen ligation, in comparison to the pentacoordination occurring in leghemoglobins, erythrocyte, and muscle hemoglobins. This hexacoordination of the heme pocket results in tighter oxygen binding, which is a basis for a conserved high fidelity NO dioxygenation mechanism (Gardner et al. 2006).

Class 1 hemoglobin is expressed in roots and other tissues within 2 h of exposure to hypoxia (Taylor et al. 1994). Strong hypoxic induction of the Hb gene, comparable to the induction of alcohol dehydrogenase, occurs in Arabidopsis root cultures in concert with the induction of enzymes of nitrogen metabolism, including nitrate reductases (Klok et al. 2002). Hb is also expressed in germinating seeds, which can become highly hypoxic at certain stages. In barley grain, Hb mRNA is detectable within 2 h of imbibition, and its expression continues to increase up to the point of radicle elongation, i.e., the completion of germination (Guy et al. 2002). Limiting oxygen availability increases the expression of the Hb gene in the embryos (Guy et al. 2002).

Oxygen deficiency by itself does not trigger Hb gene expression. A possible link between decreasing ATP levels and Hb synthesis has been observed (Nie and Hill 1997). Further investigation showed that both protein dephosphorylation and anaerobic elevation in cytosolic Ca^{2+} are factors in Hb gene expression (Nie et al. 2006). Hb induction is also observed in response to nitrate (Nie and Hill 1997), nitrite, and NO treatment (Ohwaki et al. 2005), implicating Hb expression with these nitrogenous compounds.

4
Nitric Oxide Turnover by Class 1 Hemoglobin

Class 1 hemoglobins can effectively convert NO to nitrate at nanomolar concentrations of O_2. In the course of this reaction, the ferrous form of hemoglobin is oxidized to the ferric form. Reduction back to the ferrous form to sustain the catalytic cycle can be achieved by a reductase using the appropriate reducing agent. In microorganisms, NO is scavenged by NO dioxygenase (NOD), which is a flavohemoglobin possessing two domains, one of which is hemoglobin and the other is an NAD(P)H-dependent reductase (Gardner et al. 1998). The NOD reaction is described by the equation

$$2NO + 2O_2 + NAD(P)H \rightarrow 2NO_3^- + NAD(P)^+ + H^+ .$$

When the reductase domain is lacking, the rate of reaction is exceedingly slow (Frey et al. 2002). Even leghemoglobin in its ferric form can be reduced non-enzymatically by NADH, reduced glutathione, or ascorbate, but with a much lower rate than with the help of methemoglobin reductase (Becana and Klucas 1990). Our studies with a mutant, class 1 barley Hb demonstrate that Hb alone is incapable of sustaining physiologically significant NAD(P)H-dependent NO-degrading activity (Igamberdiev et al. 2006a). We have purified and identified a cytosolic monodehydroascorbate reductase that acts in conjunction with the class 1 Hb to sustain turnover of NO (Igamberdiev et al. 2006a). Its proposed role is the removal of monodehydroascorbate formed in the reduction of methemoglobin by ascorbate. Correlations between ascorbate levels and hemoglobin expression indicate that the suggested ascorbate-dependent mechanism of methemoglobin reduction may be operative in vivo (Igamberdiev et al. 2006b).

The expression of a class 1 hemoglobin has a direct effect on the level of NO found under hypoxic conditions (Dordas et al. 2003a, 2004). Cytoplasmic ex-

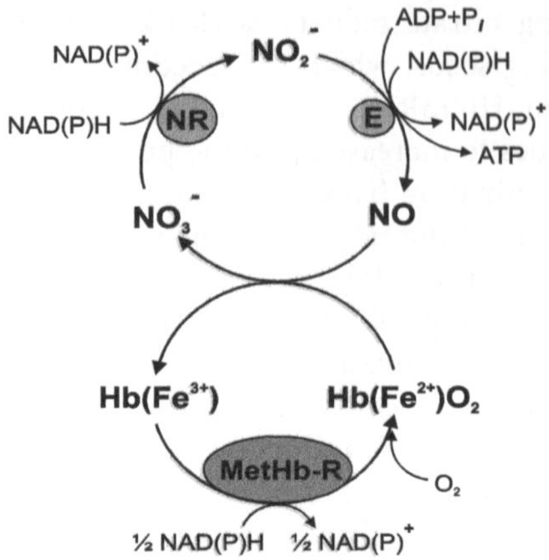

Fig. 1 Suggested scheme of hypoxic respiratory pathway involving Hb and NO (Hb/NO cycle) resulting in regeneration of NADH and anaerobic ATP synthesis. *NR* nitrate reductase. *E* nitrite:NO reductase activity of mitochondria. NO is oxidized to nitrate by oxyhemoglobin [Hb(Fe^{2+})O_2], which is converted to metHb [Hb(Fe^{3+})]. MetHb is reduced by a MetHb reductase (*MetHb-R*), such as the ascorbate/monodehydroascorbate reductase system. The high affinity of Hb(Fe^{2+}) for O_2 results in its immediate oxygenation even at very low (nanomolar) O_2 concentration. Nitrite reduction to NO in mitochondria at the sites of complex III and IV is coupled to NAD(P)H oxidation by external NADH and NADPH dehydrogenases, producing a chemiosmotic potential resulting in ATP synthesis. Adapted from Igamberdiev et al. (2005)

tracts of alfalfa root cultures have NO dioxygenase activity that is dependent on hemoglobin and NADH or NADPH. The activity is flavin- and sulfhydryl-dependent, exhibiting a broad pH optimum and a strong affinity to NADH and NADPH. The K_m is 3 μM for both nucleotides, while the maximum rate with NADH is 2.5 times higher than with NADPH (Igamberdiev et al. 2004). These properties are consistent with the reported properties of a monodehydroascorbate reductase (Hossain and Asada 1985). The sequence of reactions involving nitrate reductase-catalyzed nitrate conversion to nitrite, NO production from nitrite, and Hb-dependent NO conversion to nitrate in conjunction with the ferric Hb reduction is defined as the Hb/NO cycle (Dordas et al. 2003b; Igamberdiev and Hill 2004; Igamberdiev et al. 2005) (Fig. 1).

5
Maintenance of Redox and Energy Status by Hemoglobin During Hypoxia

The hypoxic or anoxic cell must deal with two critical metabolic conditions: a highly reductive environment due to accumulation of NAD(P)H, and a reduced efficiency of ATP synthesis due to the cessation of oxidative phosphorylation. It is generally accepted that glycolysis and fermentation are the major routes by which the cell adapts to these conditions. We suggest that the Hb/NO cycle is another route by which cells react to the situation.

Hb overexpression in anoxic maize cell cultures results in the maintenance of cell energy status, as evidenced by higher ATP levels and decreased utilization of fermentative pathways (Sowa et al. 1998). The Hb/NO cycle (Fig. 1) would predict greater turnover of NO in a reducing environment, diverting metabolism away from fermentative pathways involving alcohol dehydrogenase. NADH/NAD and NADPH/NADP ratios in plants overexpressing hemoglobin (Igamberdiev et al. 2004) are not significantly affected by hypoxia, while in plant lines downregulating Hb the ratios increase significantly under low oxygen tensions. The expression of hemoglobin in hypoxic cells, therefore, in addition to maintaining energy status, helps to maintain the redox status of the cell.

Mitochondrial localization of yeast flavohemoglobin has recently been reported (Cassanova et al. 2005). However, the nucleotide sequence of the barley class 1 Hb gene suggests that it is not imported into cell organelles (Taylor et al. 1994) and is probably localized mainly in the cytosol (Igamberdiev et al. 2004). A lack of localization within mitochondria should not be a significant limitation as rapid diffusion of NO from mitochondria to the cytosol would permit NO scavenging in the cytosolic compartment. Figure 2 shows how the expression of barley hemoglobin in alfalfa roots correlates with NO levels, ATP/ADP, and NADH/NAD ratios. These results are based on previously published data (Dordas et al. 2003a; Igamberdiev et al. 2004). The most pronounced effects of Hb expression are visible under hypoxic conditions. Hb expression strongly decreases NO levels, NADH/NAD ratios, and increases the ATP/ADP ratio, indicating an important metabolic role for Hb in the process of NO removal linked to NADH oxidation and ATP production. NAD(P)H oxidation in the Hb/NO cycle occurs at three sites in the Hb/NO cycle (Fig. 1), accounting for the observed changes in the NADH/NAD ratio. The possible implications of ATP production linked to the Hb/NO cycle are described in the next section.

6
Anoxic Mitochondrial ATP Synthesis Driven by Nitrite and NAD(P)H

The data presented in Fig. 2, showing hemoglobin participation in the maintenance of redox and energy state of hypoxic plant cells, indicate that its involvement in NO removal is linked to NADH oxidation and ATP synthesis. NO removal by itself can make mitochondria more functional via relief from the inhibitory effects of NO. Even under highly anoxic conditions, however, the energy state of cells improves (Sowa et al. 1998). There is sufficient evidence to conclude that mitochondria are capable of operating under highly anoxic conditions. Exposure to anoxia results in some changes in enzyme

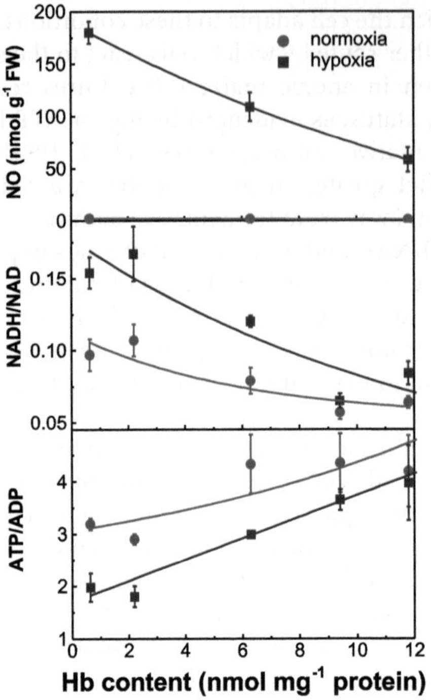

Fig. 2 Effects of differential expression of barley hemoglobin in alfalfa roots on NO levels, ATP/ADP and NADH/NAD ratios under normoxic (40 kPa O_2) and hypoxic (3 kPa O_2) conditions. Two underexpressing Hb, two overexpressing Hb, and a control line were used. Based on the data of Dordas et al. (2003a) and Igamberdiev et al. (2004)

composition in mitochondria (Couee et al. 1992; Kennedy et al. 1987), but they remain functional and preserve their ultrastructure for a prolonged time, if nitrate is present (Vartapetian and Polyakova 1999; Vartapetian et al. 2003). This evidence led to the suggestion that nitrate might serve as a terminal electron acceptor under anoxia (Polyakova and Vartapetian 2003). No evidence has been presented for nitrate reductase activity associated with plant mitochondria.

A key feature of mitochondrial operation under anoxia is the oxidation of extra-mitochondrial NAD(P)H. The high K_m, low pH optimum, and Ca^{2+} dependence of externally facing NADH and NADPH dehydrogenases (Edman et al. 1985; Møller 1997) suggests that they operate when the cytosolic NAD(P)H and Ca^{2+} concentrations are elevated and the cytosolic pH is decreased, as observed under hypoxia (Roberts et al. 1992; Subbaiah et al. 1998).

We have shown that anaerobic plant mitochondria can synthesize ATP using nitrite and NAD(P)H, forming NO and NAD(P)$^+$ (Stoimenova et al. manuscript in preparation). The reaction was insensitive to rotenone and antimycin A, partially sensitive to diphenyleneiodonium and myxothiazol, and strongly sensitive to oligomycin and uncouplers. Oxidation of cytosolic NADH and NADPH via externally facing dehydrogenases precludes the formation of a proton gradient at the site of electron transport from NAD(P)H to ubiquinone. Inhibition of ATP synthesis and NO production by myxothiazol and the absence of inhibition by antimycin A is similar to the inhibition pattern of nitrite reduction observed in mammalian mitochondria (Kozlov et al. 1999). The effect is analogous to a single electron leak to dioxygen, indicating a one-electron mechanism of nitrite reduction due to an oxidized state of ubiquinol at the cytochrome bc_1 complex. We cannot, however, exclude the possibility that nitrite reduction may also occur at a later step and involve cytochrome *c* oxidase, since the reactions of NO production and ATP synthesis are sensitive to cyanide (Fig. 3).

Anaerobic mitochondrial production of ATP is approximately 5% of the aerobic rate, yielding at a minimum 0.4–0.6 ATP per NAD(P)H oxidized, indicating that membrane potential generation is limited (Stoimenova et al. manuscript in preparation). The reaction involves the release of two electrons by one NADH and the reduction of a single nitrite by one electron. While the observed rate of ATP synthesis may be low relative to aerobic respiration, it is important to note that the rate is comparable to ATP generation by glycolytic substrate phosphorylation.

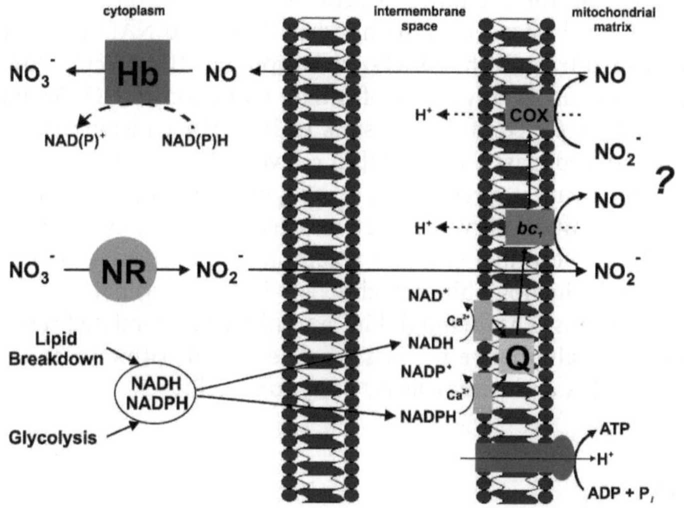

Fig. 3 Operation of plant mitochondria under oxygen deprivation. Glycolytic fermentation and lipid breakdown in hypoxia result in the increase of cytosolic NADH and NADPH. Externally facing mitochondrial dehydrogenases oxidize NADH and NADPH. At levels of oxygen well below saturation of cytochrome *c* oxidase, nitrite serves as an alternative electron acceptor at the sites of complex III and/or complex IV. Nitric oxide formed in this reaction is converted by hypoxically induced hemoglobin (*Hb*) to nitrate. The latter is reduced to nitrite by hypoxically induced nitrate reductase (*NR*)

7 Possible Link of Hemoglobin-Dependent NO Turnover to Nitrite-Dependent ATP Synthesis by Mitochondria

Theoretically, NO formed by mitochondria can be recycled back to nitrite via a corresponding side reaction with cytochrome *c* oxidase (Cooper 2002). The cytochrome *c* oxidase under low oxygen concentrations will be present in a reduced form, however, which makes it very unlikely that this reaction occurs at the copper site of the protein (Cooper 2002). When the cytochrome *c* oxidase is reduced, nitrite binds at the heme iron center of the protein and a reverse reaction of NO formation from nitrite is more likely (Cooper 2002). A plausible alternative is the operation of a hemoglobin/NO cycle (Igamberdiev and Hill 2004) converting NO to nitrate, needing only nanomolar traces of oxygen present, even in highly anaerobic conditions, to recycle NO formed by mitochondria back to nitrate (Fig. 1), while the operation of mitochondria with nitrite as a terminal acceptor does not need oxygen at all (Fig. 3).

8 Conclusions

We present a case for a pathway operating during oxygen deprivation, in which ATP is synthesized in mitochondria, linked to NAD(P)H-driven NO formation from nitrite. Both mitochondrial complex III and cytochrome *c* oxidase could serve as putative sites of nitrite reduction to NO. Accumulation of NO under hypoxic conditions results in its further metabolism in which hypoxically induced class 1 hemoglobin converts NO to nitrate. Nitrate is reduced to nitrite by nitrate reductase, an enzyme also induced under hypoxia. NO, nitrate, and nitrite serve as electron carriers in this cyclic process with oxygen being the terminal electron acceptor. The oxygen is made available through a class 1 hemoglobin, which exists in the oxygenated form at very low oxygen tensions. This hemoglobin- and nitric oxide-related respiration is considered as an alternative to classic fermentation pathways and serves to maintain the redox and energy levels of hypoxic cells.

References

Arnon DI (1937) Ammonium and nitrate nitrogen nutrition of barley and rice at different seasons in relation to hydrogen-ion concentrations, manganese, copper and oxygen supplied. Soil Science 44:91–121

Arredondo-Peter R, Hargrove MS, Sarath G, Moran JF, Lohrman J, Olson JS, Klucas RV (1997) Rice hemoglobins: gene cloning, analysis and oxygen-binding kinetics of a recombinant protein synthesized in *Escherichia coli*. Plant Physiol 115:1259–1266

Becana M, Klucas RV (1990) Enzymatic and nonenzymatic mechanisms for ferric leghemoglobin reduction in legume root nodules. Proc Natl Acad Sci USA 87:7295–7299

Bethke PC, Badger MR, Jones RL (2004) Apoplastic synthesis of nitric oxide by plant tissues. Plant Cell 16:332–341

Bogusz D, Appleby CA, Landsmann J, Dennis ES, Trinick MJ, Peacock WJ (1988) Functioning haemoglobin genes in non-nodulating plants. Nature 331:178–180

Botrel A, Kaiser WM (1997) Nitrate reductase activation state in barley roots in relation to the energy and carbohydrate status. Planta 201:496–501

Botrel A, Magne C, Kaiser WM (1996) Nitrate reduction, nitrite reduction and ammonium assimilation in barley roots in response to anoxia. Plant Physiol Biochem 34: 645–652

Brudvig GW, Stevens TH, Chan SI (1980) Reactions of nitric oxide with cytochrome *c* oxidase. Biochem 19:5275–5285

Cassanova N, O'Brien KM, Stahl BT, McClure T, Poyton RO (2005) Yeast flavohemoglobin, a nitric oxide oxidoreductase, is located in both the cytosol and the mitochondrial matrix: effects of respiration, anoxia, and the mitochondrial genome on its intracellular level and distribution. J Biol Chem 280:7645–7653

Cooper CE (2002) Nitric oxide and cytochrome oxidase: substrate, inhibitor or effector? Trends Biochem Sci 27:33–39

Couee I, Defontaine S, Carde JP, Pradet A (1992) Effects of anoxia on mitochondrial biogenesis in rice shoots – modification of in organello translation characteristics. Plant Physiol 98:411–421

Dordas C, Hasinoff BB, Igamberdiev AU, Manac'h N, Rivoal J, Hill RD (2003a) Expression of a stress-induced hemoglobin affects NO levels produced by alfalfa root cultures under hypoxic stress. Plant J 35:763–770

Dordas C, Rivoal J, Hill RD (2003b) Plant haemoglobins, nitric oxide and hypoxic stress. Ann Bot 91:173–178

Dordas C, Hasinoff BB, Rivoal J, Hill RD (2004) Class 1 hemoglobins, nitrate and NO levels in hypoxic maize cell suspension cultures. Planta 219:66–72

Duff SMG, Wittenberg JB, Hill RD (1997) Expression, purification, and properties of recombinant barley (*Hordeum* sp.) hemoglobin: Optical spectra and reactions with gaseous ligands. J Biol Chem 272:16746–16752

Edman K, Ericson I, Møller IM (1985) The regulation of exogenous NAD(P)H oxidation in spinach (*Spinacia oleracea*) leaf mitochondria by pH and cations. Biochem J 232:471–477

Fan TWN, Higashi RM, Frenkiel TA, Lane AM (1997) Anaerobic nitrate and ammonium metabolism in flood-tolerant rice coleoptiles. J Exp Bot 48:1655–1666

Finlay BJ, Span ASW, Harman JMP (1983) Nitrate respiration in primitive eukaryotes. Nature 303:333–336

Frey AD, Farres J, Bollinger CJT, Kallio PT (2002) Bacterial hemoglobins and flavohemoglobins for alleviation of nitrosative stress in *Escherichia coli*. Applied Environ Microb 68:4835–4840

Gardner PR, Gardner AM, Martin LA, Salzman AL (1998) Nitric oxide dioxygenase: an enzymic function for flavohemoglobin. Proc Natl Acad Sci USA 95:10378–10383

Gardner PR, Gardner AM, Brashear WT, Suzuki T, Hvitved AN, Setchell KDR, Olson JS (2006) Hemoglobins dioxygenate nitric oxide with high fidelity. J Inorg Biochem 100:542–550

Guo FQ, Okamoto M, Crawford NM (2003) Identification of a plant nitric oxide synthase gene involved in hormonal signalling. Science 302:100–103

Gupta KJ, Stoimenova M, Kaiser WM (2005) In higher plants, only root mitochondria, but not leaf mitochondria reduce nitrite to NO, in vitro and in situ. J Exp Bot 56:2601–2609
Guy PA, Sidaner J-P, Schroeder S, Edney M, MacGregor AW, Hill RD (2002) Embryo phytoglobin gene expression as a measure of germination in cereals. J Cereal Sci 36:147–156
Hargrove MS, Barry JK, Brucker EA, Berry MB, Phillips GN, Olson JS, Arredondo-Peter R, Dean JM, Klucas RV, Sarath G (1997) Characterization of recombinant soybean leghemoglobin a and apolar distal histidine mutants. J Mol Biol 266:1032–1042
Hill RD (1998) What are haemoglobins doing in plants? Can J Bot 76:707–712
Hossain MA, Asada K (1985) Monodehydroascorbate reductase from cucumber is a flavin adenine-dinucleotide enzyme. J Biol Chem 260:2920–2926
Huber SC, MacKintosh C, Kaiser WM (2002) Metabolic enzymes as targets for 14-3-3 proteins. Plant Mol Biol 50:1053–1063
Igamberdiev AU, Hill RD (2004) Nitrate, NO and haemoglobin in plant adaptation to hypoxia: An alternative to classic fermentation pathways. J Exp Bot 55:2473–2482
Igamberdiev AU, Seregélyes C, Manac'h N, Hill RD (2004) NADH-dependent metabolism of nitric oxide in alfalfa root cultures expressing barley hemoglobin. Planta 219:95–102
Igamberdiev AU, Baron K, Manac'h-Little N, Stoimenova M, Hill RD (2005) The haemoglobin/nitric oxide cycle: involvement in flooding stress and effects on hormone signalling. Ann Bot 96:557–564
Igamberdiev AU, Bykova NV, Hill RD (2006a) Scavenging of nitric oxide by barley hemoglobin is facilitated by monodehydroascorbate reductase-mediated ascorbate reduction of methemoglobin. Planta 223:1033–1040
Igamberdiev AU, Stoimenova M, Seregélyes C, Hill RD (2006b) Class-1 hemoglobin and antioxidant metabolism in alfalfa roots. Planta 223:1041–1046
Kaiser WM, Gupta KJ, Planchet E (2007) In: Lamattina L, Polacco JC (eds) Nitric oxide in plant growth, development and stress physiology. Plant Cell Monographs, vol 6. Springer, Berlin Heidelberg New York (in press)
Kaiser WM, Huber SC (2001) Post-translational regulation of nitrate reductase: mechanism, physiological relevance and environmental triggers. J Exp Bot 52:1981–1989
Kennedy RA, Fox TC, Siedow JN (1987) Activities of isolated mitochondria and mitochondrial enzymes from aerobically and anaerobically germinated barnyard grass (*Echinochloa*) seedlings. Plant Physiol 85:474–480
Klok EJ, Wilson IW, Wilson D, Chapman SC, Ewing RM, Somerville SC, Peacock WJ, Dolferus R, Dennis ES (2002) Expression profile analysis of the low-oxygen response in Arabidopsis root cultures. Plant Cell 14:2481–2494
Kobayashi M, Matsuo Y, Takimoto A, Suzuki S, Maruo F, Shoun H (1996) Denitrification, a novel type of respiratory metabolism in fungal mitochondrion. J Biol Chem 271:16263–16267
Kozlov AV, Staniek K, Nohl H (1999) Nitrite reductase activity is a novel function of mammalian mitochondria. FEBS Lett 454:127–130
Kozlov AV, Costantino G, Sobhian B, Szalay L, Umar F, Nohl H, Bahrami S, Redl H (2005) Mechanisms of vasodilatation induced by nitrite instillation in intestinal lumen: possible role of hemoglobin. Antiox Redox Signaling 7:515–521
Kundu S, Trent JT III, Hargrove MS (2003) Plants, humans and hemoglobins. Trends Plant Sci 8:387–393
Lacsa Z, Kozlov AV, Pankotai E, Csordás A, Wolf G, Redl H, Kollai M, Szabó C, Busija DW, Horn TFW (2006) Mitochondria produce reactive nitrogen species via an arginine-independent pathway. Free Radic Res 40:369–378

Mason MG, Nicholls P, Wilson MT, Cooper CE (2006) Nitric oxide inhibition of respiration involves both competitive (heme) and noncompetitive (copper) binding to cytochrome c oxidase. Proc Natl Acad Sci USA 103:708–713

McKenney DJ, Drury CF (1997) Nitric oxide production in agricultural soils. Global Change Biol 3:317–326

Meyerhoff PA, Fox TC, Travis RL, Huffaker RC (1994) Characterization of the association of nitrate reductase with barley (*Hordeum vulgare* L.) root membranes. Plant Physiol 104:925–936

Millar AH, Eubel H, Jansch L, Kruft V, Heazlewood JL, Braun HP (2004) Mitochondrial cytochrome c oxidase and succinate dehydrogenase complexes contain plant specific subunits. Plant Mol Biol 56:77–90

Møller IM (1997) The oxidation of cytosolic NAD(P)H by external NAD(P)H dehydrogenases in the respiratory chain of plant mitochondria. Physiol Plant 100:85–90

Nie XZ, Hill RD (1997) Mitochondrial respiration and hemoglobin gene expression in barley aleurone tissue. Plant Physiol 114:835–840

Nie XZ, Durnin DC, Igamberdiev AU, Hill RD (2006) Cytosolic calcium is involved in the regulation of barley hemoglobin gene expression. Planta 223:542–549

Nohl H, Kozlov AV, Staniek K, Gille L (2001) The multiple functions of coenzyme Q. Bioorg Chem 29:1–13

Ohwaki Y, Kawagishi-Kobayashi M, Wakasa K, Fujihara S, Yoneyama T (2005) Induction of class-1 non-symbiotic hemoglobin genes by nitrate, nitrite and nitric oxide in cultured rice cells. Plant Cell Physiol 46:324–331

Paitian NA, Markossian KA, Nalbandyan RM (1985) The effect of nitrite on cytochrome oxidase. Biochem Biophys Res Commun 133:1104–1111

Planchet E, Gupta KJ, Sonoda M, Kaiser WM (2005) Nitric oxide emission from tobacco leaves and cell suspensions: rate limiting factors and evidence for the involvement of mitochondrial electron transport. Plant J 41:732–743

Polyakova LI, Vartapetian BB (2003) Exogenous nitrate as a terminal acceptor of electrons in rice (*Oryza sativa*) coleoptiles and wheat (*Triticum aestivum*) roots under strict anoxia. Russian J Plant Physiol 50:808–812

Reggiani R, Brambilla I, Bertani A (1985) Effect of exogenous nitrate on anaerobic metabolism in excised rice roots. 2. Fermentative activity and adenylic energy charge. J Exp Bot 36:1698–1704

Reggiani R, Brambilla I, Bertani A (1986) Effect of exogenous nitrate on anaerobic metabolism in excised rice roots. 3. Glycolytic-intermediates and enzymatic activities. J Exp Bot 37:1472–1478

Roberts JKM, Hooks MA, Miaullis AP, Edwards S, Webster C (1992) Contribution of malate and amino acid metabolism to cytoplasmic pH regulation in hypoxic maize roottips studied using nuclear magnetic resonance spectroscopy. Plant Physiol 98:480–487

Rockel P, Strube F, Rockel A, Wildt J, Kaiser WM (2002) Regulation of nitric oxide (NO) production by plant nitrate reductase in vivo and in vitro. J Exp Bot 53:103–110

Saglio PH, Drew MC, Pradet A (1988) Metabolic acclimation to anoxia induced by low (2–4 kPa partial pressure) oxygen pretreatment (hypoxia) in root tips of *Zea mays*. Plant Physiol 86:61–66

Sakihama Y, Nakamura S, Yamazaki H (2002) Nitric oxide production mediated by nitrate reductase in the green alga *Chlamydomonas reinhardtii*: an alternative NO production pathway in photosynthetic organisms. Plant Cell Physiol 43:290–297

Sowa AW, Duff SMG, Guy PA, Hill RD (1998) Altering hemoglobin levels changes energy status in maize cells under hypoxia. Proc Natl Acad Sci USA 95:10317–10321

Stöhr C, Stremlau S (2006) Formation and possible roles of nitric oxide in plant roots. J Exp Bot 57:463–470
Stoimenova M, Hansch R, Mendel R, Gimmler H, Kaiser WM (2003a) The role of nitrate reduction in the anoxic metabolism of roots – I. Characterization of root morphology and normoxic metabolism of wild type tobacco and a transformant lacking root nitrate reductase. Plant Soil 253:145–153
Stoimenova M, Libourel IGL, Ratcliffe RG, Kaiser WM (2003b) The role of nitrate reduction in the anoxic metabolism of roots – II. Anoxic metabolism of tobacco roots with or without nitrate reductase activity. Plant Soil 253:155–167
Stöhr C (2007) In: Lamattina L, Polacco JC (eds) Nitric oxide in plant growth, development and stress physiology. Plant Cell Monographs, vol 6. Springer, Berlin Heidelberg New York (in press)
Subbaiah CC, Bush DS, Sachs MM (1998) Mitochondrial contribution to the anoxic Ca^{2+} signal in maize suspension-cultured cells. Plant Physiol 118:759–771
Taylor ER, Nie XZ, MacGregor AW, Hill RD (1994) A cereal haemoglobin gene is expressed in seed and root tissues under anaerobic conditions. Plant Mol Biol 24:853–862
Tielens AGM, Rotte C, van Hellemond JJ, Martin W (2002) Mitochondria as we don't know them. Trends Biochem Sci 27:564–572
Tischner R, Planchet E, Kaiser WM (2004) Mitochondrial electron transport as a source for nitric oxide in the unicellular green alga *Chlorella sorokiniana*. FEBS Lett 576:151–155
Vanin AF, Svistunenko DA, Mikoyan VD, Serezhenkov VA, Fryer MJ, Baker NR, Cooper CE (2004) Endogenous superoxide production and the nitrite/nitrate ratio control the concentration of bioavailable free nitric oxide in leaves. J Biol Chem 279:24100–24107
Vartapetian BB, Polyakova LI (1999) Protective effect of exogenous nitrate on the mitochondrial ultrastructure of *Oryza sativa* coleoptiles under strict anoxia. Protoplasma 206:163–167
Vartapetian BB, Andreeva IN, Generozova IP, Polyakova LI, Maslova IP, Dolgikh YI, Stepanova AY (2003) Functional electron microscopy in studies of plant response and adaptation to anaerobic stress. Ann Bot 91:155–172
Ward MR, Tischner R, Huffaker RC (1988) Inhibition of nitrate transport by anti-nitrate reductase IGG fragments and the identification of plasma membrane associated nitrate reductase in roots of barley seedlings. Plant Physiol 88:1141–1145
Ward MR, Grimes HD, Huffaker RC (1989) Latent nitrate reductase activity is associated with the plasma membrane of corn roots. Planta 177:470–475
Weitzberg E, Lundberg JON (1998) Nonenzymatic nitric oxide production in humans. Nitric Oxide: Biol Chem 2:1–7
Yamasaki H, Shimoji H, Ohshiro Y, Sakihama Y (2001) Inhibitory effects of nitric oxide on oxidative phosphorylation in plant mitochondria. Nitric Oxide: Biol Chem 5:261–270
Zumft WG (1997) Cell biology and molecular basis of denitrification. Microbiol Mol Biol Rev 61:533–616

Plant Cell Monogr (6)
L. Lamattina · J. C. Polacco: Nitric Oxide in Plant Growth
DOI 10.1007/7089_2006_097/Published online: 20 October 2006

Fluorometric Detection of Nitric Oxide with Diaminofluoresceins (DAFs): Applications and Limitations for Plant NO Research

Nao O. Arita[1] · Michael F. Cohen[2] · Gaku Tokuda[3] · Hideo Yamasaki[1] (✉)

[1]Faculty of Science, University of the Ryukyus, 903-0213 Okinawa, Japan
yamasaki@sci.u-ryukyu.ac.jp

[2]Department of Biology, Sonoma State University, 1801 E. Cotati Ave., Rohnert Park, CA 94928, USA

[3]Center of Molecular Biosciences (COMB), University of the Ryukyus, 903-0213 Okinawa, Japan

Abstract Nitric oxide (NO), a reactive nitrogen species, serves as a signaling molecule in plants, animals, fungi, and bacteria. In spite of its potential significance, however, the unique challenges of NO research can bring confusion to investigations, primarily due to difficulties in detecting and quantifying biological NO production. To overcome such barriers, we recommend that researchers choose a combinatorial approach for monitoring NO levels, in which multiple methods having distinct detection principles are employed. After an overview of the major methodologies for NO detection, we highlight the usefulness and application limits of the fluorescence probe diaminofluorescein (DAF), which has been preferentially applied in studies of plant NO-producing systems.

1 Introduction

Nitric oxide (NO) is a ubiquitous molecule that exists in interstellar space, the atmosphere, and soils. As one of the nitrogen oxides (NO_x) produced through the combustion of fossil fuels it is also recognized as a harmful air pollutant (Yamasaki 2000). The discovery of NO synthase (NOS), which produces NO from O_2 and arginine, combined with our growing understanding of the pivotal roles for NO in living cells has offered a new paradigm to the fields of biochemistry, cell biology, and physiology: NO as an important signaling molecule not only in plants but also in animals, fungi, and bacteria (Yamasaki 2004).

Scientific breakthroughs, in many cases, have been sustained by developments of new methods or technologies; research in NO biology and biochemistry is no exception. Reliable methods for the specific detection and quantitative determination of NO production in tissues, cells, and organelles are prerequisite for a better understanding of the complex biological functions of the radical gas molecule. It is undeniable that development of new

NO detection methodologies has promoted recent advances in NO research in living systems. Biogenic NO can be detected either by direct means or in the form of its stable oxidized products nitrite (NO_2^-) and nitrate (NO_3^-). In this chapter, we describe NO detection technologies that have been applied to plant systems. After a brief review of these methods we focus on potential problems and common errors encountered in utilizing diaminofluorescein fluorescence technique for detection of NO.

2 Griess Assay

The assay developed in the 19th century by Johann Peter Griess to detect nitrite by the formation of a colored diazo compound in an acidified solution (Ivanov 2004) has been widely adopted for use in modern NO research. The most common refinement of the Griess assay, available in over a dozen commercially available kits (Sun et al. 2003), precedes via initial reaction of nitrite (or more precisely, its acidified unstable product N_2O_3) with sulfanilamide to form a transient diazonium salt followed by addition of *N*-naphthylethylenediamine to yield a stable azo compound that absorbs strongly at 540 nm. A nitrite standard reference curve should be prepared for each assay. The practical sensitivity limit for this procedure is approximately 0.5–3 μM nitrite, depending on the matrix.

For a more accurate measurement of NO produced in a sample one must also measure the nitrate formed via oxidation of nitrite. This is most often accomplished by reducing nitrate to nitrite immediately prior to introduction of the Griess reagents. Reduction of nitrate can be carried out by treating samples with purified nitrate reductase (NR) or appropriate reductants; the official method of the American Official Analytical Chemists calls for use of cadmium, a hazardous waste. Problems with reaction efficiency, interference, and labor demands of the nitrate-to-nitrite reduction step have been described by Sun et al. (2003). To those who have the available instrumentation, ion chromatography offers a cleaner, more convenient, and sensitive alternative for end-point nitrite/nitrate detection (Dionex 1996).

In plant systems, tissue sometimes contains a high amount of nitrate (or nitrite), a condition where a low level of NO production is difficult to determine due to the high background. This is the case in green leaves and roots where nitrate assimilation takes place (Kawamura et al. 1996). It is important to be reminded that the oxidation product, nitrite, is also a substrate of NO production (Yamasaki et al. 1999; Yamasaki 2000). This complexity in the relationship between nitrite and NO could potentially result in inaccurate quantification of biological NO production in plant systems (Yamasaki 2005). Furthermore, the procedure for the conversion of nitrate to nitrite using NR also includes a potential risk because the enzyme itself catalyzes not only nitrate reduction but also NO production via nitrite reduction in a condition of excess reductants (Yamasaki and Sakihama 2000). Therefore, one should take these characteristics into consideration in quantitatively determining NO production activity in plant cells. Nevertheless, the Greiss assay is useful because it does not require specific instruments and skills; the assay is recommended for quantification of NO production in a sample that does not include a substantial amount of nitrate (or nitrite) or in a purified in vitro system.

3 Oxyhemoglobin

NO can bind to heme-proteins including hemoglobins and changes the absorption spectrum as a result. The binding of NO to oxyhemoglobin (oxyHb) can be detected photometrically in real-time as a spectral shift. Key early investigations of NO biology utilized this assay but subsequent findings have revealed several drawbacks, including interference by other compounds and NO_x that are commonly encountered in NO assays (Artz and Thatcher 1998; Privat et al. 1997; Schmidt et al. 1994). Conversely, actual NO production may be missed by this assay (Adak et al. 2002a,b). The need to purify commercially obtained oxyHb reagent (Taha 2003) is further reason to avoid this method. The obvious advantage and disadvantage of this method are similar to those of Griess method: no requirement for a specific instrument but low specificity. The method is applicable to an in vitro purified system.

4 Electron Spin Resonance

NO is a radical molecule that is paramagnetic, a characteristic that can be distinguished from other non-paramagnetic molecules. Electron spin resonance (ESR) or electron paramagnetic resonance (EPR) is applicable as another NO detection method that offers an advantage in real-time continuous monitoring of NO production. In addition, this method can be widely applied for NO detection in vivo because it can detect NO at room temperature. However, the ESR method involves two apparent difficulties: its low sensitivity and higher equipment cost (Taha 2003; Yao et al. 2004). The application of a spin trapping agent is an option to overcome the difficulty of low sensitivity. The organic radical PTIO has been applied as a spin trapping agent specific for NO (Akaike et al. 1993; Akaike and Maeda 1996). Membrane permeability of such a spin trapping agent should be considered when one applies the method to an in vivo system.

5
Electrochemical Sensors

The advantages and limitations of electrodes for detection of NO are well-summarized in a recent review by Taha (2003), who had a pivotal role in developing these technologies. Briefly, a chemically modified electrode surface is positioned behind an NO-selective permeable membrane, and oxidation of NO at the surface is detected as a change in amperage. Previously, electrodes could be applied strictly for liquid phase NO measurement but are now available for gas-phase measurement (Taha 2003).

Detection levels are as low as 0.3–0.5 nM and a linear range can be achieved at up to 25 μM NO (Mantione and Stefano 2004; Taha 2003; Zhang and Broderick 2000). Considering the short half life of NO in oxic environments, the thin 100 nm diameter of one available nanosensor is an especially desirable feature for reaching source cells (Zhang and Broderick 2000); a recent evaluation found this electrode to perform with markedly better reliability and durability than past probes (Mantione and Stefano 2004). Being electrochemical sensors, NO sensors are sensitive to temperature fluctuations but newer models are available with a temperature compensation option (Taha 2003). Experimental designs should control for the possibility of artifacts resulting from even low intensity illumination of a sample.

6
Chemiluminescence

In the chemiluminescence assay NO is reacted with ozone, producing excited-state NO_2, which upon decay to the ground state releases a photon that is detected by a photomultiplier. It is considered a useful method because of its high sensitivity and capability for real-time monitoring of NO but is limited to detection of gasiform NO (Taha 2003; Yao et al. 2004). End-point detection of NO produced in fluids requires reduction of nitrite and nitrate to NO by vanadium (III) chloride in hydrochloric acid at 90 °C. This method is ideal for monitoring NO release from plants (Rockel et al. 2002).

7
Fluorescence Methods

In an oxic environment N_2O_3, the immediate unstable product of NO oxidation, can react with various conjugated ring-containing compounds, such as diaminonaphthalene, dihydrorhodamine, or diaminofluorescein (DAF), resulting in the formation of fluorophores (just as N_2O_3 from nitrite acidification converts the Griess reagents to a stable chromophore). Near stochiometric real-time detection of NO is possible in by use of a fluorimeter for liquid samples or by fluorescent or confocal microscopy for cellular localization. Synthesis of new DAF derivatives in recent years has expanded the permissible pH range for detection and allowed for entry of the reagent into cells.

7.1 Application of DAF

Diaminofluorescein-2 (DAF-2) was the first product in the DAF series to become commercially available (Kojima et al. 1998). DAF-2 detects NO through the formation of a triazole molecule (DAF-2T) that exhibits strong green fluorescence. DAF-2DA, a diacetate analog of DAF-2, was designed in particular for imaging of NO produced in cells (Kojima et al. 1998). The diacetate form possesses an obvious advantage in efficient uptake of the fluorescent probe into the living cells; the intracellular DAF-2DA is quickly transformed to DAF-2 by endogenous esterase activities, enabling visualization of in vivo NO production. Because of high sensitivity, specificity, and simplicity, DAF-2DA has been extensively applied in many fields of NO research (Cohen and Yamasaki 2003; Lacza et al. 2003). The detection of NO with DAF fluorescence techniques is now the most widely used methodology in plant research.

Figure 1 shows an example for the measurement of NO production in *Chlamydomonas* cells with DAF-2DA (Sakihama et al. 2002). NO produc-

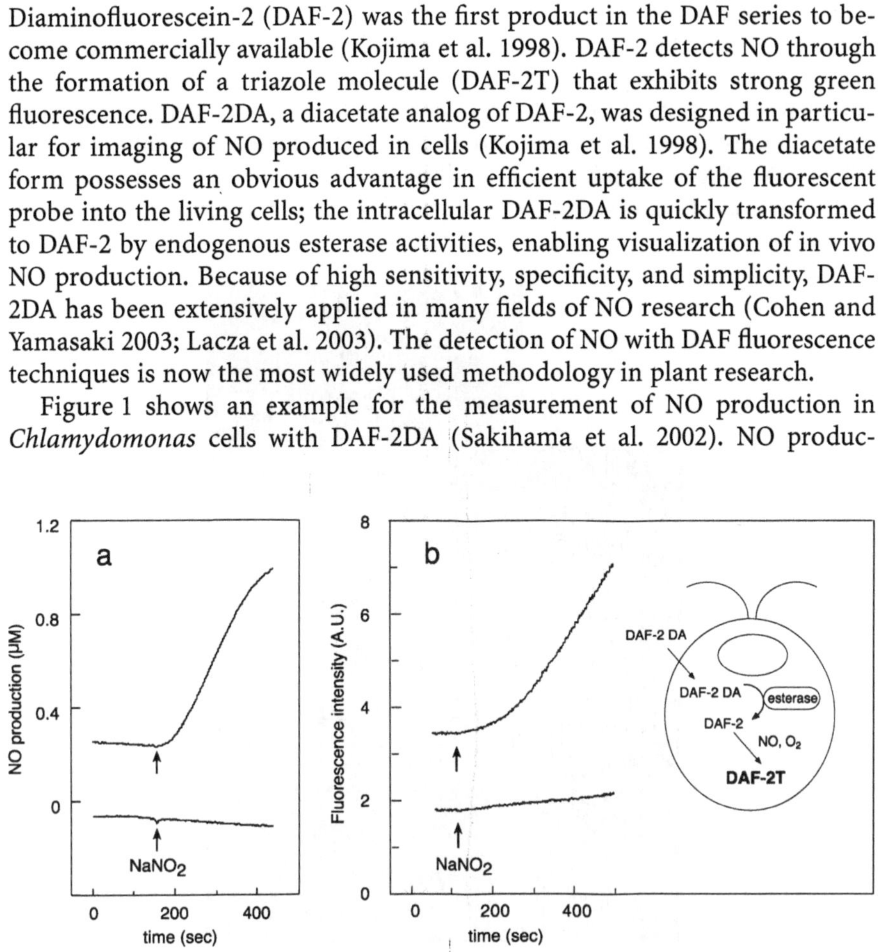

Fig. 1 Detection of extracellular and intracellular NO productions in the unicellular alga *Chlamydomonas reinhardtii*. **a** Time courses for NO production measured through NO released into an extracellular aqueous phase with an NO-specific electrode. **b** Time courses for NO production monitored with the intracellular NO indicator DAF-2DA. In both panels the *upper* and *lower traces* present NO production in the wild type and an NR-less mutant, respectively. Nitrite (5 mM sodium nitrite) was added to initiate NO production

tion in the bulk aqueous phase outside the cells was monitored with an NO electrode following addition of nitrite (Fig. 1a). Because of the membrane-permeable characteristic of DAF-2DA, one can monitor endogenous NO production within the cells in a real-time fashion. Figure 1b represent time courses for nitrite-dependent NO production within the *Chlamydomonas* cells using a fluorescence spectrophotometer. As in the NO electrode measurement, fluorescence intensity of the DAF by wild-type cells is increased in response to a nitrite supply. Because nitrate reductase (NR) is responsible for the NO production from nitrite, in both assay systems the NR-less mutant does not show any increase in the signals. Figure 2b shows visible fluorescence from DAF-2DA-incubated wild-type *Chlamydomonas* cells in the presence of nitrite. In good accordance with the time course measurements of endogenous and exogenous NO production, NR-less mutant cells do not show green fluorescence but exhibit red chlorophyll fluorescence only (Fig. 2c). When DAF-2DA is applied to a seedling of *Arabidopsis*, stomata of leaves and roots shine brightly (Fig. 2e,f), suggesting active NO production in these tissues.

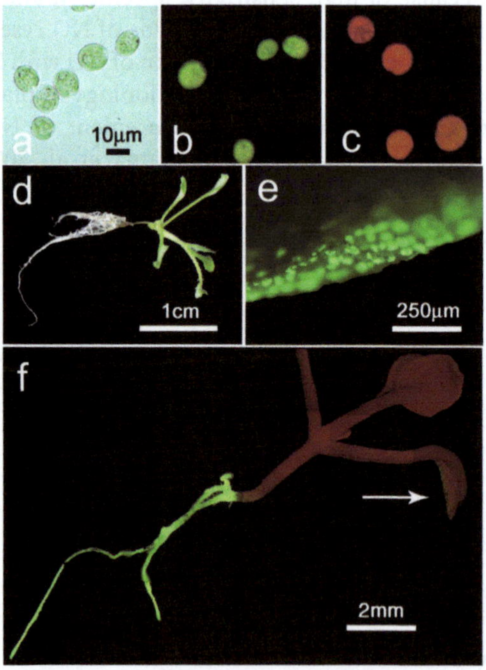

Fig. 2 Visualization of NO production with DAF-2DA. **a** *Chlamydomonas* cells (light microscope). **b** *Green* fluorescence from *Chlamydomonas* wild-type cells in the presence of nitrite. **c** *Red* chlorophyll fluorescence from *Chlamydomonas* NR-less cells in the presence of nitrite. **d** A seedling of *Arabidopsis thaliana*. **e** *Green* fluorescence from stomata of an abaxial side of a leaf. **f** Fluorescence image of a seedling of *A. thaliana*. The *arrow* indicates *green* fluorescence from stomata

7.2
Potential Errors in DAF Measurement

In addition to the development of NO detection methodology, there have been strong demands for NO scavengers. Until recently, only a few endogenous compounds such as hemoglobin and glutathione (GSH) were known to act as NO scavengers but with low specificity for NO (Herold 2003; Folkes and Wardman 2004). Development of 2-(4-carboxyphenyl)-4,5-dihydo-4,4,5,5-tetramethyl-1*H*-imidazolyl-1-oxy-3-oxide or cPTIO by Akaike and coworkers solved this problem. Now, cPTIO is widely used as an effective NO scavenger (and also as a spin trapping agent), owing to its high specificity and stoichiometric reaction with NO (Akaike and Maeda 1996). The combination of DAF and cPTIO has been frequently applied for proving in vivo NO production in animals (Pittner et al. 2003; Urno et al. 2005), bacteria (Creus et al. 2005), and plants (more than 20 papers as of 2006).

Figure 3 demonstrates the in vitro effect of cPTIO on fluorescence intensity of DAF-2T, which is formed from DAF-2. In the presence of cPTIO its fluorescence intensity is significantly increased. In principle, cPTIO acts as an NO scavenger through an oxidation reaction producing cPTI and NO_2. Contrary to the widely held presumption, the NO scavenger cPTIO does not suppress but actually enhances conversion of DAF into the DAF-2T fluorophore.

Figure 4a represents such a fluorescence enhancement effect of cPTIO on DAF fluorescence. In the presence of 0.1 mM cPTIO, visible green fluorescence is not inhibited but rather is increased. It should be noted that cPTIO itself has an intense color that poses another potential problem in

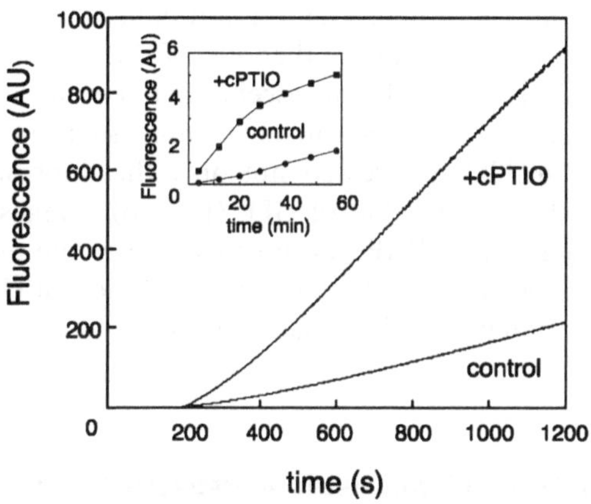

Fig. 3 Increase in DAF fluorescence in the presence of cPTIO monitored with a fluorescence spectrophotometer. The reaction medium contained 10 μM DAF-2, 250 μM SNAP, and 100 mM sodium phosphate (pH 7.4) with or without 100 μM cPTIO. *Inset* DAF fluorescence monitored with a microplate reader

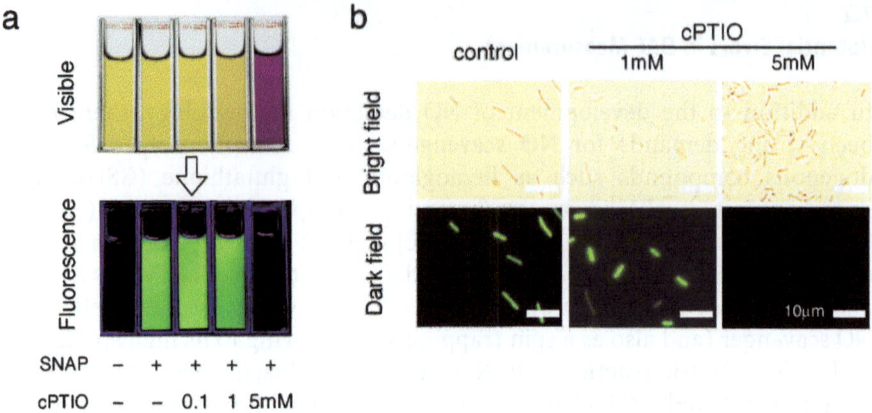

Fig. 4 Non-specific optical masking of fluorescence detection by cPTIO. **a** Different concentrations of cPTIO were added to each solution in quartz cuvettes. Black light was applied to the cuvettes for visual observation of DAF fluorescence. **b** Fluorescence microscopy images of *E. coli* cells expressing the green fluorescent protein in the presence or absence of cPTIO

fluorescence detection. The peak absorption wavelengths of cPTIO are 340 and 558 nm whereas the peak fluorescence emission of DAF-2T is 518 nm at 25 °C in 0.1 M phosphate buffer at pH 7.4. There is an obvious overlapping of the cPTIO absorption spectrum onto the DAF-2T fluorescence emission spectrum. Due to this, NO-dependent green fluorescence is not observed in the presence of a high concentration of cPTIO (Fig. 4a). To verify the masking effect of cPTIO on green fluorescence, Fig. 4b demonstrates the effect of cPTIO on green fluorescence from the green fluorescence protein (GFP) expressed in *Escherichia coli*. The excitation and emission spectrum of GFP is 475 nm and 509 nm, which is close to that of DAF-2T. It is important to be reminded that the principle of GFP fluorescence is irrelevant to NO production but its emission spectrum (emission max: 509 nm) is close to that of DAF-2T. Importantly, 5 mM cPTIO suppressed not only the fluorescence of a solution containing DAF but also that of GFP (Fig. 4b), suggesting that cPTIO could reduce NO-dependent DAF fluorescence due to the overlapping emission spectra. These results indicate that the application of cPTIO to the DAF system includes the potential risk of masking green fluorescence, irrespective of its origin.

7.3 Requirement for Oxidants in Fluorometric NO Detection by DAF

Kojima et al. have pointed out that DAF does not react with NO directly but with nitrous anhydride (N_2O_3) to form the fluorescent triazole compound (Kojima

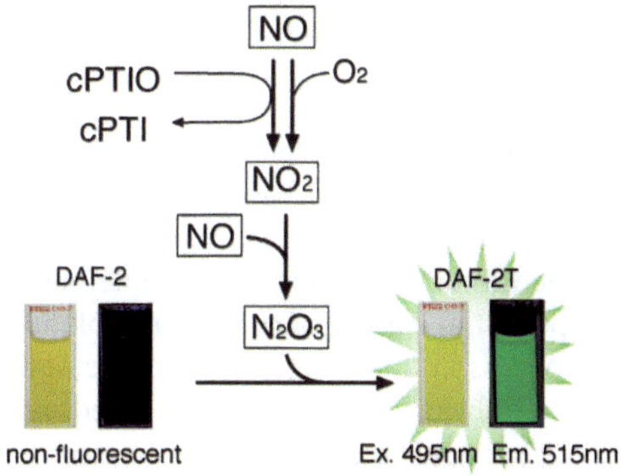

Fig. 5 Schematic diagram for the cPTIO reaction in fluorescence NO detection by DAF-2. NO is oxidized to NO_2 by either O_2 or cPTIO (without the requirement for O_2) and further changed to N_2O_3 via reaction with NO to form the highly fluorescent compound DAF-2T

et al. 1998). N_2O_3 is formed in the following scheme (Miles et al. 1996):

$$NO + O_2 \rightarrow 2NO_2 \quad (1)$$

$$NO_2 + NO \rightarrow N_2O_3 \,, \quad (2)$$

where $k_1 = 2 \times 10^6\ M^{-1}\ sec^{-1}$ and $k_2 = 2 \times 10^9\ M^{-1}\ sec^{-1}$. Under ambient conditions, the oxidation proceeds through the reaction of NO with molecular oxygen (O_2) dissolved in solution (1). As presented in Fig. 5, cPTIO can eliminate NO to produce NO_2 and subsequently N_2O_3 in the reaction series (rate constant, $k = 1.01 \times 10^4\ M^{-1}\ sec^{-1}$) (Akaike et al. 1993; Akaike and Maeda 1996; Zhang and Hogg 2002). Faster formation of fluorescent DAF-2T rather than suppression could be observed in the presence of cPTIO owing to production of N_2O_3. Similar reactions with DAF have been reported in the case of PTIO, an analog of cPTIO. Therefore, cPTIO and PTIO do not suppress but enhance NO-dependent DAF fluorescence detected under ambient conditions.

In principle, the DAF–NO detecting system requires O_2 to form the fluorescent molecule, as mentioned above (Fig. 5). It should be noted that cPTIO can mimic the oxygen effect by acting as an oxidant for NO (Fig. 5). A saturated concentration of O_2 in water can be 0.253 mM at 25 °C but this is strongly affected by conditions. There is a wide range in O_2 tension among tissues and cell compartments: high in oxygen-evolving chloroplasts of plants and low in muscles of animals. The DAF-NO detecting system does not work in anoxic conditions, but supplementing cPTIO into anoxic compartments or cells may enable us to detect NO even in the absence of O_2. This aspect could provide an opposite application for cPTIO, namely, not to scavenge NO but to visualize the radical in anoxic compartments. For example, the detection of NO from cells in anoxic or hypoxic conditions could be another potential application of cPTIO for the DAF system.

8 Concluding Remarks

In a half-century, biochemistry and molecular biology have established many new technologies that enable us to investigate proteins (enzymes), lipids, sugars, and nucleic acids (DNA and RNA). NO has unique characteristics contrary to these conventional biomolecules: simple, small, ubiquitous, and unstable. Despite the potential significance of NO in plant physiological functions, however, our progress in this field is not satisfactory. Even the source of intracellular NO production has not yet been clarified, neither in plants nor in vertebrate animals. One apparent reason can be ascribed to the difficulty of detection and quantification of NO due to its uniqueness.

To date, two enzymes have been recognized for their capacity to produce NO in plants: nitrate reductase (NR) and NO synthase (NOS). Because mammalian NOS enzymes produce NO from O_2 and arginine, plant NO-producing activity is often estimated through measurement of citrulline product formation (citrulline assay). It should be noted that such an estimate rests on many assumptions. Often overlooked in studies of NO production by eukaryotes is the potential for bacterial associates, some of which are endophytes that cannot be removed by surface sterilization, to contribute NO via nitrification, denitrification, or NOS activity (Cohen et al. 2006; Creus et al. 2005; Cohen and Yamasaki 2003). In relying on production of citrulline from arginine as a measure of NOS activity one must keep in mind that this activity is shared by arginase, an enzyme not uncommon among bacteria (Xu and Verstraete 2001). Although a bacteria-free assay can be ensured by simple passage of an extract through a 0.2 μm filter, the possibility of an ultimate prokaryotic origin for any given NO-producing activity should always be taken into consideration. Definitive evidence of plant-encoded NO production should include transgene expression and demonstrated NO-producing activity of a cloned product along with subcellular localization of the enzyme in its native condition.

The DAF and DAF-DA series are powerful tools for the exploration of plant NO biology. There is no doubt that this fluorescence technique is one of the best options for NO detection due to its high sensitivity and simple procedure. In the application of this method to living systems, however, we must be aware of potential errors that can be avoided by knowing the reaction principle. It is always true that there is no perfect method in experimental science; awareness of application limits is needed to avoid misinterpretation of data. This is particularly true for the detection of NO in living systems. Therefore, we strongly recommend employment of at least two methods having distinct detection principles when initiating a study of an NO-producing activity in an in vivo system.

Acknowledgements Due to space limitations we were not able to cite many brilliant works on plant NO research that have applied the methods described in this chapter. Please refer to other chapters for such investigations. This work was supported by the grants (Grant-in-Aid for Scientific Research (B) and (C) to HY, and the 21st Century COE program of the University of the Ryukyus) from the Japanese Ministry of Education, Science, Culture and Sports.

References

Adak S, Aulak KS, Stuehr DJ (2002a) Direct evidence for nitric oxide production by a nitric-oxide synthase-like protein from *Bacillus subtilis*. J Biol Chem 277:16167–16171

Adak S, Bilwes AM, Panda K, Hosfield D, Aulak KS, McDonald JF, Tainer JA, Getzoff ED, Crane BR, Stuehr DJ (2002b) Cloning, expression, and characterization of a nitric oxide synthase protein from *Deinococcus radiodurans*. Proc Natl Acad Sci USA 99:107–112

Akaike T, Yoshida M, Miyamoto Y, Sato K, Kohno M, Sasamoto K, Miyazaki K, Ueda S, Maeda H (1993) Antagonistic action of imidazolineoxyl *N*-oxides against endothelium-derived relaxing factor/NO though a radical reaction. Biochemistry 32:827–832

Akaike T, Maeda H (1996) Quantitation of nitric oxide using 2-phenyl-4,4,5,5-tetramethylimidazoline-1-oxyl 3-oxide (PTIO). Method Enzymol 268:211–221

Artz JD, Thatcher GR (1998) NO release from NO donors and nitrovasodilators: comparisons between oxyhemoglobin and potentiometric assays. Chem Res Toxicol 11:1393–1397

Cohen MF, Yamasaki H (2003) Involvement of nitric oxide synthase in sucrose-enhanced hydrogen peroxide tolerance of *Rhodococcus* sp. strain APG1, a plant-colonizing bacterium. Nitric Oxide 9:1–9

Cohen MF, Mazzola M, Yamasaki H (2006) Nitric oxide research in agriculture: bridging the plant and bacterial realms. In: Rai AK, Takabe T (eds) Abiotic stress tolerance in plants: toward the improvement of global environment and food. Springer, Berlin Heidelberg New York, pp 71–90

Creus CM, Graziano M, Casanovas EM, Pereyra MA, Simontacchi M, Puntarulo S, Barassi CA, Lamattina L (2005) Nitric oxide is involved in the *Azospirillum brasilense*-induced lateral root formation in tomato. Planta 221:297–303

Dionex Coorporation (1996) Application note 107: Ions in physiological fluids. Salt Lake City, UT

Folkes LK, Wardman P (2004) Kinetics of the reaction between nitric oxide and glutathione: Implications for thiol depletion in cells. Free Radic Biol Med 37:549–556

Herold S (2003) Interaction of nitrogen monoxide with hemoglobin and the artefactual production of S-nitroso-hemoglobin. Comptes Rendus Biologies 326:533–541

Ivanov VM (2004) The 125th anniversary of the Greiss reagent. J Anal Chem 59:1109–1112

Kawamura Y, Takahashi M, Arimura G, Isayama T, Irifune K, Goshima N, Morikawa H (1996) Determination of level of NO_3^-, NO_2^- and NH_4^+ ions in leaves of various plants by capillary electrophoresis. Plant Cell Physiol 37:878–880

Kojima H, Nakatsubo N, Kikuchi K, Kawahara S, Kirino Y, Nagoshi H, Hirata Y, Nagano T (1998) Detection and imaging of nitric oxide with novel fluorescent indicators: diaminofluoresceins. Anal Chem 70:2446–2453

Lacza Z, Snipes JA, Zhang J, Horvath EM, Figueroa JP, Szabo C, Busija DW (2003) Mitochondrial nitric oxide synthase is not eNOS, nNOS or iNOS. Free Radic Biol Med 35:1217–1228

Mantione KJ, Stefano GB (2004) A sub-nanomolar real-time nitric oxide probe: in vivo nitric oxide release in heart. Med Sci Monit 10:MT47–49
Miles AM, Wink DA, Cook JC, Grisham MB (1996) Determination of nitric oxide using fluorescence spectroscopy. Method Enzymol 268:105–120
Pittner J, Liu R, Brown R, Wolgast M, Persson AEG (2003) Visualization of nitric oxide production and intracellular calcium in juxtamedullary afferent arteriolar endothelial cells. Acta Physiol Scand 179:309–317
Privat C, Lantoine F, Bedioui F, Van Brussel EM, Devynck J, Devynck M-A (1997) Nitric oxide production by endothelial cells: comparison of three methods of quantification. Life Sci 61:1193–1202
Rockel P, Strube F, Rockel A, Wildt J, Kaiser W (2002) Regulation of nitric oxide (NO) production by plant nitrate reductase in vivo and in vitro. J Exp Bot 53:103–110
Uruno A, Sugawara A, Kanatsuka H, Kagechika H, Saito A, Sato K, Kudo M, Takeuchi K, Ito S (2005) Upregulation of nitric oxide production in vascular endothelial cells by all-trans retinoic acid through the phosphoinositide 3-kinase/Akt pathway. Circulation 112:727–736
Sakihama Y, Nakamura S, Yamasaki H (2002) Nitric oxide production mediated by nitrate reductase in the green alga *Chlamydomonas reinhardtii*: an alternative NO production pathway in photosynthetic organisms. Plant Cell Physiol 43:290–297
Schmidt K, Klatt P, Mayer B (1994) Reaction of peroxynitrite with oxyhaemoglobin: interference with photometrical determination of nitric oxide. Biochem J 301:645–647
Sun J, Zhang X, Broderick M, Fein H (2003) Measurement of nitric oxide production in biological systems by using Griess reaction assay. Sensors 3:276–284
Taha ZH (2003) Nitric oxide measurements in biological samples. Talanta 61:3–10
Xu J, Verstraete W (2001) Evaluation of nitric oxide production by lactobacilli. Appl Microbiol Biotechnol 56:504–507
Yamasaki H (2000) Nitrite-dependent nitric oxide production pathway: implications for involvement of active nitrogen species in photoinhibition in vivo. Roy Soc Lond Phil Trans B 355:1477–1488
Yamasaki H (2004) Nitric oxide research in plant biology: its past and future. In: Magalhaes JR, Singh RP, Passos LP (eds) Nitric oxide signaling in higher plants. Studium, Houston, pp 1–23
Yamasaki H (2005) The NO world for plants: achieving balance in an open system. Plant Cell Environ 28:78–84
Yamasaki H, Sakihama Y, Takahashi S (1999) Alternative pathway for nitric oxide production in plants: new features of an old enzyme. Trends Plant Sci 4:128–129
Yamasaki H, Sakihama Y (2000) Simultaneous production of nitric oxide and peroxynitrite by plant nitrate reductase: in vitro evidence for the NR-dependent formation of active nitrogen species. FEBS Lett 468:89–92
Yao D, Vlessidis AG, Evmiridis NP (2004) Determination of nitric oxide in biological samples. Microchim Acta 147:1–20
Zhang X, Broderick M (2000) Amperometric detection of nitric oxide. Mod Asp Immunobiol 1:160–165
Zhang Y, Hogg N (2002) Mixing artifacts from the bolus addition of nitric oxide to oxymyoglobin: implications for S-nitrosothiol formation. Free Radic Biol Med 32:1212–1219

Subject Index

MIX
Papier aus verantwortungsvollen Quellen
Paper from responsible sources
FSC® C105338

If you have any concerns about our products,
you can contact us on
ProductSafety@springernature.com

In case Publisher is established outside the EU,
the EU authorized representative is:
Springer Nature Customer Service Center GmbH
Europaplatz 3, 69115 Heidelberg, Germany

Printed by Libri Plureos GmbH
in Hamburg, Germany